Systematics and Conservation Evaluation

The Systematics Association
Special Volume No. 50

Systematics and Conservation Evaluation

Edited by

P. L. FOREY,
C. J. HUMPHRIES, and
R. I. VANE-WRIGHT
*The Natural History Museum,
London*

THE
Systematics
ASSOCIATION

Published for THE SYSTEMATICS ASSOCIATION by
CLARENDON PRESS · OXFORD
1994

Oxford University Press, Walton Street, Oxford OX2 6DP
Oxford New York Toronto
Delhi Bombay Calcutta Madras Karachi
Kuala Lumpur Singapore Hong Kong Tokyo
Nairobi Dar es Salaam Cape Town
Melbourne Auckland Madrid
and associated companies in
Berlin Ibadan

Oxford is a trade mark of Oxford University Press

Published in the United States
by Oxford University Press Inc., New York

A catalogue record for this book is available from the British Library

Library of Congress Cataloging in Publication Data
Systematics and conservation evaluation/edited by P. L. Forey, C. J. Humphries, R. I. Vane-Wright.
(The Systematics Association special volume: no. 50)
Papers presented during a symposium held in London, June 17–20, 1992.
Includes bibliographical references.
1. Biological diversity conservation—Congresses. 2. Biology—Classification—Congresses. I. Forey, Peter L. II. Humphries, C. J. (Christopher John), 1947–. III. Vane-Wright, Richard Irwin. IV. Systematics Association. V. Series.
QH75.A1S97 1994 333.95'16—dc20 93-38147
ISBN 0 19 857771 0

Typeset by Cambrian Typesetters, Frimley, Surrey
Printed in Great Britain
on acid-free paper by
St Edmundsbury Press, Bury St Edmunds, Suffolk

Preface

This special volume is based on papers presented during a symposium, with the same title, held in London, 17–20 June 1992. The meeting was sponsored by the The Natural History Museum (where the meeting took place), The Linnean Society of London, and The Systematics Association. Over 150 people gathered to debate what connections could be made between our understanding of the classification, taxonomy, and phylogenetic relationships of the millions of organisms that occur on Earth, and the growing awareness of the need to set priorities for their conservation. The meeting was timely: during June 1992 over 150 countries, represented at the United Nations Conference on Environment and Development held in Rio de Janeiro, signed the Biodiversity Convention. The same month saw publication of *Global biodiversity: status of the Earth's living resources* (Groombridge 1992), following closely on the *Global biodiversity strategy* (Reid *et al.* 1992).

Systematics and conservation evaluation divides into two halves. The first 12 chapters set limits and context to the debate over systematic methods, covering the issues of biodiversity loss and the socio-political agenda (Chapters 1–3), ecological and land form insights into biodiversity conservation (Chapters 4 and 5), and basic questions about how many taxa exist, and how we can gather and present data about them (Chapters 6–12). In the second part, Chapters 13–24 are concerned with different systematic approaches, based on empirical data about the numbers, distributions, and interrelationships of organisms, to recognizing and setting priorities for biodiversity conservation. These contributions concern assessment of 'hotspots' (Chapters 13–15), the relative value of different organisms based on systematic principles (Chapters 16 and 17), and the development of systematic methods for conservation evaluation based on entire groups, or on what Williams and Gaston (in press) have called 'wholesale biodiversity' (Chapters 18–24). Finally, Chapter 25 represents an attempt to bridge the gap between ecology and systematics. In sequence, we look at each of these 25 papers.

Objectives, perspectives, and empirical problems

Purpose, policy, and philosophy

Walter Reid (Chapter 1) challenges conventional views on the objectives of conservation, particularly those concerned with the preservation of natural ecosystems. Present goals for ecological management were developed at a time when ecologists believed that natural communities consisted of stable assemblages of species, population sizes tended to equilibrium levels, and that humans had little effect on the structure of such communities. Although recent advances have brought into question the very meaning of 'natural' in this context, the old ideas remain embedded in management policies. Reid argues that the primary goal of ecological management should reflect a social purpose, to maximize human capacity to adapt to change, and thus sees the maintenance of biodiversity as a prerequisite for maximizing that human ability to adapt to future conditions.

That ecological conditions will alter radically in the coming decades—due to increasing effects of human population pressure and suffering, as well as 'systemic' changes brought about by global warming, ozone depletion (whatever their final magnitude), and pollution—seems inevitable. Reid's outspoken contribution serves to establish the need for a new (and we believe, in part, systematic) approach to conservation evaluation: the conservation of biodiversity is not the same as traditional approaches to conservation, exactly because it serves different goals and different purposes.

In some contrast, Vernon Heywood (Chapter 2) takes a more pluralist standpoint, at least amongst the professional conservation community. He points out that biodiversity is viewed differently by ecologists, managers, plant breeders, population biologists, agriculturists, taxonomists, and so on. Based on various ways of looking at the hierarchical organization of biodiversity—ecosystem, species, and genetic levels, or as compositional, structural, and functional systems—Heywood argues that, to achieve a variety of conservation objectives, a multiple-level strategy for the measurement of biodiversity will be required. The question then arises, will this continue to be the case if we subsume all our conservation goals within a single overall objective, as proposed by Reid?

Bryan Norton (Chapter 3) offers philosophical and sociological insights into these problems, and questions whether there is, or can be, an 'objective' scientific definition of biodiversity. He concludes that there cannot be such a definition, both for epistemological reasons, and because of the irreducible complexity of natural systems. Natural systems are seen as dynamic, with diversity being maintained when natural populations adapt to the dangers and opportunities created by change. According to Norton, we can only understand biodiversity if we interpret nature as a hierarchy in which the smaller subsystems tend to change more rapidly than larger units or the whole system. In the absence of one scientific definition of biodiversity, other means are needed to establish conservation targets. Norton argues that there is an essential link between human culture, landscape, and the maintenance of biodiversity. Our goal then becomes the maintenance of diversity on an inter-generational scale sufficient to permit sustainablity of both human culture and of nature—on which we are dependent, and of which we also form part.

During the meeting the apparent incompatibility between these views and the more reductionist approach which followed later (notably in Chapters 13–24) was not addressed. Whether or not a philosophical approach can lead to practical and scientifically defensible proposals for the systematic conservation of biological diversity remains to be seen. Also, we need to know if the systematic approach, which at this stage explores, and to some extent revolves around a search for a single system of measurement for biodiversity (Chapters 18–24), can mature to the point where it can be integrated with other measures of biodiversity (Heywood, Chapter 2), and with any overriding concern for cultural values (Norton, Chapter 3).

At present the systematic approach to the conservation of biodiversity sits most readily amongst those general views expressed here, with the position adopted by Walter Reid: the purpose of preserving as much biological diversity as possible is to maximize the ability of succeeding generations of humans to adapt to change. However, as most systematists who are faced with the enormous number of different organisms also believe, most biodiversity conservation will have to be carried out locally, *in situ*, in real and dynamic landscapes, then the apparent differences in viewpoint may be more of style than substance—and also give a natural link to Chapters 4 and 5 concerned with land forms and biodiversity. This link is made explicit in Chapter 25.

Ecological perspectives

Chapters 4 and 5 examine, at widely differing scales, ecological perspectives on the problems of conserving biodiversity. Although this book is primarily about systematic perspectives, most conservation action ultimately represents a problem in applied ecology, particularly at local scales, and so ecological insights are essential.

In Chapter 4, Sam McNaughton examines the wide variation and hierarchical complexity seen in grassland biodiversity of the Serengeti National Park, Tanzania. This variety has direct consequences for grazing mammals, notably in terms of food availability at critical stages of their annual cycles. McNaughton concludes that conservation evaluation of ecosystems supporting large ungulates needs to incorporate an understanding of the functional effects of multi-level biodiversity patterns and how they relate to the carrying capacity of an ecosystem. This paper is thus a reminder that landscape scale, and the patchiness of a wide range of different local habitats, may be crucial for the ability of a region to support, in the long-term, its full range of biodiversity, especially macro-fauna.

Ian Atkinson (Chapter 5) assesses ecological and taxonomic criteria for conserving terrestrial biodiversity, based on work carried out in New

Zealand, a biotically and geologically diverse country. Atkinson examines an imaginary scenario, a retrospective attempt to protect biodiversity in New Zealand as it was in the 1840s, including steps necessary to avoid some of the mistakes that have since been made. These steps would include the need to give greater priority to lowland dry-climate areas, identifying 'hot-spots' of high species diversity by correlation with areas of high geological and landform diversity, locating centres of endemism, and focusing on areas thought to have long-term, inherent sustainability. This analysis is summarized as a general strategy for reducing future biodiversity losses, elements of which may have wider application to the biodiversity crisis now affecting so many parts of the world.

In the absence of better data such indirect approaches to biotic evaluation may have much to commend them. However, the historical dimension, which is also a major factor in the repartition of diversity, can only be taken into limited account by such methods, as through the identification of areas of endemism—itself a fundamentally biogeographic and systematic problem, rather than an ecological one.

Problems with data

In Chapter 6, Nigel Stork examines the question of how many millions of species exist, and the problems that these vast numbers create for making biotic inventories. Stork suggests that efforts should be focused on particular sites, such as biosphere reserves. Certain sites should be intensively documented, but for the others it will only be possible to make extensive surveys, involving the selection and use of indicator groups (there may, however, be alternatives to indicator groups, such as the use of higher taxa; Williams and Gaston, in press). Stork makes a number of recommendations regarding surveys, including the use of standardized sampling procedures.

But modern sampling techniques (including those frequently used by Nigel Stork himself), tend to produce an embarrassment of riches, especially when it comes to their documentation, identification, and permanent preservation. In Chapter 7, Ebbe Nielsen and Judy West examine the continuing challenge to museums to preserve and make available for study the mountains of existing biological material, a problem exacerbated by the onset of an increasingly 'industrialized' approach to biodiversity sampling and monitoring. Nielsen and West first comment on the question of the number of species, particularly in underworked groups such as fungi and insects (cf. Chapter 6), and how it affects national efforts to develop systematic information for all groups. A frequent mismatch between the deployment of the existing taxonomic workforce and our state of knowledge of different taxonomic groups is noted (e.g. in botanical institutions, with their low input to the hyper-diverse fungi). Nielsen and West then focus on the problems of taxonomic information generation, access, and transfer. With respect to original research, the utility and cost-effectiveness of a checklist is contrasted with the basic, but inevitably expensive full taxonomic revision. Methods for information transfer are similarly compared, notably of direct physical access to curated collections, by conventional printed form, or through electronic media. The costs of obtaining, processing, and maintaining natural history specimens are analysed, and it is demonstrated that only a fraction of the entire cost is added if the data they represent are stored and made available by means of an appropriate electronic database. They argue that it is the duty of collections managers in major museums and herbaria to move towards comprehensive databasing, so as to make far more effective use of the invaluable but costly investment collections already represented. The authors conclude with a discussion of the need for database standards, and development of rational protocols for carrying out primary biotic surveys in the first place, to ensure that the needs for the subsequent primary research, identification, specimen storage, and information transfer are properly planned from the beginning.

This comprehensive approach is pursued in Chapter 8 by Barry Richardson, with respect to the data itself. As Richardson clearly points out, national and international biological information systems require much careful planning. Such systems can easily fail if they do not meet the (often unreal) expectations of users, or because of failure to anticipate all likely, or even unexpected user requirements. A particular problem, regarding the validation of data gathered over different periods and to different standards, is examined in the light of his work for Australia's Environmental Information Resources Network (ERIN).

Data quality is a perennial problem, often ignored or marginalized, and the unique experience within ERIN and other Australian organizations for creating automated checking or validation procedures (e.g. BIOCLIM: Busby 1991) is of great value. In light of this, the contribution by Ian Oliver and Andrew Beattie (Chapter 9) is bound to cause debate. They describe a rapid method for assessing biodiversity in limited areas based on use of taxa such as invertebrates and cryptogams, evaluated as Recognizable Taxonomic Units (RTUs) by 'biodiversity technicians' rather than trained specialists. Cranston and Hillman (1992) have criticized this approach, based on direct attempts to replicate the procedure. The use of RTUs also poses strong limitations for comparisons between sites where complementarity (e.g. Chapter 19) or irreplaceability (Chapter 23) are seen as particularly potent factors for conservation evaluation, rather than just taxon richness alone. A more traditional approach to survey work is examined in the next chapter, by Sudhendu Mandal, on the conservation status of Indian plants.

Mapping approaches

Once we have all this data, more or less systematically collected, how can it be used for conservation evaluation? One of the most time-honoured and practical approaches is through the use of mapping. In Chapter 11 Ghillean Prance explores phytogeographic data for the New World tropics, using a variety of mapping techniques, to demonstrate how such data presentation can guide conservation planning at different spatial scales, ranging from analyses across the entire neotropical region down to investigations of a 23-ha plot in Brazil. In Chapter 12, Ron Miller, Bob Press, Marian West, and Ricardo Baldini describe a computerized system for mobilizing museum data and monitoring threatened Italian plants. By linking such a database to GIS technology, it is possible to analyse distributions in a wide variety of ways, and to produce high quality maps.

Although graphical mapping techniques are undoubtedly valuable for alerting us to general patterns, including high diversity areas, fundamentally simple methods lack analytical power. These problems are reworked in Chapters 18–24, which concentrate on iterative procedures for analysing raw data.

Systematic analyses

The second half of the book develops more detailed arguments about how analyses of patterns of biodiversity should be undertaken and how reserve networks should be designed. The papers can be divided into three sections: those concerned, essentially, with the question of generalized 'hot-spots' (Chapters 13–15), those with the conservation value of particular taxa based on systematic assessments (Chapters 16 and 17), and a final group of papers developing and refining iterative approaches to conservation evaluation (Chapters 18–24).

'Hot-spots'—fact or fiction?

John Lawton, John Prendergast, and Brian Eversham (Chapter 13) make an assessment of a variety of taxa from the British Isles, using data from the UK Biological Records Centre. By scoring presence and absence of species in each of the 10 × 10 km squares of the national grid, they produce and analyse a data set to ask a number of pertinent questions which have previously remained as assumptions in the minds of theoretical ecologists and conservationists alike. Where are the most species-rich areas for particular taxa? Do diversity 'hot-spots' coincide for different taxa? Exactly how much taxon turnover is shown by different groups along geographic transects? What proportion of the national pool of species in a particular taxonomic group might be conserved by setting aside reserves in a limited number of locations? They find, superimposed on north–south and east–west trends in alpha diversity, significant departures in local species richness. Some 'hot' and 'cool' spots are relatively easy to explain, but others are not. Furthermore, 'hot-spots' for the different taxa examined appear not to be coincident. The hope that a reserve in one area that is rich for one or two well-known groups will also be rich in more poorly-known taxa is not supported. Beta-diversity is low in the UK, however, and so even a few well-selected reserves should be capable of preserving many taxa.

However, this is not necessarily true for tropical taxa, as demonstrated in Jeremy Holloway's contribution (Chapter 14). Using biogeographic and ecological data for about 1200 Bornean moth species, relative vulnerabilities of these groups to habitat change are calculated. Holloway demonstrates a spectrum of vulnerability amongst the groups to different types of habitat change, with those most at risk having a high proportion of endemic species dependent on primary lowland rainforest. The implications for environmental management and conservation policy are discussed. Here, however, it is obvious that beta-diversity in the Bornean moth fauna is high. Thus, for taxonomic groups in tropical regions it is likely that many species-rich areas will have to be conserved to encompass the total biodiversity.

The International Council for Bird Preservation (now BirdLife International) recently published their global scheme for assessing conservation priorities based on areas of endemism for birds (ICBP 1992). Assuming that centres of endemism would be priorities for the conservation of bird diversity they considered distributions of all species with breeding ranges restricted to 50 000 km^2, or less. This study, reviewed in Chapter 15 by Simon Thirgood and Melanie Heath, reveals that 2497 species (26 per cent of all birds) are restricted to 221 endemic bird areas (EBAs). In total, these high-priority areas cover only about 5 per cent of the world's land surface. Thirgood and Heath then raise two very important questions: is there congruence of patterns of endemism between different animal and plant groups, and are endemic birds good indicators for areas of endemism in other such groups? As already noted, the idea of 'hot-spot' congruence can be seriously questioned (Chapter 13), even though this appears only to be true (so far) for UK data. Through a regional assessment of the 221 EBAs throughout the world, Thirgood and Heath demonstrate good evidence of congruence between global patterns of endemism in birds and other taxa.

Validation and importance of individual taxa

Vernon Heywood (Chapter 2) makes the plea that no particular sector of the conservation movement should hold the key to conservation policy. A united front between all sectoral interests will be necessary if we are to succeed. Success will also require that we agree about what the units of conservation should be. Tim Crowe, Peter Ryan, Faadiel Essop, Richard Brooke, Philip Hockey, and Roy Siegfried (Chapter 16) assume that species are knowable and definable. They go far beyond Mayr's biological species concept, taking a multifaceted approach using morphological, behavioural, ecological, and genetic information, to recognize species amongst a complex of larks in south-western Africa. On this basis they propose a justifiable conservation policy for the preservation of these birds.

Several other papers consider that our energies should be channelled into conserving critical taxa other than simply worrying about species *en masse* (the so-called megadiversity or list-envy syndrome). For ecologists this translates, for example, into ideas such as keystone species. From a systematic standpoint, Melanie Stiassny and Mário de Pinna consider the problem of basal taxa (plesiomorphic members of systematic groups) which also have extremely restricted geographic distributions. These two features render them particularly vulnerable to environmental pressures. This phenomenon is strikingly prevalent among freshwater fishes, an assemblage representing upward of 25 per cent of all living vertebrates, yet largely ignored in regional biodiversity estimates and conservation evaluation programmes. Stiassny and de Pinna provide a spirited argument to conclude that until stable, well-resolved phylogenies are available for most taxonomic groups, basal taxa assume a key importance in phylogenetic analysis. Furthermore, and particularly for those groups in the tropics where new species are constantly being found, basal taxa will continue to play an important role in determining the taxonomic placement of these new species. They feel that basal taxa should be assigned far higher conservation priority than has so far been the case.

Biodiversity evaluation based on systematic criteria

The theme of systematics as fundamental to biodiversity and its conservation is dominant for the remainder of the volume, with technical details strongly in evidence. Dan Faith (Chapter 18) provides an account of a new measure, 'phylogenetic diversity', designed to identify priorities by maximizing the feature diversity that selected taxa represent. In conservation evaluation, giving priority to species subsets that maximize represented feature diversity is justified. As part of a larger study of the diversity patterns over several groups of invertebrates in Tasmanian rainforest, his preliminary results for amphipods provide a case study where phylogeny reveals diversity information presently unavailable in the existing species classification.

Paul Williams and Chris Humphries (Chapter 19) continue within this theme by describing practical approaches to measuring biodiversity. Within any defined geographical area, biodiversity is considered to be a measure of two main variables: the number of taxa and the degrees of difference between them. Using the computer program WORLDMAP, comparative trials of four different measures of cladistic diversity (root-weight, higher taxon richness, spanning-tree length, and cladistic dispersion) have been undertaken. The taxonomic measures can be used, together with a measure of endemism and in conjunction with complementarity, to determine priority sequences of areas for conservation. They conclude that the greatest contribution to conservation and representation of biodiversity is from complementarity, and differences between organisms can be most readily represented by using available information of group membership from taxonomic classifications. Endemism is seen not as a measure of biodiversity but of range size, and provides a relatively ineffectual basis for calculating priorities for areas for biodiversity conservation if only a few areas can be conserved.

The next two papers (Chapters 20 and 21) provide empirical applications of WORLDMAP, both dealing with the very thorny issue of combining data from many different taxonomic groups as one way to deal with the notion of wholesale biodiversity (Williams and Gaston, in press). Pjotr Oosterbroek (Chapter 20) concentrates on combined evaluations of Mediterranean groups, looking at diversity in birds, mammals, reptiles, amphibians, and three species-rich groups of insects. He appraises the critical issue of simple species richness *versus* taxonomic measures and concludes that, although species richness is a strong guide to biodiversity, the historical components of areas, i.e. source areas and relictual taxa, are best identified using taxonomic weighting procedures. Consequently, priority sequences vary according to which measure is employed.

Dick Vane-Wright, Campbell Smith, and Ian Kitching (Chapter 21) concentrate on the issue of how to combine data from many different taxonomic groups—perhaps up to 20 000 terminal taxa or more will be necessary if meaningful priority areas for conservation are to be recognized. They note that analysing data sets of more than 1000 taxa using measures more complex than spanning-subtree weighting is expensive on systems based on personal computers, and that coping with 20 000 terminal taxa is beyond present capability. Using data derived from owls, hawk-moths, milkweed butterflies, and tiger-beetles, analysed with the WORLDMAP program at regional (SE Asia) and national (Thailand) levels, they describe initial attempts to find methods that will combine the results of separate analyses to reproduce, or to approximate the results of all-inclusive analyses.

The three remaining papers (Chapters 22–24) describe other methods that utilize the principle of complementarity in the actual design of nature reserve networks, on a local scale. In Chapter 22, Chris Margules, Ian Cresswell, and Nick Nicholls describe a method for identifying networks of protected areas to maximize the likelihood that they sample biological diversity fully. After determining environmental regions throughout Australia, iterative heuristic algorithms were used to determine where protected areas should be located within the regions, so that all known species, communities, or environments are represented. The method is powerful because it emphasizes flexibility, efficiency, and explicitness. Margules, Creswell, and Nicholls also make an important plea by pointing out that the most urgent scientific challenge for all of us is to improve the quality of the databases upon which the algorithms operate and conservation issues are decided (thus referring back to issues raised in Chapters 6–12).

Bob Pressey, Michael Bedward, and David Keith (Chapter 23) also criticise the *ad hoc* nature of environmental policy making and procedures. Approaching the problem with attitudes and methods similar to those outlined by Margules *et al.* (Chapter 22), they question what is required to achieve representative reserve systems, and make a call for more systematic procedures for selecting reserves. They review opportunism against a variety of more systematic approaches for reserve selection, in relation to three key principles of efficiency, flexibility, and irreplaceability, to determine practical reserve networks capable of maximizing the conservation of biodiversity.

Tony Rebelo (Chapter 24) reviews conservation issues in the Cape Floristic Region (CFR), an area which covers only 90 000 km^2 yet comprises one of the world's six great floral kingdoms. With 8600 species, of which 68 per cent are endemic, it ranks amongst the richest of temperate and tropical floras (although most of the diversity is concentrated in a few species rich taxa). Some 19 per cent of the area of the CFR is in nature reserves, but the great majority of the preserved area comprises mountain Fynbos. Utilizing an iterative selection procedure to identify optimal reserve configurations, based on Proteaceae species richness, suggests that 95 per cent of all vascular plant species in Fynbos can be preserved in 16 per cent of the area. The results suggest that the ideal approach to designing a reserve configuration is to identify areas of high endemism and richness in distinctive vegetation types within particular biogeographical regions.

Systematics and ecology—conflict or empathy?

The final paper, by Steven Cousins, bears a provocative subtitle, 'will the Ark work?' Accepting that systematics and systematic approaches to conservation evaluation may be able to make a real contribution, he then questions the extent to which a highly efficient network of carefully selected sites will be effective in preserving biodiversity. This is not due to doubts over the selection procedures themselves, but because of a potential breakdown in ecosystem function, and the services they provide, if an insufficient quantity of life is preserved in a 'natural' or 'semi-natural state'. His ecotrophic module concept echoes some of the ideas in Sam McNaughton's paper (Chapter 4).

In Cousins' view, concentrating only on the quality (taxic diversity) of life may be self-defeating if the lack of sheer quantity of fundamental ecological types renders the biosphere unstable. This relates to the energy-flow problem raised by Paul Ehrlich and colleagues (Vitousek *et al.* 1986): man now appropriates about 40 per cent of the net energy fixed by photosynthesis available to support all terrestrial biodiversity, and this hugely disproportionate quantity looks set to go on rising.

Among several other points raised by Cousins is an old question in a new context: do we want many small reserves, or fewer, larger ones? Undoubtedly, to create the sort of *in situ* zoo- or plant-parks envisaged by systematists (e.g. Humphries *et al.* 1991) to preserve the widest taxonomic variety of life, then more smaller rather than a few larger areas might, at first sight, seem better. But if we take an ecosystem function (e.g. Walker 1992) or viability approach (e.g. Lamont and Klinkhamer 1993), then larger parks or reserves probably make more sense. However, as pointed out by Westman (1990), making such a stark dichotomy is almost undoubtedly a mistake: most biodiversity will be conserved, if at all, outside strict nature reserves. We need a more rational environmental policy for the whole surface of the Earth, including the oceans, not just wilderness areas or 'biodiversity' reserves.

The apparent differences between ecological and systematic approaches largely relate to different emphases and different time-frames, rather than basic differences over substance.

The development of the systematic approach to site selection (Pressey *et al.* 1993) has largely been driven by a need to respond to impending mass extinction, in an effort to develop a rational strategy for conserving truly representative stocks of the Earth's genetic, taxic, and ecosystem diversity, before it is too late. The ecosystem approach, faced with the equally unwelcome prospects of ozone depletion, climate change, pollution, and general environmental collapse, is concerned with the goal of sustaining life (one is tempted to say any life) on the planet, in the long-term. Our willingness to divorce these two approaches reflects little more than the oddly compartmentalized and reductionist heritage of the natural sciences. In attempting to meet the 'truly overwhelming task of global biodiversity conservation', Cousins forces us to acknowledge that we will have to sustain both the quantity and the quality of different life forms if we are are to succeed with this literally vital goal.

If we are going to conserve biodiversity reliably enough to span beyond our own generation, we need to be able to combine scientific solutions with local aspirations and international politics, to guarantee survival of the necessary nature reserves and conservation areas into the twenty-first century and beyond.

Acknowledgments

In addition to The Systematics Association, The Linnean Society of London, and The Natural History Museum, a number of individuals helped in a variety of ways, with organization and running of the meeting, or subsequent editing of this volume: Sasha Barrow, John Benfield, Sam Bhattacharyya, Ray Hobson, Ian Kitching, John

Marsden, Elaine Shaughnessy, David Shuker, Campbell Smith, Jacqui Staerck, and Sally Young. To the supporting organizations, and our colleagues, we offer our most grateful thanks.

References

Busby, J. R. (1991). BIOCLIM—a bioclimate analysis and prediction system. In *Nature conservation: cost effective biological surveys and data analysis* (ed. C. R. Margules and M. P. Austin), pp. 64–8. CSIRO, Melbourne.

Cranston, P. and Hillman, T. (1992). Rapid assessment of biodiversity using 'Biological Diversity Technicians'. *Australian Biologist*, **5**, 144–54.

Groombridge, B. (ed.) (1992). *Global biodiversity: status of the Earth's living resources.* Chapman and Hall, London.

ICBP (1992). *Putting biodiversity on the map: priority areas for global conservation.* ICBP, Cambridge.

Humphries, C. J., Vane-Wright, R. I., and Williams, P. H. (1991). Biodiversity reserves: setting new priorities for the conservation of wildlife. *Parks (IUCN)*, **2**, 34–8.

Lamont, B. B. and Klinkhamer, P. G. L. (1993). Population size and viability. *Nature*, **362**, 211.

Pressey, R. L., Humphries, C. J., Margules, C. R., Vane-Wright, R. I., and Williams, P. H. (1993). Beyond opportunism: key principles for systematic reserve selection. *Trends in Ecology & Evolution*, **8**, 124–8.

Reid, W., Barber, C., and Miller, K. (1992). *Global biodiversity strategy: guidelines for action to save, study and use Earth's biotic wealth sustainably and equitably.* WRI, IUCN and UNEP, Washington, DC.

Vitousek, P. M., Ehrlich, P. R., Ehrlich, A. H., and Matson, P. M. (1986). Human appropriation of the products of photosynthesis. *BioScience*, **36**, 368–73.

Walker, B. (1992). Biodiversity and ecological redundancy. *Conservation Biology*, **6**, 18–23.

Westman, W. E. (1990). Managing for biodiversity. *BioScience*, **40**, 26–33.

Williams, P. H. and Gaston, K. J. (in press). Measuring more of biodiversity: can higher-taxon richness predict wholesale species richness? *Biological Conservation.*

London
1994

P. L. F.
C. J. H.
R. I. V.-W.

Contents

Contributors

I. A. E. ATKINSON
Landcare Research New Zealand, Private Bag 31902, Lower Hutt, New Zealand.

R. M. BALDINI
Museo Botanica, University of Florence, via La Pira 4, Firenza, Italy.

A. J. BEATTIE
School of Biological Sciences, Macquarie University, NSW 2109, Australia.

M. BEDWARD
New South Wales National Parks and Wildlife Service, PO Box 1967, Hurstville, NSW 2220, Australia.

R. K. BROOKE
Percy FitzPatrick Institute for African Ornithology, University of Cape Town, Rondebosch 7700, South Africa.

STEVEN H. COUSINS
International Ecotechnology Research Centre, Cranfield Institute of Technology, Cranfield, Bedfordshire MK43 0AL, UK.

I. D. CRESSWELL
ERIN (Environmental Resources Information Network), GPO Box 636, Canberra, ACT 2601, Australia.

T. M. CROWE
Percy FitzPatrick Institute for African Ornithology, University of Cape Town, Rondebosch 7700, South Africa.

M. F. ESSOP
Percy FitzPatrick Institute for African Ornithology, University of Cape Town, Rondebosch 7700, South Africa.

B. C. EVERSHAM
Institute of Terrestrial Ecology, Monks Wood, Abbots Ripton, Huntingdon, Cambridgeshire PE17 2LS, UK.

DANIEL P. FAITH
Division of Wildlife and Ecology, CSIRO, P.O. Box 84, Lyneham, ACT 2602, Australia.

P. L. FOREY
Department of Palaeontology, The Natural History Museum, London SW7 5BD, UK.

M. F. HEATH
BirdLife International (formerly International Council for Bird Preservation), Wellbrook Court, Girton Road, Girton, Cambridge CB3 0NA, UK.

V. H. HEYWOOD
IUCN—The World Conservation Union, Plant Conservation Office, Richmond TW9 3BW, UK.

P. A. R. HOCKEY
Percy FitzPatrick Institute for African Ornithology, University of Cape Town, Rondebosch 7700, South Africa.

J. D. HOLLOWAY
International Institute of Entomology, Department of Entomology, The Natural History Museum, Cromwell Road, London SW7 5BD, UK.

C. J. HUMPHRIES
Biogeography and Conservation Laboratory, Department of Botany, The Natural History Museum, London SW7 5BD, UK.

D. A. KEITH
New South Wales National Parks and Wildlife Service, PO Box 1967, Hurstville, NSW 2220, Australia.

I. J. KITCHING
Biogeography and Conservation Laboratory, Department of Entomology, The Natural History Museum, Cromwell Road, London SW7 5BD, UK.

J. H. LAWTON
NERC Centre for Population Biology, Imperial College at Silwood Park, Ascot, Berkshire SL5 7PY, UK.

S. J. McNAUGHTON
Biological Research Laboratories, Syracuse University, Syracuse, NY 13244–1220, USA.

S. MANDAL
Department of Botany, Visva-Bharati University, Santiniketan—731235, India.

C. R. MARGULES
CSIRO Division of Wildlife and Ecology, P.O. Box 84, Lyneham, ACT 2602, Australia.

R. I. MILLER
P.O. Box 2731, Cambridge, Massachusetts 02138, USA.

A. O. NICHOLLS
CSIRO Division of Wildlife and Ecology, P.O. Box 84, Lyneham, ACT 2602, Australia.

EBBE S. NIELSEN
Australian National Insect Collection, CSIRO Division of Entomology, GPO Box 1700, Canberra, ACT 2601, Australia.

BRYAN G. NORTON
School of Public Policy, Georgia Institute of Technology, Atlanta, Georgia 30332, USA.

I. OLIVER
School of Biological Sciences, Macquarie University, NSW 2109, Australia.

PJOTR OOSTERBROEK
Institute of Taxonomic Zoology, University of Amsterdam, Plantage Middenlaan 64, 1018 DH Amsterdam, The Netherlands.

MÁRIO C. C. DE PINNA
Departamento de Zoologia, Universidade de São Paulo, Caisca Postal 11461, São Paulo-SP 0522–970, Brazil.

G. T. PRANCE
Royal Botanic Gardens, Kew, Richmond, Surrey TW9 3AB, UK.

J. R. PRENDERGAST
NERC Centre for Population Biology, Imperial College at Silwood Park, Ascot, Berkshire SL5 7PY, UK.

J. R. PRESS
European Plant Information Centre, Department of Botany, The Natural History Museum, London SW7 5BD, UK.

R. L. PRESSEY
New South Wales National Parks and Wildlife Service, PO Box 1967, Hurstville, NSW 2220, Australia.

A. G. REBELO
Conservation Biology Research Unit, National Botanical Institute, Kirstenbosch, Claremont 7735, South Africa.

WALTER V. REID
World Resources Institute, 1709 New York Avenue, NW, Suite 700, Washington, DC 20006, USA.

B. J. RICHARDSON
Faculty of Science and Technology, University of Western Sydney, Hawkesbury, Bourke St, Richmond, NSW 2753, Australia.

P. G. RYAN
Percy FitzPatrick Institute for African Ornithology, University of Cape Town, Rondebosch 7700, South Africa.

W. R. SIEGFRIED
Percy FitzPatrick Institute for African Ornithology, University of Cape Town, Rondebosch 7700, South Africa.

C. R. SMITH
Biogeography and Conservation Laboratory, Department of Entomology, The Natural History Museum, Cromwell Road, London SW7 5BD, UK.

MELANIE L. J. STIASSNY
Department of Herpetology and Ichthyology, American Museum of Natural History, New York, NY 10024, USA.

N. E. STORK
Biodiversity Division, Department of Entomology, The Natural History Museum, London SW7 5BD, UK.

S. J. THIRGOOD
BirdLife International (formerly International Council for Bird Preservation), Wellbrook Court, Girton Road, Girton, Cambridge CB3 0NA, UK.

R. I. VANE-WRIGHT
Biogeography and Conservation Laboratory, Department of Entomology, The Natural History Museum, Cromwell Road, London SW7 5BD, UK.

JUDY G. WEST
Australian National Herbarium, CSIRO Division of Plant Industry, GPO Box 1600, Canberra ACT 2601, Australia.

M. WEST
European Plant Information Centre, Department of Botany, The Natural History Museum, London SW7 5BD, UK.

P. H. WILLIAMS
Biogeography and Conservation Laboratory, Department of Botany, The Natural History Museum, London SW7 5BD, UK.

Abbreviations

ABIS	Australian Biogeographic Interchange Standards
ABRS	Australian Biological Resources Study
APNI	Australian Plant Names Index
BGCI	Botanical Gardens Conservation International
BGCS	Botanical Gardens Conservation Secretariat (now BGCI)
BIOCLIM	Bioclimatic Modelling Software (Nix, H. A. 1986. In: Atlas of Elapid Snakes (ed. R. Longmore). Pp. 4–15. Australian Flora and Fauna Series No. 7. Australian Government Publishing Service, Canberra
BioNET	Biological Resources Network, promoted by CAB International
BRC	Biological Records Centre
CAB International	Centre for Agriculture and Biosciences
CAP	Census of Australian Plants
CAUP	Census of Australian Plants
CAVS	Census of Australian Vertebrate Species
CCA/IRF	Caribbean Conservation Association/Island Resources Foundation
CGIAR	Council of Genetic Agricultural Research
CODA	Conservation Options and Decision Analysis
CSIRO	Commonwealth Scientific and Industrial Research Organisation
DIVERSITAS	IUBS-SCOPE-UNESCO Programme on Biological Diversity
ERIN	Environmental Resources Information Network
ETI	Expert-Centre for Taxonomic Information
FAO	Food and Agriculture Organisation
FI	herbarium acronym for the Florence herbarium
GIS	Global Information System
HISPID	Herbarium Information Standards and Protocol for Interchange of Data
IBPGR	International Board for Plant Genetic Resources
ICBP	International Council for Bird Preservation (now BirdLife International)

INBio	Institutio Nacional de Biodiversidad
INTKEY	Identification Programme in DELTA (DEscription Language for TAxonomy)
IOPI	International Organization for Plant Information
IGBP	International Geosphere Biosphere Program
IUCN	International Union for the Conservation of Nature
IUBS	International Union of Biological Sciences
MAB	Man And the Biosphere
NHM	The Natural History Museum, London
ONC	Ordinance Navigational Chart
PNAP	Protected Natural Areas Programme
ROTAP	Committee on Rare Or Threatened Australian Plants
SCOPE	UNESCO Scientific Committee On Problems of the Environment
SIS	Spatial Information System
TDWG	Taxonomic Databasing Working Group for Plant Sciences
UNEP	United Nations Educational Programme
UNESCO	United Nations Educational, Scientific and Cultural Programme
WCMC	World Conservation Monitoring Centre
WRI	World Resources Institute
WWF	World Wildlife Fund

1. Setting objectives for conservation evaluation

WALTER V. REID
World Resources Institute, 1709 New York Avenue, NW, Suite 700, Washington, DC 20006, USA

Abstract

This paper questions whether we have the means of measuring the objective of biodiversity conservation. For decades, the continuum from 'natural' to 'disturbed' ecosystems has underlain ecological management goals. Now, advances in ecology, paleobiology, and conservation biology are questioning the very meaning of a 'natural' ecosystem. Today's goals of ecological management were developed by ecologists who believed that communities were comprised of relatively stable assemblages of species, that population sizes tended towards equilibrium levels, and that humans played a relatively minor role in shaping the structure of communities. Scientists have now modified or abandoned these assumptions, yet their policy implications remain enshrined in management philosophy. It is time to rethink the fundamental concerns of ecological management. It will be shown that although objectives such as 'ecological health' and 'biological integrity' serve important political and educational roles, they cannot be formulated into scientifically defensible biological indicators. More generally, the unpredictability of the behaviour of ecological systems and our lack of knowledge of the predictable aspects suggests that our goals fall short of the basic concern of ensuring that the planet maintains its capacity to support life, including humanity. The principal goal of ecological management should be social, to maximize human capacity to adapt to changing ecological conditions. Specific objectives can then be defined in biological terms. In order to adapt to change, humanity needs both the diversity from which new innovations can be created, and the productive ecological systems that provide biological and economic capital to invest in those innovations. Thus, maintaining biodiversity is a prerequisite for maximizing human's ability to respond to changing conditions, as is maintaining the productivity of agricultural systems, the yield from forests and fisheries, clean water, and clean air.

Systematics and Conservation Evaluation (ed. P. L. Forey, C. J. Humphries, and R. I. Vane-Wright), Systematics Association Special Volume No. 50, pp. 1–13, Clarendon Press, Oxford, 1994.

Introduction

At the Earth Summit in 1992, the more than 150 nations that signed the International Convention on Biological Diversity signalled their commitment to the conservation of biodiversity and their obligation to monitor its status. But do we yet have the means of measuring whether the objective of biodiversity conservation is being achieved? Indeed, do we even know what the objective of biodiversity conservation should be?

With the growing awareness of the importance of biodiversity conservation and the increased willingness of governments to take conservation action, it is more urgent than ever that conservation objectives are well-defined and the means exist to evaluate progress toward those objectives. But the determination of conservation objectives is fraught with difficulty. The objectives of conservation are inextricably linked to our knowledge of ecological systems, and that knowledge is undergoing profound change.

For decades, the continuum from 'natural' to 'disturbed' ecosystems has underlain ecological management objectives. At one end of the spectrum, human disturbance is minimized or allowed only when necessary to recreate natural conditions (e.g. controlled burning). At the other end, ecosystems are managed for extractive uses. Management objectives for natural or semi-natural ecosystems are to maintain biodiversity, maintain ecological services such as flood control and pollution filtration, and to meet demands for recreation or tourism. For extractive protected areas, primary objectives include the production of timber, crops, or livestock.

But advances in the sciences of ecology, palaeobiology, and conservation biology are calling into question the very meaning of a 'natural' ecosystem—traditionally considered to be an ecosystem undisturbed by humanity. In eastern North America, for example, Native Americans' use of fire had modified ecosystems for millennia before Europeans arrived. Excluding human intervention today might produce a different unnatural state, but it certainly could not produce a 'natural' state. And while it might be possible to recreate historic disturbance regimes, how could it be determined whether the resulting system was 'natural'? Ecological systems are dynamic—changing continually through time. Thus, we can be sure that whatever community existed 500 years ago is not what would exist today under 'natural' conditions, but we have no objective means of determining what today's 'natural' state would be.

Many of today's objectives for ecological management were developed at a time when ecologists believed that communities were comprised of relatively stable assemblages of species, that population sizes tended toward equilibrium levels, and that humans played a relatively minor role in shaping the structure of communities. Today, scientists have modified or abandoned these assumptions; yet their policy implications remain enshrined in management practices.

Clearly it is time to rethink the fundamental concerns of ecological management. This chapter explores the heritage of our current ecological management concerns as they are typically applied to public land management in the United States, and examines how well those concerns hold up in view of changes in both ecological science and conservation philosophy. It briefly reviews several other possible management objectives, including 'ecological health' and 'biological integrity', and proposes a different objective more in keeping with current understanding of ecology and evolutionary biology.

A linguistic and management legacy

The roots of current land management policies in the United States trace back to two schools of conservation thought formed in the late 1800s and early 1900s. On the one hand, such founders of the wilderness movement as George Perkins Marsh, Henry David Thoreau, and John Muir saw conservation as an alternative to development (Worster 1977). To them, the value of conservation was spiritual, not economic. When Muir lost the battle for preserving Hetch Hetchy valley in California, for example, he wrote:

> These temple destroyers, devotees of ravaging commercialism, seem to have a perfect contempt for Nature, and instead of lifting their eyes to the God of the mountains, lift them to the Almighty Dollar (Fox 1981, p. 142).

In contrast, Gifford Pinchot and Theodore Roosevelt saw conservation in utilitarian terms. For Pinchot, conservation ensured the continued production of valuable biological resources. Foreshadowing current debates over sustainable forestry and the value of non-timber forest products, Pinchot wrote:

> The purpose of Forestry, then, is to make the forest produce the largest possible amount of whatever crop or service will be most useful, and keep on producing it for generation after generation of men and trees (Pinchot 1947, p. 31).

Two objectives of public land management grew out of these contrasting views of conservation. In the tradition of Thoreau and Muir, the US National Park system was established to 'conserve the scenery and the natural and historic objects and the wildlife therein, so that they can be enjoyed by this and future generations' (Congress of the United States 1916, p. 73). Pinchot's philosophy was expressed in the establishment of the Forest Reserve system, designed to maximize the productivity of those major resources in which people had a clear, direct, and immediate interest (Worster 1977). In this same tradition, wildlife refuges were created to maximize the production of game species for hunters, often through the use of intensive predator-control programmes.

By the middle of the century, however, the sharp distinction between utilitarian and non-utilitarian conservation goals began to fade with the growth of the science of ecology. Leopold's book *Game Management* (1933), for example, provided a scientific underpinning for early wildlife management efforts. But the 1940s and 1950s saw the field drop its natural history orientation and become a more empirical and experimental science. Led by the detailed empirical field studies of Elton (1924), Lack (1954), and others, and MacArthur's (1955, 1972) application of basic mathematical models to these field observations, ecology became a deductive science.

One of the first achievements of the revitalized science of ecology was to provide strong empirical support to the notion of a 'balance of nature'. Both population models and long-term datasets supported the idea that population sizes tended toward equilibrium values, and suggested that communities exhibited characteristics like 'stability' and 'resilience' (Odum 1953; MacArthur 1955). Guided by a mechanical metaphor of ecological dynamics (Botkin 1990), human disturbance could be seen to upset the stability of an ecological system. Thus, a 'natural' state could be achieved simply by removing human disturbance. Without today's palaeobiological evidence, which demonstrates the dynamic nature of community composition, the notion that biological communities represented stable associations of co-evolved species was a strong one.

If nature existed in a balance, then the consequences of disturbing that balance were obviously of concern. Conservationists and scientists argued that the disturbance of an ecosystem must be kept within bounds or the system could collapse (Holling 1973). Moreover, the preservation of species diversity emerged as one of the key elements thought to maintain ecosystem stability. Backing up intuition with a simple mathematical model, MacArthur showed that the appealing concept of a 'web of life' made scientific sense (MacArthur 1955). The greater the diversity of species in an ecosystem, the more stable the system was likely to be. [Although ecologists had discarded this idea of a direct relationship between diversity and stability by the 1970s (May 1973; Goodman 1975), it is probably too useful for environmentalists ever to reject.]

The combination of scientific support for the 'balance of nature' and the apparent risks of destabilizing ecosystems thus cast the preservation of natural ecosystems and wilderness in a very different light. Now, protecting habitats and ecosystems from human disturbance was not only in keeping with the spiritual values of Muir, but made sound ecological sense as well. Where predators had been the target of extensive control programmes in the early part of the century, their protection could now be justified by their critical role in ecosystem function (Worster 1977). Wilderness areas became more than recreation areas; they now were seen as some of the 'healthiest' of ecosystems and thus best able to provide such ecological services as clean water, flood control, and habitat for commercially valuable species.

This emergence of what Worster (1977, p. 284) refers to as the 'Age of Ecology' helped coalesce the utilitarian and non-utilitarian conservation rationales. This evolution of conservation philosophy is well-reflected in both the life and writings of Aldo Leopold. Raised in Pinchot's school of utilitarian conservation, by 1947 Leopold had changed his own beliefs to a biocentric view of conservation that saw value in nature for nature's sake. But Leopold stressed the ecological underpinnings of both utilitarian and biocentric views of conservation:

> a system of conservation based solely on economic self-interest is hopelessly lopsided. It tends to ignore, and thus eventually to eliminate, many elements in the land community that lack commercial value, but that are (as far as we know) essential to its healthy functioning. It assumes, falsely, I think, that the economic parts of the biotic clock will function without the uneconomic parts (Leopold 1966, p. 251).

Conserving natural systems and maintaining natural ecological processes was good not only for the spirit; it was also an essential means of avoiding ecological degradation that would undermine human development. The arrival of the Age of Ecology set the stage for the development of the concept of 'protected areas'. Instead of viewing National Parks as an alternative to development, they could be seen as one of many different types of public land management serving development needs. At one end of the continuum 'strictly protected areas' like National Parks were kept free of all human disturbance and maintained in a natural state to meet one set of conservation objectives (recreation, species protection, etc.), while at the other end protected areas were managed for such extractive uses as timber harvest. At the Second World Conference on National Parks, held in 1972, this new view of National Parks serving development, rather than simply protecting the 'heritage' of a nation was still the minority (Hales 1989). But by the time of the third conference in 1982 (now called the World National Parks Congress) the Protected Area concept was moving into the mainstream. The pioneering work of Dasmann (1973) and Miller (1975) had provided the conceptual basis for establishing protected area 'categories', and the publication of the World Conservation Strategy (IUCN, UNEP, WWF 1980) placed protected area conservation squarely in the mainstream of 'sustainable development'. The Third World National Parks Congress, entitled 'Parks for Sustainable Development', led to the adoption of an eight-category system of protected areas by IUCN (IUCN Commission on National Parks and Protected Areas 1984), and by 1992 the Congress was renamed the World Congress on National Parks and Protected Areas.

Implicit in the concept of protected areas is the idea that the continuum of management objectives represents differing levels of permissible human intervention and, thus, differing approximations to a 'natural' ecosystem. Yet, where the definition of a 'natural ecosystem' seemed straightforward given assumptions of equilibrium co-evolved communities, that assumption was being overturned by ecologists just as managers were succeeding in translating it into policy.

Thirty years ago science supported the notion that a natural ecosystem could be defined on the basis of its appearance. This is clearly seen in the 'Leopold Report' published in 1963 (Advisory Board on Wildlife Management 1963). This report, requested by Secretary of the Interior of the United States, Stewart Udall, and developed by a committee chaired by A. Starker Leopold—Aldo Leopold's son—provided recommendations for game management in National Parks. The report recommended that the primary goal of wildlife management should be to maintain or recreate the biotic associations in each park 'as nearly as possible in the condition that prevailed when the area was first visited by the white man' (Advisory Board on Wildlife Management 1963, p. 4). The report stated that 'A national park should represent a vignette of primitive America', and that 'Above all other policies, the maintenance of naturalness should prevail' (Advisory Board on Wildlife Management 1963, pp. 4, 9).

Prior to this report's release, Park management had focussed largely on protecting species from human disturbance, including hunting and fire. In contrast, the Leopold report argued that this management practice created unnatural conditions since fire was common under 'natural' conditions and since ecosystems were already disturbed through the extinction of large predators or the fragmentation of habitat. The report drew on the assumption that a balance of nature did exist and proposed that management interventions should be designed to restore that balance. It suggested that the state of the system when Europeans first arrived could be considered 'natural' and adopted as a target.

Although maintaining an ecosystem in the condition that existed when Europeans first arrived might still represent a rational management objective, we now know that it is anything but natural. In the case of National Parks, the choice of the reference point for recreating a 'natural' system could not have been worse. In advance of the white man's first visit to the regions that became our national parks diseases, that swept across the continent after the first contact with early explorers, had devastated Native American populations (Crosby 1986). Thus, neither the ecological communities nor the apparent interaction between humans and these communities could have been considered natural when Europeans first settled.

The ecological justification for recreating a previous set of ecological conditions has been eroded by the findings of the dynamic changes in community composition (Hunter *et al.* 1988) and the 'non-equilibrium' state of communities. Thus, even if it were possible to define the 'natural' community that existed in 1600, it is a safe bet that it would be quite different today, with or without humans on the continent. This consideration has led to revised National Park Service policies under which 'change [will] be recognized as an integral part of the functioning of natural systems. The National Park Service will not seek to preserve natural systems in natural zones as though frozen at a given point in time' (NPS 1988).

Objectives for ecological management

The fundamental changes that have taken place in our understanding of ecological systems over the four decades since Leopold wrote *A Sand County Almanac* have made the objective of conservation more elusive. The Age of Ecology had put the strength of science behind Leopold's (1966, p. 262) credo that 'a thing is right when it tends to preserve the integrity, stability, and beauty of the biotic community. It is wrong when it tends to do otherwise.' Given the assumption of the balance of nature and the reliance on mechanical metaphors of stability and integrity, it was relatively easy to see the human modification of ecosystems as necessary evils that had to be kept to a minimum lest the stability of these systems be disrupted.

But if communities are not in balance and no 'natural' state can be defined, then what should the primary concern of ecological management be? Intuition suggests that there must be some fundamental concern that can guide the relationship between humanity and ecological systems, some basic objective that addresses our underlying sense that the biosphere's capacity to support life, including humanity, must be maintained and that it must contribute to sustainable development.

Increasingly, that underlying concern is being seen to be the concept of 'ecosystem health' or 'biological integrity' (Noss 1990). Biological integrity is defined as the ability to support and maintain 'a balanced, integrated, adaptive community of organisms having a species composition, diversity, and functional organization comparable to that of natural habitat of the region' (Karr 1991). An ecosystem can be considered healthy when its capacity for self-repair when perturbed is preserved and minimal external support for management is needed (Karr 1991).

Ecosystem health and biological integrity are macro-descriptors that define a state of nature against which to measure degradation. They appear to open the way to developing biological indicators that can reveal the basic condition of our ecological systems. And they appear to provide a basic objective of ecological management that will subsume other conservation needs. For example, Walker (1992, p. 18) suggests that 'The best way to minimize species loss is to maintain the integrity of ecosystem function.'

Unfortunately, the conceptualization of biological integrity and ecosystem health as ecological management objectives fits poorly with ecological reality. While it is entirely appropriate for political or educational reasons to seek to maintain a 'healthy ecosystem', just as we seek a 'strong economy', the former provides no more guidance to the biological manager than the latter provides to the economist. Ecologists seem to be bent on exhausting every possible mechanical and medical metaphor before recognizing the unique nature of ecological systems. Concepts such as inertia, elasticity, amplitude, hysteresis, and stability (Westman 1978) have no doubt helped ecologists frame questions about ecological processes, but none has ever stood the test of time as a biological indicator. Yet these concepts still live on and guide policy, even though their meaning is unclear. Walker (1992, p. 22), for example, notes that 'The worrisome cost of decline in biodiversity, particularly to politicians who may be held accountable, is the threat of a collapse in the "stability" of ecosystems (whatever that means).'

Just as ecologists sought measures of stability and failed, it seems that the metaphors of health and integrity are also doomed to failure. Establishing when an ecosystem is 'sick' or disintegrating is entirely subjective. Is the eastern deciduous forest of North America more healthy today than it was before climate change led to the migration of oak and hemlock into the community 10 000 years ago? Did the loss of the chestnut from those forests decrease the biological integrity of the system?

A sense of objectivity can be achieved if we assume that any change from the 'natural condition' is evidence of declining health or loss of biological integrity, but this founders on the same shoals as traditional management objectives. How would we objectively define a 'natural' state? Recognizing the difficulty of defining a natural ecosystem by its composition, several attempts have been made to redefine it based on 'process'. Prescott-Allen (1991) defines a natural ecosystem as one where since the industrial revolution (1750) human impact (a) has been no greater than that of any other native species, and (b) has not affected the ecosystem's structure. Botkin (1990, p. 157), stresses that we must appreciate that change is 'natural, desirable, and acceptable', implying that natural ecosystems are those undergoing natural rates of change. But in either case, by recognizing man's role in 'nature' and in particular by giving man a role in maintaining 'natural' ecosystems, the very definition of 'nature'—that part of the material world unaffected by man—is contradicted.

The continuing pursuit of ecological management objectives related to ecosystem health and biological integrity reflects an undying faith among ecologists that we can ultimately determine what 'makes the system tick'. Guided throughout our careers by Leopold's admonition 'to keep every cog and wheel is the first precaution of intelligent tinkering' (Leopold 1966, p. 190), we continue to search for the mainspring even though it is now clear that the system does not tick at all—it undergoes both sudden and gradual change through both deterministic and chaotic processes.

Clearly, efforts to prescribe objectives for ecological management in terms of the 'state' of ecological systems will not succeed simply because ecological systems are neither static nor balanced. But what are we left with? Though we may not be able to define the state of the system, we must pursue some management objective that will ensure that the regional and global biota maintains its capacity to support life. And we can be relatively certain of the targets of ecological management even if we have difficulty defining the objective. We clearly want to maintain ecological services important to humanity, and we want to maintain biodiversity to keep options open for future generations.

One key objective could be to preserve the status quo. Given that we have little ability to predict how a change in an ecological system will affect ecological services and biodiversity conservation, we could simply acknowledge that any change carries with it a risk, and a good risk-averse strategy is to do our best not to change the existing system. In a sense, this approach would be functionally no different than traditional objectives of maintaining natural ecosystems. We would accept whatever now exists as what we should maintain, since it is providing valuable services.

The problem with the 'if it ain't broke, don't fix it' approach is, of course, that the system is already broken. Declining biodiversity, declining soil fertility, water pollution, air pollution, and so forth all suggest that the status quo is insufficient.

Alternatively, we could state the management objective in terms of maintaining ecological processes rather than maintaining particular ecological states. For instance, we could seek to establish 'evolutionary preserves'—areas in which natural evolutionary processes continue unabated. A process approach would also frame objectives in terms of ensuring that valuable ecological services are not degraded. But again, how would we judge what is an impairment of the process and what is not? Certainly, evolution would be more rapid in a fragmented habitat: so should an evolutionary preserve be large in order to slow species extinction, or small to promote speciation? And how should the goal of maintaining ecological services address the impacts of climate change, which will presumably fundamentally alter the location of various services from ecological systems.

The pursuit of a fundamental ecological management objective is doomed to failure because of the unpredictability of the behaviour of ecological systems, and because of our lack of knowledge of those aspects of the behaviour of ecological systems that are predictable. These two characteristics ensure that objectives related to ecological states, ecological processes, or macrodescriptors combining states and processes, will fall short of the more basic concern. We know enough to be sure that there is nothing ecologically profound about what we have called 'natural systems' (and they cannot be defined anyway), but we will never know enough to be sure what the 'best' state is or what the impacts of change will be. This suggests that the objective of ecological management should not be ecological, but social: the overarching objective of ecological management should be to maximize human capacity to adapt to changing ecological conditions. This may seem overly-anthropocentric, but it is entirely consistent with a biocentric outlook. To maximize human capacity to adapt to changing conditions we must also maximize life's (or, the biosphere's) capacity to adapt to change.

In practice, the adoption of such a management objective would have relatively little effect on current management targets. To adapt to change, humanity needs the diversity from which new innovations can be created, and needs productive ecological systems that provide the biological and economic capital to invest in making those innovations. Thus, maintaining biodiversity is a prerequisite for maximizing humans' ability to respond to changing conditions, as is maintaining the productivity of agricultural systems, the yield from forests and fisheries, clean water, and clean air. Similarly, the three requirements of biodiversity conservation stated in the Global Biodiversity Strategy—save biodiversity, study it, and use it sustainably and equitably—are all key elements of a strategy to maximize human adaptability (WRI, IUCN, UNEP 1992).

But, the adoption of such a management objective would have two effects. First, it would change the justification used for management goals. Rather than having recourse to concepts of biological integrity, ecosystem health, or ecological stability, justifications would be cast in terms of economic, social, and ethical needs and responsibilities. We would justify the maintenance of species within a protected river system because we have an ethical responsibility to future generations, not to maintain the system's 'health'. Or, we might seek to maintain those species because we know that their loss would impair important ecological services (or, because we do not know enough to judge whether it would), not because we are worried about the biological 'integrity' of the system.

Second, such an objective would place well-deserved emphasis on the fundamental importance of change as a characteristic of ecological systems and as a focus of management concern. The objective implies that a premium should be placed on policies geared toward flexibility and adaptation to change, rather than direct regulation geared toward static targets. For example, why not make greater use of development fees to reduce the rate of loss of habitats that are valuable for maintaining biodiversity and ecological services? Perhaps all development within five meters of sea level, or within 'ecologically sensitive areas' as identified through Gap Analysis, should be assessed a fee (which could be used to subsidize development in areas with less value for biodiversity or ecological services). Similarly, greater use of public funds for purchasing conservation easements—and defraying lost property taxes for local communities—can help management move beyond protected areas.

Conclusions

For over three decades, a coalescence of utilitarian and non-utilitarian conservation philosophy—guided by now-dated ecological theory—has helped push conservation goals into the mainstream of development planning. But this coalescence was based in part on ecological misunderstandings. Using ecological theories based on ecosystem stability, biological integrity, and the assumption that natural ecosystems provide higher quality services, wilderness preservation could be justified on the very same economic grounds as resource exploitation.

But both economic reality and ecological knowledge changed dramatically over those decades. Today, with the economic value of US National Forests to wilderness and recreational users some nine times the aggregate value of round wood timber harvest, an economic case for wilderness conservation hardly needs to rest on ecological theories (Bergstrom and Cordell 1991). At the same time, the ecological support for preservationist goals has weakened considerably.

Abandoned by ecologists who can no longer provide firm scientific support for strict preservation, 'deep ecologists' have reclaimed the moral high ground and rightfully reasserted the spiritual value of wilderness and the value of wilderness for its own sake. There is indeed value in leaving portions of the earth surface untouched by deliberate human management. This value can be defended with science—providing a base line against which the effects of human intervention can be measured (Botkin 1990)—or philosophy: 'in wildness is the preservation of the world' (Thoreau, cited in Chase 1987). There is no need to justify such areas as 'natural'. (Indeed, they are likely to become among the least 'natural' as exotics invade, and habitat fragmentation and changes in disturbance regimes take their toll on native species.)

Recognizing these changes, it is time to adopt a new objective for ecosystem management: maximizing the human capacity to adapt to changing ecological conditions. There is no point in continuing to pursue a 'correct' state of ecological systems as an objective of management when it is clear that ecosystems are both dynamic and subject to random changes. There is every reason, however, to seek to manage ecological systems—including areas left untouched by humanity—to ensure that humanity will be capable of responding to change.

Acknowledgments

This article is adapted from the paper 'Beyond protected areas: setting objectives for ecological management' prepared for the *Symposium on Biodiversity in Managed Landscapes*, held in Sacramento, California in July 1992. I would like to thank Jeffrey McNeely for providing helpful comments on an earlier draft of this manuscript, and Kathleen Courrier for her help in editing the manuscript.

References

Advisory Board on Wildlife Management (1963). *Wildlife management in the national parks*. A. S. Leopold (chairman). Appointed by Secretary of the Interior Udall, March 4, 1963.

Bergstrom, J. C. and Cordell, H. K. (1991). An analysis of the demand for and value of outdoor recreation in the United States. *Journal of Leisure Research*, **23**, 67–86.

Botkin, D. B. (1990). *Discordant harmonies*. Oxford University Press, New York.

Chase, A. (1987). *Playing God in Yellowstone: the destruction of America's first national park*. Harcourt Brace Jovanovich, New York.

Congress of the United States (1916). The National Park Service Act. *US Statutes at Large*, **28**, 73.

Crosby, A. W. (1986). *Ecological imperialism*. Cambridge University Press.

Dasmann, R. F. (1973). A system for defining and classifying natural regions for purposes of conservation. *IUCN Occasional Paper*, **7**, 1–18.

Elton, C. (1924). Periodic fluctuations in the numbers of animals: their causes and effects. *British Journal of Experimental Biology*, **2**, 119–63.

Fox, S. (1981). *John Muir and his legacy: the American conservation movement*. Little, Brown, Boston.

Goodman, D. (1975). The theory of diversity-stability relationships in ecology. *Quarterly Review of Biology*, **50**, 237–66.

Hales, D. (1989). Changing concepts of National Parks. In: *Conservation for the twenty-first century* (ed. D. Western and M. Pearl), pp. 139–44. Oxford University Press, New York.

Holling, C. S. (1973). Resilience and stability of ecological systems. *Annual Reviews of Ecology and Systematics*, **3**, 33–50.

Hunter, M. L., Jacobson, G. L., and Webb, T. (1983). Paleoecology and the coarse-filter approach to maintaining biological diversity. *Conservation Biology*, **2**, 375–85.

IUCN Commission on National Parks and Protected Areas. (1984). In *National parks, conservation, and development: the role of protected areas in sustaining society* (ed. J. A. McNeely and K. R. Miller), pp. 47–55. Smithsonian Institution Press, Washington, DC.

IUCN, UNEP, WWF (1980). *World conservation strategy*. Morges, Switzerland.

Karr, James R. (1991). Biological integrity: a long-neglected aspect of water resource management. *Ecological Applications*, **1**, 66–84.

Lack, D. (1954). *The natural regulation of animal numbers*. Oxford University Press.

Leopold, A. (1933). *Game management*. Scribner, New York.

Leopold, A. (1966) [first published in 1949 by Oxford University Press]. *A Sand*

County almanac: with essays on conservation from Round River. Ballantine Books, New York.

MacArthur, R. H. (1955). Fluctuations of animal populations, and a measure of community stability. *Ecology*, **36**, 533–6.

MacArthur, R. H. (1972). *Geographical ecology*. Harper and Row, New York.

May, R. M. (1973). *Stability and complexity in model ecosystems*. Princeton University Press.

Miller, K. R. (1975). Guidelines for the management and development of national parks and reserves in the American humid tropics. In *Proceedings of a meeting on the use of ecological guidelines for development in the American humid tropics, Caracas 20–22 February, 1974*, pp. 94–105. IUCN, Switzerland.

NPS [USDI National Park Service]. (1988). *Management policies*. US Department of Interior, National Park Service, Washington, DC.

Noss, R. F. (1990). Can we maintain biological and ecological integrity? *Conservation Biology*, **4**, 241–43.

Odum, E. P. (1953). *Fundamentals of ecology*. Saunders, Philadelphia.

Pinchot, G. (1947). *Breaking new ground*. Harcourt Brace Jovanovich, New York.

Prescott-Allen, R. (1991). Classification of ecosystem conditions. In *IUCN, UNEP, WWF 1991. Caring for the Earth: a strategy for sustainable living*, p. 34. IUCN, Gland, Switzerland.

Walker, B. H. (1992). Biodiversity and ecological redundancy. *Conservation Biology*, **6**, 18–23.

Westman, W. E. (1978). Measuring the inertia and resilience of ecosystems. *Bioscience*, **28**, 705–10.

Worster, D. (1977). *Nature's Economy*. Cambridge University Press.

WRI, IUCN, UNEP. (1992). *Global biodiversity strategy: guidelines for action to save, study and use Earth's biotic wealth sustainably and equitably*. World Resources Institute, Washington, DC.

2. The measurement of biodiversity and the politics of implementation

V. H. HEYWOOD
IUCN — The World Conservation Union, Plant Conservation Office, Richmond TW9 3BW, UK

Abstract

The reason for failure on the part of many agencies and governments to recognize biodiversity as a serious issue stems from the difficulties of defining it so that it can be applied to real problems of management, regulation, and monitoring. Biodiversity is viewed differently by community ecologists, park managers, plant breeders, population geneticists, agriculturalists, conservation biologists, and taxonomists. This is not a weakness provided that no one group insists on pre-eminence of their sectoral interest: on the contrary, having broad constituency strengthens both the concept and the likelihood of support. On the other hand it is critical for biodiversity conservation to appreciate that different objectives will require the use of different measures. This paper will examine some of the problems in defining biodiversity at its various levels. It is conventional to consider biodiversity at ecosystem, species, and genetic levels—although an alternative approach recognizes compositional, structural, and functional biodiversity, each encompassing multiple levels of organization. The problems of measuring biodiversity at each of the three main levels are discussed particularly with respect to the difficulties of measuring taxonomic diversity and its application to conservation. It will be argued that to achieve many of the declared conservation objectives, a multiple level strategy of measuring biodiversity is required.

Introduction

There is a broad consensus today that biodiversity should be protected, but it is not at all clear what this means in policy and practical terms. This

Systematics and Conservation Evaluation (ed. P. L. Forey, C. J. Humphries, and R. I. Vane-Wright), Systematics Association Special Volume No. 50, pp. 15–22, Clarendon Press, Oxford, 1994.

uncertainty is in part a consequence of divergent views as to the exact meaning of biological diversity. Indeed the public might be forgiven for thinking that it refers to tropical forests full of myriad insects, colourful birds, bizarre animals, and plants containing potential new wonder drugs! Part of the reason for the failure on the part of many agencies and governments to recognize biodiversity as a serious issue stems from the difficulties of defining it in such a way that is both measurable and operational so that it can be applied to real world problems of management, regulation, and monitoring. Moreover, it is viewed differently by different interest groups such as community ecologists, park managers, plant breeders, population geneticists, agriculturalists, conservation biologists, taxonomists, and so on. This is not necessarily a weakness provided that no one group insists on the pre-eminence of their sectoral interest: on the contrary, having a broad constituency strengthens both the concept and the likelihood of support. On the other hand it is critical for the conservation of biodiversity to appreciate that different objectives will require the use of different measures as Reid (this volume) has stressed. It is important that we recognize that while there is a broad general agreement that biodiversity exists at three levels—ecosystem, species, and genetic—this is a gross oversimplification and makes a highly complex situation seem deceptively simple (cf. McNaughton, this volume). Indeed at each level there are myriad problems that only now are we beginning to understand.

This debate also comes at a time when an increasing number of authors are questioning the wisdom or practicality of focussing attention on species. Walker (1992), for example, suggests that conservationists should spend less time worrying about the persistence of particular plant or animal species and begin to think instead about maintaining the nature and diversity of ecosystem processes. Yet the effective functioning of ecosystems depends on the successful maintenance of their component species, especially those that play a critical role (so-called keystone species), assuming that one can determine which these are. Moreover, species are one of the commonest measures of biodiversity and the most frequently voiced unit of concern. They are also the focus of much conservation action and policy as witness, for example, the Endangered Species Act of the USA, the Species Survival Commission (one of the most successful operations of IUCN), and the Bern Convention of the Council of Europe. Much of the clamour of concern about biodiversity has been about the large numbers of species that many reckon to be threatened with extinction (often ignoring the even more widespread occurrence of genetic erosion within species). The conservation movement has been very adept at producing lists but, as John Lawton has said, most ecologists have so far produced telephone directories; what we need now is yellow pages. On the other hand, yellow pages in the form of the protected areas system have been developed worldwide, but often in the absence of any knowledge of the species actually being protected in the parks or reserves concerned.

Norton and Ulanowicz (1992) have recently suggested that this conflict can be resolved by adopting a hierarchical approach to biodiversity conservation policy: we can, they say, save species both because we value them directly (at least in many cases) and because of their roles in ecosystem processes.

Another important dimension that can be overlooked when considering the problems of biodiversity protection in global terms is, as Common and Norton (1992) have recently pointed out, that addressing the conservation of biological diversity is essentially a political problem with the relevant political unit being the nation state. However, there are obviously global dimensions and these require agreements between nations, as indeed the signature of the Convention on Biological Diversity agreed at Rio de Janeiro in June 1992 by many governments indicated. All these issues and problems will have to be addressed in the wake of the meeting of the United Nations Commission on Environment and Development at Rio and solutions found as the agreements worked out there come to be implemented.

Definitions of biological diversity

The measurement of biodiversity is highly complex. It is conventional to consider biodiversity at three basic levels—ecosystem, species, and genetic—although there is not always agreement as to the composition of these levels. An alternative approach is proposed by Noss (1990), who recognizes compositional, structural, and functional biodiversity, each encompassing multiple levels of organization. Another useful distinction is between alpha-diversity and beta-diversity. The problems of measuring biodiversity at each of the three main levels are discussed below and in particular the difficulties of measuring taxonomic diversity. From a conservation point of view, there are serious practical and political problems which bear on the measurement of biodiversity. For example, since resources do not permit all species to be conserved, how should priorities be established? Various approaches will be considered: one is to identify keystone, flagship, umbrella, or indicator species; another is to focus on taxonomic or phyletic distinctiveness; yet another is to focus on degree of threat—Endangered, Rare, Vulnerable, etc according to the IUCN categories (although these are currently under review). The question of redundancy and how to measure it has also to be addressed—by ecological, taxonomic, phyletic, economic, social, and other criteria. To achieve many of the declared conservation objectives, a multiple level strategy of measuring biodiversity is needed.

The three conventionally recognized levels of biodiversity will be considered in more detail.

1. *Level 1: landscape, ecosystem, and habitat diversity*

Landscape is sometimes recognized as separate and more comprehensive than habitat or ecosystem diversity. In both cases there are problems of definition, whether in terms of composition, structure, or function. Indeed different methods of classification are based on which of these aspects is given priority. Thus landscapes and ecosystems can be classified in terms of physiognomy (e.g. the Clements or Tansley schools of vegetation ecology), phytosociology (various systems including the much-debated Zurich-Montpellier school of vegetation classification), or function, which itself can be ecophysiological or in terms of land use (as is often preferred by geographers, such as Dansereau 1957).

Also, biodiversity in terms of species will also vary according to the successional status of the ecosystem concerned. As Kemp (1992) has noted:

> genetic diversity is a product not only of the number of species in an area but of successional change, from the colonization of gaps by pioneer species to the mature climax forest. The most species-rich areas are likely to be those including secondary forest in various stages of recovery.

In fact, maintenance of an ecosystem at a pre-climax stage may be a necessary managment tool to conserve particular species populations *in situ*.

Recognition of the landscape as a separate level is desirable in that it allows emphasis to be given to socio-economic and human population factors. For example, the conversion of tropical forests in the Amazon for ranching or agriculture, or of oak/pine-dominated landscapes in Central Portugal for eucalyptus plantations leads not only to the loss of ecological and taxonomic diversity, but to serious alterations in the diversity of the life style and socio-economic structure of the inhabitants of the areas concerned.

The impression should not be given that our knowledge of vegetation structure is complete, even at the gross level. Detailed composition of ecosystems and communities is known for only a small percentage, and phytosociological classification of communities has been largely restricted to the North Temperate region and the Mediterranean Basin.

2. *Level 2: Taxonomic diversity, or learning to live with ignorance*

At the taxonomic level the commonest measure is in terms of species, but there are other valid measures such as the number of genera or even families. There are serious difficulties in using species as a measure, and it should be remembered that it is a fact of systematic biology that the vast majority of the species recognized are equivalent by designation only, not through their degree of evolutionary, genetic, ecological, or phenetic differentiation (Heywood 1958). It is therefore not particularly meaningful to compare species from different phyletic groups: an insect with a flowering plant, for example. Even within a taxonomic group, such as the flowering plants, an ephemeral herb cannot be equated, except nominally, with a forest tree. Also, a polytypic species containing several differentiated races or subspecies represents a different situation from a narrowly endemic relict species. A major area of concern is the typological approach to taxonomy that is adopted by many conservation bodies, especially in matters of environmental legislation. Species are treated as though they are invariant entities in many Red Lists and Databooks, databases, legislative instruments, etc., almost in a Linnaean philosophical sense. We need to recognize that species are usually made up of variable populations showing morphological, chemical, genetical, and physiological diversity and data on these aspects need to be recorded for effective conservation action to be carried out.

Then there are major regional differences in the levels of differentiation which are recognized by taxonomists as meriting species or generic status. Do these represent different patterns of evolution and differentiation in the Neotropics, Africa, and the tropics of Asia, as Gentry (1990) has recently suggested? Or is it that different standards are applied in different regions? Certainly different taxonomic standards have been applied regionally for certain groups. The 'Flora Malesiana', for example, followed a philosophy of taxonomic lumping which led to fewer species being recognized than in other regions such as the Neotropics, and this has been criticized by Gentry on the grounds that it masks variation by not recognizing it at species level. On the other hand, Van Steenis, the founding father of 'Flora Malesiana' used to say that the aim was to determine how few species there were in an area, not how many. Again, the 'Flora USSR' did not recognize any subspecies, a decision justified on the grounds of dialectical materialism, rather than on biological facts.

The great discrepancies in species numbers between different continents may be partly a reflection of their different geological histories, partly the consequence of different taxonomic attitudes and philosophies—the so-called species problem. There are no easy solutions to these problems which are inherent in the evolutionary and taxonomic processes, and it has been suggested that no single species concept will meet the needs of all biologists.

Another approach is to weight taxa according to their taxonomic or phyletic distinctiveness based on the topology of a hierarchical taxonomic/phyletic classification such as those that have been proposed by Williams *et al.* (1993), and these may well be valuable as an additional approach in helping identify priorities for conservation, depending on the conservation circumstances concerned. This in turn raises the question of species redundancy in ecosystems, identification of keystone, critical or umbrella species. Not all species are equally important, however defined (Heywood 1978).

Arguments have been advanced that the genus is a more useful indicator of taxonomic diversity in tropical countries where insufficient work has been done at the species level. What is becoming clear is that no one measure of taxonomic diversity is universally applicable.

Even within the present unsatisfactory framework there are serious problems in estimating the size of floras and faunas in different countries. This is partly due to incomplete inventory for many groups such as soil fungi, insects, and other invertebrates, especially in the tropics. We can only make an intelligent guess at the number of flowering plants in Venezuela, Ecuador, or Colombia and the total figure quoted for Latin America as a whole varies from 90 000–150 000, although work by the Missouri Botanical Garden in producing catalogues of the floras of these countries will give us clearer estimates. It is in practice very difficult to add together the floras or faunas of different countries because of the incompleteness of the inventory, overlap in species, and inadequate taxonomy: one approach is to take the country with the largest known number as a baseline and add to this the number of endemics found in the other countries. Even if the inventory were complete, this method is fallible in that some species are country-endemics, others are two- or three-country endemics, regional endemics, and so on and these are not included customarily in the calculations. On the other hand, algorithms such as those employed by Margules (this volume), Pressey (this volume), and others are designed to resolve these problems of complementarity, as is implemented in the WORLDMAP program (Williams *et al.* 1993).

There is a need for a critical look at endemism which is frequently used as a qualifier of biodiversity although not a measure (nor an indicator) of taxonomic diversity as such. Vane-Wright (pers. comm.) points out that all taxic measures (including simple species richness) usually perform in a very similar manner when coupled with complementarity to find optimal priority sequences. In marked contrast, endemism, he notes, usually gives a very different, and far less efficient result, except that it is usually at least as efficient in finding the critical set (a minimal or near-minimal set which is fully representative), even though it usually performs very poorly over the first few steps (Williams *et al.* 1993).

Many comments have been made recently on our ignorance of the total number of species of organisms in existence today. Even if one accepts an elevated figure for the arthropods (and the basis for the recent estimates of this group has been critically analysed, suggesting a much lower total than the often quoted figure of 30 million; cf. Stork this volume), it is quite clear that our knowledge of biodiversity in terms of taxonomy is highly skewed in favour of certain groups: mammals, birds, flowering plants, rather than invertebrates, fungi, and microorganisms, except when the latter are of vital importance to humankind.

Our current taxonomic knowledge is largely demand-led in that we have tended to describe only those groups of organisms that are known or believed to be of some value in some way to humankind, or that impinge on our lives in some way. This is why flowering plants have been the subject of fairly intensive study over the last two centuries—up to 100 000 of an estimated 250 000 have been used for one purpose or another by humans and currently 20–25 000 are in current use (Heywood 1992). For other groups, such as soil fungi in the tropics which may well play a key role in the functioning of ecosystems, we are largely ignorant and in general we have learned to live with this ignorance. We cannot convincingly argue that enormous amounts of effort and resources be put into collecting, housing specimens of, describing, and naming the millions of as yet unknown species that it is believed exist, especialy in the light of the many other priorities. Nor is it feasible to describe only the important ones since we have no means of determining this in advance of their recognition and description.

In practice, our handling of biological diversity at the taxonomic level requires a great deal of selectivity and priority making. Our priorities can be set in terms of keynote species, umbrella species, flagship species, charismatic species, economic species, endemics, species modules, or those that are ecologically significant for management or monitoring. Yet all of these pose almost as many problems as they solve.

3. *Level 3: genetic diversity*

Genetic diversity exists itself at different levels—population, genomic, chromosomal, genic, allelic, amino acid sequence—and this is perhaps the least understood set of components of biological diversity. Even chromosome number and karyotype morphology is unknown for the vast majority of organisms. In terms of conservation, most interest in this level has been shown by the genetic resources sector, especially for plants as witness the achievements of the CGIAR Crop Centres, FAO, IBPGR, and about 150 major Gene Banks around the world which have collectively brought together major *ex situ* collections of germplasm for crop, forage, and selected tree species. The network of botanic gardens coordinated by BGCI is attempting to do the same for genetic collections of wild species. Here the emphasis is on the extent and pattern of genetic variation within populations of species, but for the vast majority of species this is largely unknown and represents another gap in our knowledge.

In conclusion, we need to foster a much closer cooperation between taxonomists, conservation biologists, and conservation practitioners in addressing these problems. As a first step we need to understand each other's aims, aspirations, and limitations. Time is too short for us to work in isolation and ignorance.

References

Common, M. S. and Norton, T.W. (1992). Biodiversity: its conservation in Australia. *Ambio*, **21**, 258–65.

Gentry, A. H. (1990). Herbarium taxonomy versus field knowledge. *Flora Malesiana Bulletin*, Special Vol. 1, 31–5.

Heywood, V. H. (ed.) (1958). *The presentation of taxonomic information: a short guide for contributors to Flora Europaea*. Leicester University Press.

Heywood, V. H. (1978). European floristics: past, present and future. In *Essays in plant taxonomy* (ed. H. Street) pp. 275–89. Academic Press, London.

Heywood, V. H. (1992). Conservation of germplasm of wild species. In *Conservation of biodiversity for sustainable development* (ed. O. T. Sandlund, K. Hindar, and A. H. D. Brown), pp. 189–203. Scandinavian University Press, Oslo.

Kemp, R. H. (1992). The conservation of genetic resources in managed tropical forests. *Unasylva 169*, **43**, 34–40.

Norton, B. G. and Ulanowicz, R. E. (1992). Scale and biodiversity policy: a hierarchical approach. *Ambio*, **21**, 244–9.

Noss, R. F. (1990). Can we maintain biological and ecological integrity? *Conservation Biology*, **4**, 241–4.

Walker, B. H. (1992). Biodiversity and ecological redundancy. *Conservation Biology*, **6**, 18–23.

Williams, P. H., Vane-Wright, R. I., and Humphries, C. J. (1993). Measuring biodiversity for choosing conservation areas. In *Hymenoptera and biodiversity* (ed. J. LaSalle and I. D. Gauld), pp. 309–28. CAB International, Wallingford, UK..

3. On what we should save: the role of culture in determining conservation targets

BRYAN G. NORTON
School of Public Policy, Georgia Institute of Technology, Atlanta, Georgia 30332, USA

Abstract

This paper examines the role of culture in the protection of biological diversity. The present situation exists that while a few commentators question the urgency sometimes asserted for biodiversity protection, at the same time there is no organized, scientifically respectable, and political influential group that advocates destruction of species, nor suggests that we do nothing. The cultural basis of this enquiry is, that whilst it is not seriously questioned that there is an incumbent moral obligation for us to sustain biological diversity for the benefit of future generations, we still have to find the rationale for consensus and to articulate it in a specific and operational manner. The consensus we seek implicitly sets out the temporal frame over which humans will 'manage' natural systems, provides a basis for strengthening and making more specific the agreed protocol to protect biodiversity, and provides the means to accomplish this task. First, I discuss the view that there is an 'objective' scientific definition of biodiversity. I conclude that there is not for two reasons: one of epistemology, which allows us to realize that all knowledge is theory bound, and undermines any attempts to aggregate multiple levels of biodiversity into a single measure; and secondly, physical theory which admits the irreducible complexity of natural systems. Second I introduce 'hierarchy theory' as a useful technique to identify conservation targets in the absence of a single scientific definition of diversity. Hierarchy theory is used to explain new explorations by many ecologists and ecological economists of methods of minimizing human alteration of natural cycles, shifting attention to 'ecosystem health' and the structures and processes that perpetuate natural systems. I develop three assumptions: that natural systems are dynamic; biotic diversity is maintained when natural populations adapt creatively to dangers and opportunities; and that we can only understand biodiversity if we interpret the dynamics of nature as a natural hierarchy in which the smaller sub-systems tend to change more rapidly than larger units or the whole system. Human

Systematics and Conservation Evaluation (ed. P. L. Forey, C. J. Humphries, and R. I. Vane-Wright), Systematics Association Special Volume No. 50, pp. 23–9, Clarendon Press, Oxford, 1994.

culture and maintenance of biodiversity must go hand in hand. I will argue that ecological economics is the bridging discipline that links ecology and culture, a link that is forged in the crucible of policy formation to create a dialectical dynamic between culture and nature which drives landscape development and attendant diversity on the intergenerational scale relevant to sustainablity goals.

Introduction

I have been asked to discuss the question, what is the role of culture in the protection of biological diversity? This is a good, but extremely difficult and complex question to answer.

Loss of biodiversity is one of the 'global' environmental problems which, along with ozone depletion and greenhouse warming of the earth's atmosphere, have come to dominate environmental policy discussions. While scientific uncertainties remain about the exact dimensions of the problem, loss of biodiversity, these uncertainties have not stood in the way of a scientific and policy consensus that the problem requires a strong and decisive response. A commitment to protect biological resources has therefore become the centrepiece of the renewed commitment to environmental protectionism that has been intensified by the recent political and economic realignment of Eastern Europe. There may be an emerging opportunity to unify the world behind an environmentalist banner with protection of biodiversity as a centrepiece. While there are still a few commentators who question the urgency sometimes asserted for biodiversity protection (see, for example, Kahn 1982; Simon and Kahn 1984), there is no organized, scientifically respectable, and politically influential group that advocates destruction of species, nor are there even serious advocates of a do-nothing approach to the subject. We can therefore base our exploration of the role of culture in biodiversity policy on a very broad and strong consensus: it is not seriously questioned that the present generation has an important moral obligation to sustain biological diversity for the benefit of all future generations. Our task is mainly to provide a rationale for this consensus, and to articulate this goal in more specific and operational terms.

It is important that this consensus for sustainability introduces an implicit, multi-generational time frame. If we are seriously committed to protecting biological diversity over many human generations, we are committed to understanding and modelling our impacts on populations of plants and animals on scales of at least 150–200 years (Norton and Ulanowicz 1992). This commitment starts to give shape to a rational and comprehensive policy to protect biodiversity. So the consensus in favour of long-term biodiversity protection is no vacuous consensus. It implicitly sets the temporal frame over which humans will 'manage' natural systems and provides a basis for strengthening and making more specific the consensus to protect biodiversity and how to accomplish this urgent task. It should be noted that 'management' of ecological systems more often than not means 'managing' human impacts on ecological systems. We usually do not know how to manage complex, dynamic, creative, and inherently unpredictable (see Faber *et al.* 1992) systems to produce desired outcomes without undesirable consequences.

From the outset we must recognize that there is much less agreement as to what exactly we mean by 'biodiversity'. What exactly are we trying to save? I call this the question of 'conservation targets'. Should we be trying to save genes, individuals, species, communities, or ecosystems? It is tempting to say that we should save all of these to the fullest extent possible; but this would surely require more resources than will be available, and we must face the question of targets and set conservation priorities. The important question of conservation targets can be illuminated by a careful analysis of the relationship of human cultures with surrounding biotic communities. Accordingly, I consider two questions on the relationship between culture and the natural systems that form their context: (1) should cultural factors influence our characterization and description of biological diversity? and (2) is it possible to save the biological resources in an area without saving, also, important aspects of indigenous cultures?

First I will examine the view that there is an 'objective' scientific definition of biodiversity that can guide conservation efforts. I conclude that there is no objective scientific definition. Second, a theoretical approach, called 'hierarchy theory', is introduced to provide a useful technique for identifying conservation targets in the absence of a scientific definition of diversity. From this technique I can answer two questions by explaining the crucial role that local cultures should play in our selection of conservation targets. I can then begin to address the interaction of cultures with the biological resources they depend upon in the last part, and to sketch a general answer to the second question. Drawing upon insights from environmental history, I will argue that any definition of biodiversity which is potentially capable of guiding an intelligent, multi-generational policy must accept that nature is dynamic and that changes are driven by local interactions of human cultures and biotic communities. Identifying conservation targets—which is necessary to articulate a rational biodiversity policy—consequently requires that we understand how cultural practices, and the aspirations of the institutions that sustain them, are expressions of local values which have evolved in response to local ecological conditions.

Measuring diversity scientifically

Most advocates of biodiversity protection assume that we have—or will have—a precise scientific measure of biological diversity and that the measure, once articulated, will tell us which elements of nature to 'target' in our conservation efforts. The goal of protecting biological diversity would then be precisely stated as the goal of maximizing diversity so measured. This optimistic view is, unfortunately, untenable. Considerable progress has been made in understanding diversity scientifically; but that progress makes it less, not more, likely that a single, scientifically objective measure of diversity will emerge to guide policy.

Several scientific definitions of diversity, applicable at differing levels of natural systems, have been proposed (Whittaker 1960; McNaughton, this volume). Within-habitat diversity is the species diversity that occurs within a particular habitat; but the diversity represented across an entire landscape will also depend heavily on the diversity of types of habitats that occur within the landscape. At first glance these multi-scalar definitions may seem to cause no serious problems—why not simply choose a spatially synoptic concept, such as McNaughton's δ-, or 'landscape' diversity as the concept most descriptive of diversity and use delta diversity as the measure of how well we are doing?

While this step is a useful one, and may help us to explain in general terms what is sought in biodiversity protection, it does not provide a satisfactory theoretical solution to our problem, because the idea of delta diversity is only as 'objective' and 'descriptive' of nature as are the various boundaries and partitions that have been introduced to describe the organization of the landscape as a hierarchy of 'microhabitats' (subsystems), 'habitats' (systems), and 'landscapes' (complexes of interacting systems). If these boundaries could be set based simply on objective data describing ecological systems, landscape diversity would also be a descriptive concept since, on this assumption, it would represent a synoptic indicator of diversity at all levels of the biotic hierarchy. If delta diversity were to reflect unique descriptions of diversity on smaller levels, it could claim to represent an objectively measurable and computable synoptic concept of system diversity, and delta diversity might provide an objective policy beacon.

But hierarchies defining ecological systems are not objective in this way (Allen and Starr 1982). The hierarchical organization is created by the choices of biologists and ecologists who embody scalar and perspectival decisions in the models they produce for understanding systems. Even among ecologists there are significant choices as to the scale and level at which any system is described. Since nature can be modelled on many different levels, there is therefore a real question as to which model to choose and what perspective to take when one 'measures' diversity.

There are actually two powerful arguments to support the conclusion that there exists no single, correct descriptive model for any natural system. First, there is a philosophical, or 'epistemological', argument (Quine 1960; Kuhn 1962; Allen and Starr 1982, p. 6). According to this very abstract argument, any amount of data gathered to describe a system will, in and of itself, underdetermine the choice of theoretical principles that describe and explain that system. Scientific perception is therefore theory-bound to some degree. Two different, but equally 'objective', scientists can observe the same natural system and, depending on the language they choose, the assumptions they make, and the goals that motivate them, provide nonequivalent descriptions of the systems they both observe. In particular, scientific describers can interpret the world as composed of many, quite different hierarchies, with alternative and conflicting boundaries.

While this argument is admittedly abstract it has surprisingly direct applications to the measurement of diversity; it implies that diversity measures are constructs by human individuals who undertake their studies for many different motives. Each motive gives rise to at least one 'useful' scale for interpreting and predicting the events around us. But differing motives, and attention to events occurring on multiple levels exist on many scales, some of which are not commensurate with others. The argument also implies that the choices that descriptive scientists make in deciding to characterize observed events—the models they construct and the descriptive concepts they employ—cannot, in principle, be executed independently from theory construction or of motivational pressures. It follows that the choices we make in describing natural events cannot be understood independently from our theories, our attitudes, and our goals.

There is also another argument for the same view of descriptive scientific concepts. This is based on physical theory rather than on logic and philosophical reasoning. Contemporary physicists, in the aftermath of the stunning discoveries of relativity theory and quantum mechanics earlier this century, have been forced to accept the irreducible complexity of natural systems. This acceptance is marked by the introduction of physical constants, such as Planck's constant, into physical descriptions and calculations. But these constants are indicative of a fundamental truth about physical systems: these systems are irreducibly complex (Prigogene and Stengers 1984; Gleick 1987). The proposal to use delta, or landscape diversity, as a synoptic definition of diversity implicitly assumes that landscape diversity does reflect a unique product of the appropriate diversity measures that are applicable on each level. But the irreducible complexity of natural ecosystems undermines any attempt to aggregate multiple levels of diversity into a single measure that is both synoptic and objectively descriptive of the system.

Consider an example that illustrates this point. Suppose we assumed that there exists a synoptic measure of biodiversity over a large geographic area, and that we can use the total number of species that exist in that large area as a rough guide to its 'diversity'. Looked at naively it would be possible to reason as follows: since the area contains *n* species now, we would increase its diversity by adding *m* additional species from other ecosystems. For example, we would have reasoned, in the 1920s, that adding Melaleuca (*Melaleuca quinquinervia*) and Brazilian Pepper (*Schinus terebrinthinus*) to Florida's landscape would increase species diversity there. In fact this effect was only achieved in the very short run because these exotic species quickly got out of control and destroyed previously existing habitats. These introductions initiated, on a longer scale of time and on a different geographic scale, a reduction in the number of types of habitats and microhabitats in south-west Florida. Addition of exotics to the system can therefore have a negative impact on diversity in the long run. Simple examples such as this, and other, more complex ones, show that we cannot simply assume that a synoptic concept of diversity (such as the total number of species in the landscape) is a useful guide to policy. Any synoptic concept of diversity must be applied with a careful understanding of the irreducible complexity of ecological systems and the importance of not just total numbers of species, but also of the complex ways that species interact.

We must be very cautious of any definition that is put forward as a purely descriptive measure of biodiversity. It is not quite correct to say that, in post-modern, post-positivistic science there are no correct descriptions of actual events. It is more accurate to say that there are *too many* correct descriptions of reality. The different scales and categories that we use to describe nature in all of its multi-levelled complexity cannot all be 'reduced' to a common vernacular, nor can they be plotted in an ideal space inhabited by irreducible atoms. It follows that, even though scientists can provide 'descriptive' accounts of natural systems, these descriptive accounts gain a special nonreducible content by virtue of the conceptual and value assumptions incorporated in them. They are in this sense both true—they are consistent within a given frame of reference—and at the same time relative—they cannot be verified in frames of reference that do not embody the essential assumptions of their own frame.

If our choice of concepts and models to describe a natural object or process depends essentially on the purposes we have in describing that object or process then certain consequences follow. One is that the concepts embodied in the analysis of population biology are not necessarily the most useful concepts to use in characterizing the goals of conservation. They may be, but we need an open and probing discussion of this question before we assume any descriptive model of the diversity we are trying to protect. Indeed, since biologists and ecologists study natural systems on many different levels, we can expect that they will develop many different concepts of diversity that will be applicable at very different scales in the natural world. The task, in designing a rational policy is to choose from among these many scientific concepts a particular concept of diversity that provides the most useful understanding of policy, *from a conservationists' perspective*.

The search for a unique scientific measure of diversity to guide policy is based, ultimately, on the principle that fact is separable from value, that nature can be fully described prior to being evaluated. If this were true, we could use pure science to guide biodiversity policy. Instead, I believe that the definition of biodiversity used by policy makers is *both* scientific and normatively political in a very specific sense. Science defines a whole spectrum of possible meanings of 'diversity'. Choosing a scientifically acceptable public policy concept that we wish to foster is a decision that can only be determined politically.

To summarize, it appears that scientists can offer a very large number of possible 'diversity measures', but that these measures cannot be aggregated into a unique measure of *the* diversity of the system. In the next section, we will see how hierarchy theory can, despite this perplexing theoretical impossibility, provide guidance in choosing a useful definition for guiding biodiversity policy from among the many possible 'scientific' definitions of diversity.

Hierarchy and biological diversity

My theoretical considerations have so far not contributed to the practical question what should be the main targets of conservation efforts? I have only, so far, argued that no fully satisfactory answer is forthcoming from 'objective' science. United States policy has emphasized species protection, under the influence of the Endangered Species Act of 1973. However, this approach is extremely expensive because it institutes protection only after a species is at the brink of extinction. It is also entirely inappropriate in tropical ecosystems suffering deforestation, where many species are lost before they are even identified.

In this volume, the related but more sophisticated idea that conservation efforts should concentrate on 'biodiversity hotspots' is advocated (see Heywood this volume). This policy approach is no doubt justified, at least as a stopgap measure, while we try to stop the haemorrhage of species lost as a result of runaway deforestation in the tropics. Speaking theoretically, however, it should be noted that the hotspots approach remains very species-oriented—it still sees diversity mainly in terms of elements, as an inventory of items, and see Lawton *et al.* this volume for difficulties with the 'hotspots' approach.

Many ecologists and especially, ecological economists, are exploring a more radical alternative to single-species management, one that attempts to minimize human alteration of natural cycles, and shifts attention from the elements of a system to the structures and processes that perpetuate that system. This approach, the 'ecosystem health' approach as it is sometimes called (Costanza *et al.* 1992), advocates a post-positivist science of conservation in which it is readily admitted that environmental management, and also scientific fields such as conservation biology, are explicitly value-laden scientific disciplines. In this respect, they resemble human medicine and veterinarian medicine more than they resemble the 'ideal' of scientific objectivity as it was understood in Newtonian physics.

Conservation biology, like medicine, studies and acts in the service of an ideal; if health is to be pursued as a matter of public policy then this ideal must be articulated in the crucible of public policy formation. While science is extremely important in defining what is meant by health, whether of humans or of ecosystems, our ideal of health goes far beyond 'pure' science to encompass, among other values, our aspirations for our children and our culture. The models chosen and the goals defined, which are inevitably expressions of cultural values and concerns, are also an essential element in the process of policy formation. The question is: how can we identify which cultural values and which natural dynamics should receive priority?

A hierarchical approach to the relationship between cultural values and biodiversity has been proposed (Norton and Ulanowicz 1992). The hierarchical approach is built firmly on three assumptions: (1) natural systems are dynamic and constantly changing on many levels and on many irreducibly complex scales simultaneously; (2) biotic diversity is maintained and enhanced when species, reacting to varied local conditions, adapt creatively to different dangers and opportunities in many varied and constantly evolving communities; variations in gene pools accompanying these adaptations are the markers representing diversity; and (3) we can only understand the complex dynamic of nature if we interpret it as a hierarchy of systems in which smaller subsystems tend to change more rapidly than do the larger systems. On each level one can identify systems that are *both* 'holons' (independently functioning wholes) and 'partons' (individual elements that play a functional role on a larger scale on which they exist as parts) (Koestler 1967). Thus, on the hierarchical approach, nature is seen as a complex of systems in which smaller, nested subsystems change and adapt against a backdrop of 'stability'. Even though every level of the system is dynamic, development of genetic and cultural adaptations is possible because the slow-changing environment provides relative stability over many generations of individuals.

At first glance, this approach may seem to plunge us into subjectivism, because the systems approach apparently leaves us with so many unguided choices regarding which scale of natural systems to focus our attention upon. It is important, however, to distinguish between 'subjectivism' (the view that the multiple descriptions of nature by scientists are all equally valid) and 'relativism' (the view that scientific descriptions are 'relative' to a particular frame of reference). Once it is recognized that a strict separation of description from evaluation is impossible we move past the myth of a value-neutral science and can propose models of nature that are designed to describe natural systems in ways that are appropriate, *given certain purposes*. If purposes are accepted, such as the need to understand and protect biodiversity, then we have a basis on which to develop models that are descriptive of nature.

The distinctive theoretical assumption of hierarchy theory is (3), that smaller sub-systems change more rapidly than do the larger systems they compose. Reasoning, then, from the endorsement of biodiversity protection for many human generations—setting a temporal policy horizon of at least 150–200 years—we can begin to zero in on the ecological scale on which to model the systems that create and sustain biodiversity, and to concentrate on the scales at which to seek the most important conservation targets. We can also begin to eliminate certain candidates as the most central targets of protection. Individuals and also populations of many species will be ephemeral over the period of concern even in systems undisturbed by human activities. Indeed, since species themselves may become extinct for reasons unrelated to human activities, the hierarchical approach directs us to emphasize not species but those particular processes that support energy flows. As long as we maintain 'healthy', self-organizing systems—ones that maintain their varied layers of structure through complex adaptive processes—individuals will be reproduced and populations that lose habitats will invade and colonize new areas as they become appropriate for them in the dynamic of ecosystem development and succession. Species will therefore be maintained by healthy processes, provided we model nature at the larger scale on which the concept of health is applied to a landscape-sized ecological system. A patchy landscape with multiple levels of constancies and change is most likely to protect the most biological diversity possible over many generations of individuals, and as populations wax and wane. The problem is to specify the crucial processes and scales on which to characterize, model, and evaluate landscape-level changes.

Emphasis on processes that maintain patchiness must therefore be supplemented with a coherent, scientifically adequate, publicly understandable conception of scale to guide the choice of patterns of patchiness. It seems certain that any landscape capable of maintaining complex and healthy systems must include undisturbed areas, buffer areas (including migration corridors), and areas of heavier use. But since predation and grazing are important ecosystem functions, it follows that there must be sufficient undisturbed areas in the landscape matrix to maintain viable populations of the largest, free-ranging predators and grazers remaining in the system (Harris 1984). It is possible then, using hierarchy theory in tandem with our assumption that biodiversity protection has at least a 150–200 year horizon, to begin to set ecologically informed boundaries as the basis for an appropriate model for formulating biodiversity policy.

Drawing on our hierarchical assumptions we now have a basis for characterizing crucial scales, as understood from a conservation viewpoint: the concrete goals of management will be dictated by the need to protect processes that support culturally important and politically defined values. In the next section, we will explore the local scale at which values in nature must originate.

How do cultural values shape conservation goals?

The idea I develop here is at least implicit in the ideas and work of Aldo Leopold, the US forester, amateur philosopher, and theoretician of environmental management (Norton 1991). Leopold (1949) realized that we naturally focus on dynamics that are relevant on the human, perceptual scale—this is the scale on which humans act economically, seeking food and shelter. But humans are also part of a human and a natural *community*, a community that has value on an intergenerational scale. Some of these values are expressed on the day-to-day scale of economics; others are expressed in the intergenerational scale in which our culture is passed from generation to generation and in which our species is perpetuated. The need to integrate these various levels is, in my view, the central idea of Leopold (1949) who recognized that, in addition to perceiving and valuing the world in economic time, we must also think in the multi-generational ('ecological') time of mountains and wetlands.

Learning a new way to perceive can only be accomplished in democratic countries through dialogue between scientists, politicians, and the public. But what, exactly, can we expect from this. We have seen above that there is no *objective* scale provided by science. The policy dialogue must therefore originate from many local viewpoints, and flow upward through the hierarchy, integrating individual, economic, and environmental concerns on all levels. Creativity in natural hierarchies occurs when an organism or population responds in a new way to the natural dynamic they live within. New adaptations are given meaning and significance by virtue of their response to constraints implicit in larger, slower-moving systems. It is this dynamic that creates environmental values (Page 1992). As Leopold realized, objectivity cannot be judged in the short run. Drawing upon the American pragmatists' idea that truth is that which survives many observations, experiments, and applications, Leopold argued that a culture has found the 'truth' only if it has developed a set of cultural ideas and ideals that result in behaviours that can be sustained for many generations (Leopold 1923; Norton 1988). If there is rapid alteration of the larger context in which those behaviours evolved and gained meaning, today's genetic and cultural adaptations become irrelevant more quickly than new ones can evolve. Leopold illustrated this point by noting that, in the American south-west, the Pre-Columbian Pueblo cultures survived and reproduced themselves for many generations in the fragile arid ecosystems of the area; but the arrival of European settlers and their grazing culture led to drastic declines in the quality of the environment in just one or two generations, because agricultural methods adapted to Europe were applied in an arid ecosystem. Leopold saw survival in conjunction with a slowly evolving environment as the 'objective' test of whether a culture has found 'the truth'. Note that, on this approach to objectivity, the truth lies in adaptation to *local* conditions.

This insight is expressed on the hierarchical approach by saying that values toward nature are shaped at the local level. Systems that model nature for the purpose of protecting it therefore must be based on an articulation of local values as expressed within descriptive/managerial models for describing nature. See Fig. 3.1 for a flow diagram for policy formation that takes into account the interaction of values and science in formulation of an environmental problem. Note that the formulation of the problem implicitly requires the identification of the correct scale on which to formulate goals and measure progress toward them.

It is now possible to reformulate the question of identifying conservation targets. On the hierarchical approach it is reasonable to equate the problem of conservation targets with scalar problems in conservation ecology. Conservation ecology differs from 'pure' biological sciences (if there are such), in unabashedly choosing a perceptual, observational perspective, and a framework and a scale on which to understand natural systems so as to help us to protect their essential features. We have just seen that science can provide us with many different, equally measurable criteria of diversity—it is the task of policy makers, with the guidance and cooperation of scientists, to choose among the possible models for describing a natural system. It has been argued (Norton and Ulanowicz 1992) that the proper scale at which to model biodiversity decisions can be set rather precisely, provided one can decide on a goal for management that embraces a time scale. Determination of an important social value allows us to articulate goals because the social value allows us to choose a relevant *perspective* and *associated* scale for modelling a system that produces the desired goals (e.g. maintenance of biological diversity). But the recognition of the multi-generational scale of biodiversity policy guides us away from a static, elements-oriented plan to protect diversity. Instead, it pushes us toward a more dynamic, process-oriented view, and it also directs our attention toward processes that are expressed at the landscape level.

Consider an example of a difficult policy problem now facing environmental managers in south-west Florida. The Everglades ecosystem has been in serious deterioration for decades because of reductions in the flow of water due to draining and removals of water for agricultural and residential purposes. Recently, there has emerged a political commitment to restore the

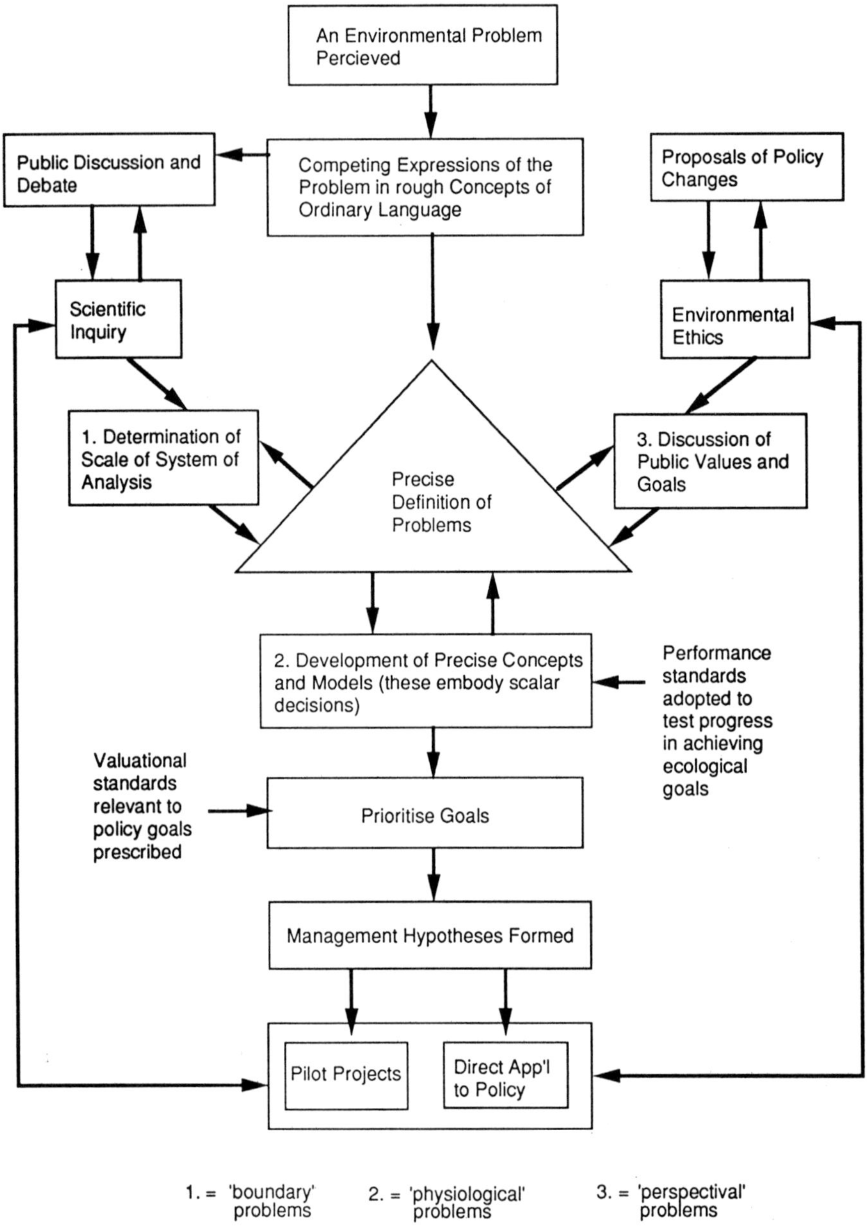

Fig. 3.1 Public policy formation does not begin with clearly worked out 'problems'. Definition of a problem requires development of a model for understanding changes in natural systems. This flow diagram shows how values, science, and public policy discussion all contribute to the development of a model, and an associated perspective and scale for modelling a system that requires management. From Norton and Ulanowicz (1992).

Everglades ecosystem by increasing water flow through it, but the restoration is now being held up because of a conflict between ecosystem restoration and responsibilities implied by the Endangered Species Act (Alper 1992). Increasing water flow would, in addition to re-establishing the temporary marshes that flood in the wet season and dry out gradually through the year, flood the hunting grounds of the Everglades snail kite (an endangered species), which feeds on one species of snail that is exposed and easily available when water is low. The conflict is clearly stated by David J. Wesley, Florida supervisor for the US Fish and Wildlife Service:

> In the Everglades we have over a dozen endangered or threatened species, each with varying needs that aren't necessarily compatible, within a changing ecosystem. As an ecologist, I would argue that the best solution is to restore the Everglades as a functioning ecosystem, and in the long-run, that will benefit all the wildlife there. But as an official of the agency that oversees the Endangered Species Act, I have to consider the effects these changes will have on those species that are endangered at this moment (Alper 1992).

The shift to the more dynamic, long-term view of Wesley-the-ecologist does not, of course, mean that species are unimportant. Species are essential elements in ongoing processes; and for a long time to come, rough species counts over specified areas including the trends in their numbers, will remain our best general indicator of ecosystem health. But species counts will not be valued so much for their own sakes as for their ability to indicate that the system remains in sufficient health to support and sustain the ecosystem structures (such as forest cover and patchy dynamics at many levels) necessary to support healthy processes. This change in emphasis will affect policy by making explicit the extent to which our understanding of social values, such as the goal of sustainability, shape the models we construct for understanding and modelling nature. The social goals we emphasize serve to focus our attention, and our descriptive prowess, on certain levels and scales of complex natural systems that form our environment. The idea of landscape, or delta diversity (see MacNaughton, this volume), may now be able to be revived as an important indicator of system health through time. But δ-diversity should not be considered a measure of elements that occur at many levels; δ-diversity must be related more directly to the dynamic models we choose to track, understand and manage, and to achieve our goal of protecting biodiversity for many generations into the future.

Why cultures and habitats must be saved together

I believe that the theoretical structure and set of assumptions outlined above provides a basis to address our second question: is it possible to save the biological resources in an area without saving, also, the local indigenous cultures that have developed there? Hierarchy theory helps us to explain the importance of local valuations. Another way to understand the importance of culture in biodiversity protection is to examine a well-documented case history of how cultural changes alter landscapes in dynamic ways.

Perhaps the most intellectually influential book published on environmentalism and environmental studies in the last decade is William Cronon's (1983) *Changes in the Land.* Cronon challenges on historical grounds the 'wilderness' ideal which has guided so much of our thinking (especially in the US, but elsewhere as well) about how nature preserves should be managed (see also Callicott 1991). This ideal assumes that natural systems are most healthy, and most diverse, in the absence of human disturbance. He challenges this idea, which apparently assumes the already-discredited idea that each landscape will tend toward a single 'climax' ecosystem type, by showing that, for one landscape, the New England area of the North American continent, the existence of a 'wilderness' at the time of the arrival of European settlers, is based entirely on myth.

Cronon (1983) shows that the New England landscape had been intentionally managed by New England Indian tribes for millennia prior to the arrival of Europeans, and that it is meaningless to imply any ideal of a 'pristine' wilderness as the ideal state that we should try to perpetuate or recreate in natural preserves. He argues that:

> The replacement of Indians by predominantly European populations in New England was as much an ecological as a cultural revolution, and the human side of that revolution cannot be fully understood until it is embedded in the ecological one (Cronon 1983, p. 6).

Admitting the difficulties involved in reconstructing that ecology, Cronon painstakingly uses multiple methods to arrive at a general account of changes taking place between 1650 and 1800. To summarize, he found that the Indians had maintained a diverse landscape, employing such tools as fire and cultivation. This diversity was made possible by the Indian's lifestyle, which was to collect few possessions and to migrate seasonally from one productive resource-producing subsystem to another. The possession of loose, but recognized rights to use areas such as shellfish beds or reed ponds, rather than fixed ownership of surveyed plots, allowed many family bands to use shared resources. European settlers, who brought with them a much more rigid concept of private ownership of land, and different concepts and institutions of property ownership characteristic of developed capitalist economies, disrupted these patterns and, simultaneously, reduced much of the diversity of the landscape as private owners 'improved' their land, by clearing it, draining it if necessary, tilling it, and eventually fencing it. Since settlers considered only these intrusive improvements true 'use' of the land, they justified forcing the Indians off more and more of their lands which had been dedicated to diverse uses such as hunting, reed-gathering, and so forth.

Since improvement was often the only recognized requirement for establishing ownership of 'open' land, the Indian lifestyle became increasingly untenable as more and more areas were 'improved' and fenced. The result was a shift from a relatively diverse, managed landscape to an increasingly monotonic and homogenized one, with attendant losses in landscape diversity:

> The colonial interaction of forests, furbearers, hunters, axes, grazing animals, plows, crops, [—introduced old-world] weeds—and the rival way of owning and selling these things—all contributed to a redrawn map of New England. It was a map that, over the course of European settlement, more and more traced, not the earlier world of movement between hunt and harvest, but the new world of cropland and pasture, of agricultural cycles entrapped within the fixed boundaries of individual possession. In the hands of the colonists, New England had become a world of fields and fences (Cronon 1983, p. 156).

Cronon's elegant ecological history is premised on the central, hierarchical assumption of ecological economics, that economies are faster-changing systems that can be more or less well 'adapted' to their spatially larger, ecological context. Besides forcing us to re-think the nature of 'pristine' reserves as a management ideal in most situations, Cronon's work points toward two further important consequences for biodiversity policy. First, he supports the importance of *dynamic* models which trace processes. Second, he shows how social values and institutions shape, and are shaped by, the interaction of human populations with their *local* environments; as well as how external demands, such as the need to produce goods tradable in world markets to obtain money to buy the various necessities Indians could gather from their more diverse environment. Since Indians managed and shared a richer landscape, they had no need to accumulate, and no need to exchange money for necessities. But the clearing and destruction of those local habitats to satisfy international demand for a few cash crops set up a positive feedback system ensuring deforestation in a short period. The point is not that we should try to return to a presettlement landscape, which would now be impossible, but that we should analyse large historical changes in the landscape on these local scales if we wish to understand how human populations affect biodiversity. Using sensitive interpretation of the interplay of culture and landscape may provide the best guidance in future development activities both in undisturbed areas of developed countries and also in less developed countries.

Conclusion

Ecological economics is a 'bridge' discipline—it examines the link between ecology, economics, and culture, a link that is forged in the crucible of public policy formation. This new discipline, which seeks to create an integrated conceptual framework, represents an essential forum in which to discuss crucial issues in biodiversity protection, including the location of conservation targets (Costanza 1991).

Ecological economics understands environmental management as taking place on multiple scales and in pursuit of many different values. The problem is to integrate these values and to reduce conflicts among them. For example, we value economic development to increase the standard of living of current human populations, especially for the poor. But we also hope to pass on to the future a healthy and well-functioning planet. If it is assumed that every type of development is equally destructive of natural systems, then development for human welfare necessarily entails destruction of natural systems. If, however, it is hypothesized that there are multiple scales on which management takes place—economic events occur at a more rapid-fire pace than do ecological changes—and that we must articulate goals on multiple levels, it becomes possible to seek integration among levels, by seeking policies that have positive impacts on all the multiple levels that are identified by scientists and the public as socially and culturally important.

The goals of ecological economics, in other words, is not simply to maximize productivity, but to find approaches to development that are consistent with protecting natural systems on the intergenerational, social scale on which we hope to sustain our culture *and* the natural context that gives it meaning over many generations. It is that dialectical dynamic between culture and nature which drives landscape development and attendant diversity on the intergenerational scale relevant to sustainability goals; this is the scale on which human populations depend upon, and either support or destroy the diversity essential to a rich interplay of human and natural forces.

References

Allen, T. F. H. and Starr, T. B. (1982). *Hierarchy: perspectives for ecological complexity.* University of Chicago Press.

Alper, J. (1992). Everglades rebound from Andrew. *Science*, **257**, 1852–4.

Callicott, J. B. (1991). The wilderness idea revisited: the sustainable development alternative. *Environmental Professional*, **13**, 235–47.

Costanza, R. (1991). *Ecological economics: the science and management of sustainability.* Columbia University Press, New York.

Costanza, R., Norton, B. G., and Haskell, B. D. (ed.) (1992). *Ecosystem health: new goals for environmental management.* Island Press, Washington, DC.

Cronon, W. (1983). *Changes in the land.* Hill and Wang, New York.

Faber, M., Manstetten, R., and Proops, J. (1992). Toward an open future: ignorance, novelty, and evolution. In *Ecosystem health: new goals for environmental management* (ed. R. Costanza, B. G. Norton, and B. D. Haskell), pp. 72–96. Island Press, Washington, DC.

Gleick, J. (1987). *Chaos: making a new science.* Penguin Books, New York.
Harris, L. (1984). *The fragmented forest.* University of Chicago Press.
Kahn, H. (1982). *The coming boom: economic, political, and social.* Simon and Schuster, New York.
Koestler, A. (1967). *The ghost in the machine.* Macmillan, New York.
Kuhn, T. (1962). *The structure of scientific revolutions.* The University of Chicago Press.
Leopold, A. S. (1923). Some fundamentals of conservation in the Southwest. In *The river of the mother of God and other essays by Aldo Leopold* (ed. S. L. Flader and J. B. Callicott), pp. 86–97, (reprinted 1991). The University of Wisconsin Press, Madison.
Leopold, A. S. (1949). *A Sand County almanac.* Oxford University Press.
Norton, B. G. (1988). The constancy of Leopold's land ethic. *Conservation Biology*, **2**, 93–102.
Norton, B. G. (1991). *Towards unity among environmentalists.* Oxford University Press, New York.
Norton, B. G. and Ulanowicz, R. E. (1992). Scale and biodiversity policy: a hierarchial approach. *Ambio*, **21**, 244–9.
Page, T. (1992). Environmental existentialism. In *Ecosystem health: new goals for environmental management* (ed. R. Costanza, B. G. Norton, and B. D. Haskell), pp. 97–123. Island Press, Washington, DC.
Prigogene, I. and Stengers, I. (1984). *Order out of chaos: Man's new dialogue with nature.* Bantam Books, Toronto.
Quine, W. V. O. (1960). *Word and object.* MIT University Press, Cambridge, MA.
Simon, J. L. and Kahn, H., (ed.) (1984). *The resourceful Earth: a response to global 2000.* Blackwell, Oxford.
Whittaker, R. H. (1960). Vegetation of the Siskiyou mountains, Oregon and California. *Ecological Monographs*, **30**.

4. Conservation goals and the configuration of biodiversity

S. J. McNAUGHTON
Biological Research Laboratories, Syracuse University, Syracuse, NY 13244-1220, USA

We need to be clear what kind of conservation is meant when it is talked about . . I believe that conservation should mean the keeping or putting in the landscape of the greatest possible ecological variety . . . Charles S. Elton, 1958, p.155.

Abstract

Biodiversity in grasslands in the Serengeti National Park varies widely. Consequently, the variety of food available to grazers also varies widely. Diversity varies hierarchically. The lowest spatial level is point diversity within sampling areas of small extent. The intermediate level is diversity of grassland swards or stands which contribute to diversity at two sub-levels: the communities of which they are members due to similarity of species composition, and the landscapes within which they occur. In the Serengeti grasslands, pattern diversity contributes to biodiversity at all levels. Within a sward about half of the diversity is due to pattern above the point level. At the level of point diversity species richness is almost five times as important as evenness. About half the diversity is variation in one place as compared to others, about a third variety of surrounding species, and about a tenth is due to how evenly species are represented. At the level of communities, classified by similarity of species composition of different swards, stand diversity is the overriding determinant. Integrity of species composition within communities means that there are similar grassland types throughout the ecosystem although they occur in different places. At the level of landscape regions, pattern diversity is an important element of vegetation diversity. However, less of the total variation in vegetation diversity is explained at this level than at any of the other levels, because it cannot be explained by either average diversity of stands or degrees of difference between stands. This is due to intergradation of different communities across a landscape. These levels of biodiversity are

Systematics and Conservation Evaluation (ed. P. L. Forey, C. J. Humphries, and R. I. Vane-Wright), Systematics Association Special Volume No. 50, pp. 41–62, Clarendon Press, Oxford, 1994.

accompanied by dietary diversity due to differences in the nutritional properties of grasses. Conservation evaluation of ecosystems supporting large ungulates should incorporate an understanding of the role of pattern diversity in biodiversity, and how that is related to the carrying capacity of the ecosystem as influenced by nutritional sufficiency.

Introduction

Biodiversity is variety in the properties and organization of organic nature. It resides at many levels in organisms, extending across the genetic variation within an individual (heterozygosity), between individuals within a species, the variety of species and their relative abundances in ecological communities, and the immense, but very incompletely known, assortment of species on Earth (May 1988). The theme of this chapter is the configuration of biodiversity in communities, organisms that co-occur in time and space. I believe the general patterns relating biodiversity, spatial heterogeneity, and ecosystem functional properties presented have widespread validity beyond the specific ecosystem considered here (e.g. Frank and McNaughton 1991, 1992).

This chapter is concerned with the landscape organization of ecological variety. It emphasizes wilderness grasslands as a key to understanding biodiversity in relatively undisturbed landscapes. As representative data, it considers grasslands in grazing ecosystems where the fauna is dominated by large mammalian herbivores. It appraises biodiversity as both a systematic and a functional property of ecosystems, focusing on levels of expression of biodiversity, interactions between herbivores and their food sources that influence it, and some of its functional consequences for the grazing food web.

Two means of considering the configuration of biodiversity are emphasized. First, an explicit approach considers biodiversity as observed in nature: its spatial distribution as manifested in hierarchical structure from very small to very large spatial scales. Second, a synthetic approach considers how biodiversity is related to other ecosystem variables, some that are properties of the grassland community itself, such as peak biomass, and others beyond the community, such as soil properties and grazer effects on the community.

Systematics and biodiversity

Systematics is the essential gateway to biodiversity and, therefore, a fundamental tool of conservation policy. Without the foundation of knowledge that systematics creates, biodiversity would still exist, but it would be only an unorganized phenomenon to us. The applications of cladistic methods of classification and molecular methods of evaluating genetic relationships are currently creating a revolution in modern systematics. That revolution is creating a redefinition of our understanding of biodiversity because it employs cladistics to reveal quantitative relationships and documents variation at organizational levels below those accessible to morphological or physiological resolution, unveiling genetic diversity at much more finely subdivided levels of biological organization.

Much of this symposium volume considers conservation from the species standpoint, emphasizing centres of species richness, endemism, and cladistic relationships. I consider a related, complementary, ecological viewpoint: does the diversity of communities of co-existing species vary in a predictable manner? If the answer is 'yes', how can that knowledge assist conservation evaluation?

Conservation policy, because it is ultimately site-specific, must recognize the configuration of biodiversity. Capturing biodiversity at different spatial scales will, of necessity, capture biodiversity with different configurations. In addition, it also captures other ecological attributes that are associated with biodiversity in different ways. These are the themes of this chapter.

A migratory grazer's view of biodiversity

The tangible focus of this chapter is grasslands of the Serengeti National Park, Tanzania. In that ecosystem, the mammalian grazing fauna combines migratory herbivores that dominate animal numbers and biomass with resident herbivore species that are less common and more sparsely distributed, but account for much of mammalian species diversity.

The Serengeti ecosystem occupies approximately 25 000 km^2 in Tanzania and Kenya just east of Lake Victoria at 1°15′ to 3°30′ S (McNaughton 1983). The first protected area was a lion reserve established in 1929. The present core areas of Serengeti National Park and surrounding multiple-use zones in Tanzania and Masai Mara Game Reserve in Kenya gradually developed over the succeeding 50 years. It is principally designed to protect Earth's largest herds of migratory big game animals, and one of its most diverse large mammal concentrations. There is an annual rainfall gradient from 350 mm in the SE to over 1 m in the NW (Norton-Griffiths *et al.* 1975). This is superimposed on substantial geological heterogeneity from young volcanic soils on the Serengeti Plains to portions of ancient Precambrian shield in the north (McNaughton 1983).

An individual within the Serengeti's 1.5 million head (Campbell 1989) migratory wildebeest (*Connochaetes taurinus*) herd encounters grassland diversity at many different hierarchical levels and many different types of vegetation as it moves through its annual migratory cycle (Fig. 4.1). It will usually be on the Serengeti Plains in the SE during the wet season, nominally from November through May (Norton-Griffiths *et al.* 1975; McNaughton

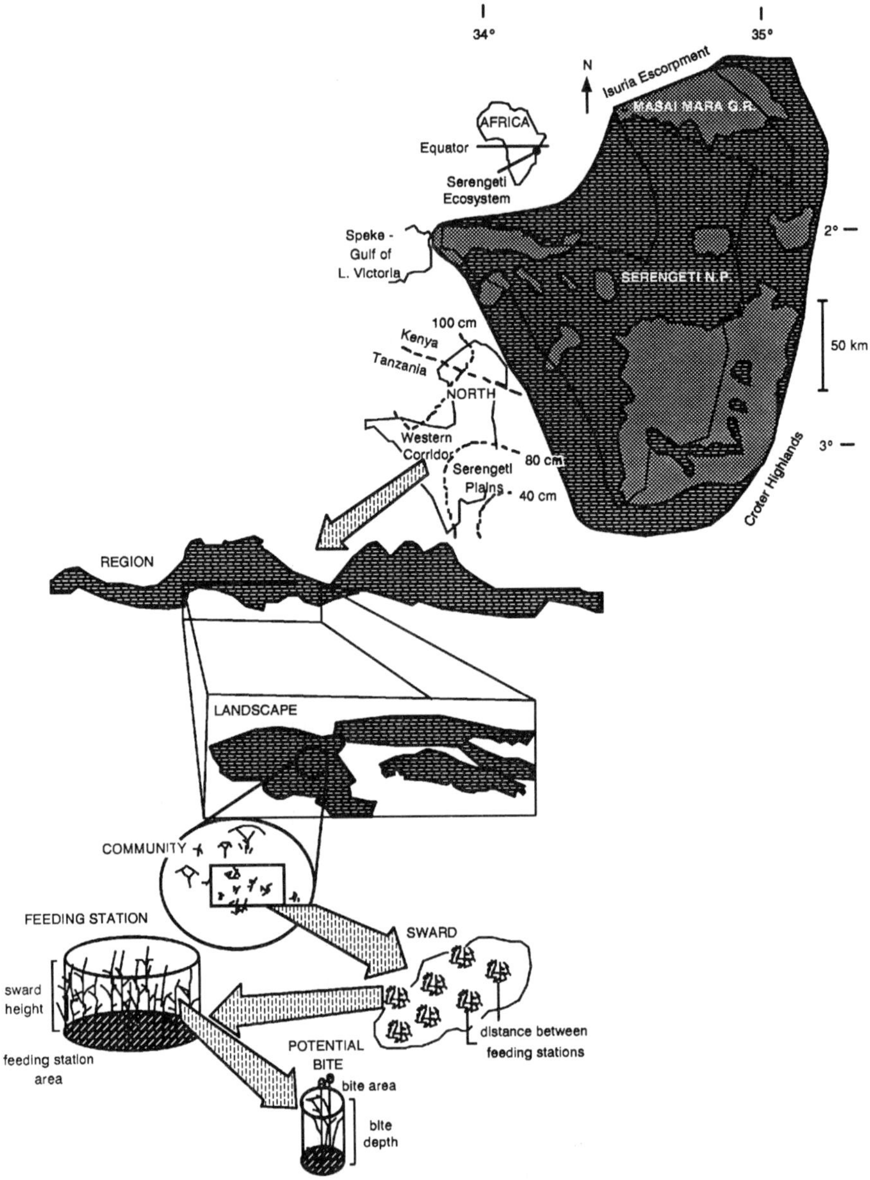

Fig. 4.1 Annual foraging hierarchy for a large, migratory grazing ungulate characterizing the levels at which grazers interact with their habitat in the Serengeti ecosystem. In the ecosystem map at the upper right, dark areas are savannah woodlands and light areas are open grasslands. Isolines on the small ecosystem inset are mean annual rainfall in cm. Both temporal and spatial scales decrease from top to bottom in the figure, with the whole ecosystem exploited on an annual basis and bites taken on scales of seconds. McNaughton (1989*a*).

1985). At seasonal transitions it will move into the Western Corridor, extending toward Lake Victoria, and it will move to the far NW during the dry season peak. The grasslands encountered vary in every conceivable way: species composition, stature, productivity, susceptibility to damage by grazing, frequency of fire, and nutritional suitability (McNaughton 1983, 1985, 1988, 1990).

On the Serengeti Plains in the SE, where the gravid wildebeest females calve, are short grasslands maintained as grazing lawns (McNaughton 1984) by the hordes of wildebeest, 250 000 plains zebras (*Equus burchelli*), and 300 000 Thomson's gazelles (*Gazella thomsoni*), as well as lesser numbers of other grazers (Campbell 1989). As the dry season progresses, the migrants move upward on a rainfall and grassland-stature gradient to the wettest portions of the ecosystem, occupied by mosaics of tall and mid-grasslands in the far north.

The moving wildebeest and its companions in the herd may choose to feed intensively in a given grassland, or to move through it without feeding. Species diversity of the grasslands encountered by herbivores varies widely. The quality of food available to grazers also varies substantially (McNaughton 1988, 1989*a*). Together with choices based on plant communities, the travelling wildebeeste has many other choices to make as it moves from locale to locale, region to region. Choices are influenced by food quantity (McNaughton 1985), foraging efficiency (McNaughton 1984), food quality (Sinclair 1974, 1977; McNaughton 1985, 1988, 1990), the availability of drinking water (McNaughton and Georgiadis 1986), and avoidance of habitats likely to harbor predators (Longland 1991; Deutsch and Nefdt 1992).

Accompanying the hierarchy of spatial scales is a hierarchy of time (Senft *et al.* 1987). That temporal hierarchy reflects the operational interaction between a grazer and its food sources. Bites take place on a scale of seconds, movements from feeding station to feeding station on fractions of minutes, and the migrants entire range is defined by their annual movements.

The Configuration of Biodiversity

Diversity also varies in a hierarchical fashion (Whittaker 1965) with spatial scale (Fig. 4.2). At the lowest spatial level is point diversity: the diversity within some sampling area of sufficient extent that it is completely censused. This is the only level that is fully known and all levels above are estimated from it. Moreover, that knowledge is time-specific, commonly limited to a certain sampling time, and point diversity will vary over both seasonal and longer time scales (McNaughton 1985).

At an intermediate level, there is the diversity of a specific stand, in this case, a grassland sward, called α-diversity (Whittaker 1965). Stands

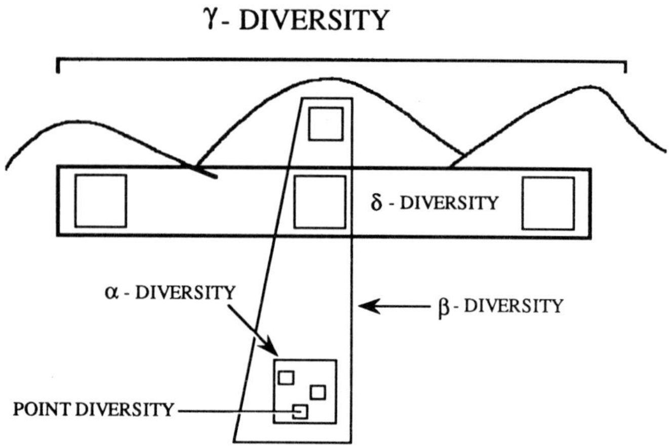

Fig. 4.2 The spatial hierarchy of diversity in vegetation. Point diversity is a unit completely censused, for the taxa studied, at a given time or several times, alpha-diversity is derived from point diversity in a local community, beta-diversity is the differentiation of communities along habitat gradients, delta-diversity is the diversity of similar communities occuping similar habitats, and gamma-diversity is the inclusive diversity within a landscape region.

contribute to diversity at higher organizational levels: β-diversity, the differences between communities along habitat gradients; and γ-diversity, variation at the landscape level (both recognized by Whittaker). In addition, we can recognize δ-diversity, the aggregate variety of communities in similar habitats, i.e. community-types (McNaughton 1983).

This hierarchical approach considers the configuration of biodiversity in a spatially explicit manner. From the small scale of point (sample unit) diversity, it aggregates lower levels into increasingly inclusive categories up to the regional diversity of landscapes.

The configuration of biodiversity can also be considered in a synthetic manner by emphasizing properties of the community and, more broadly, the ecosystems that are associated with biodiversity. Succeeding sections of this chapter treat biodiversity of Serengeti grasslands in both ways.

Pattern diversity

Pattern diversity, the degree of difference from place to place, can contribute to species diversity at all levels above point diversity (McNaughton 1983). At the point level, some very localized area that is entirely sampled, diversity is influenced by only two community properties: the number of species present, and the similarity of their relative abundances (called evenness or equitability). That is, diversity increases with the number of species per unit area or fixed sample size. And, for a given number of species in that standardized sample diversity also increases as the relative abundances of species become more similar, reaching a maximum under the theoretical condition of all species having exactly the same abundances.

The difference in species composition along habitat gradients within a landscape, β-diversity, also contributes to the vegetation biodiversity encountered by a herbivore. As herbivores move along soil catenas from hilltops to lowlands, they encounter different vegetation types (Bell 1970; McNaughton 1983), often with different nutritional properties (McNaughton 1989*a*; Seagle and McNaughton 1992). Variations in available soil moisture and resultant duration of the growing season, soil texture, and soil fertility along catenas will be reflected in differences in community properties at different catena positions (McNaughton 1983).

Partitioning diversity

For Serengeti grasslands, samples from 105 stands throughout the ecosystem (McNaughton 1983) are used here to explore diversity patterns in more detail, and in a more synthetic manner, than in previous analyses. Pattern diversity was evaluated as the proportional difference in species composition (Beals 1969), 1-C, where C, or proportional similarity, is $2w/(a+b)$, with w the sum of species' abundances for the species two samples have in common, and a and b are the sums of species' abundances in samples a and b. Of course, this simplifies to the sum of the minimum relative abundances of species in common between two samples. The dissimilarity index, 1-C, ranges from zero for samples containing exactly the same species with exactly the same relative abundances, to zero for samples with no species in common. This index was applied to samples within stands to evaluate pattern diversity, between stands in a landscape to evaluate landscape diversity, and between stands along habitat gradients to evaluate habitat diversity. At all levels, diversity was evaluated as the Shannon-Wiener index, $H' = -\Sigma p_i * \ln p_i$, an index with well-known properties (Whittaker 1972).

In the Serengeti grasslands, pattern diversity contributes to species diversity at all levels, although its importance varies conspicuously for different levels of biodiversity (Fig. 4.3). This result indicates that patchiness is a major contributor to the diversity of Serengeti grasslands at several spatial scales.

1. α-*Diversity*

At the level of α-diversity, within a sward, about half of the diversity is due to pattern at levels above the point level, in this case plots 0.5 × 1 m in size, and about 40 per cent is due to point diversity (Fig. 4.3). There is an evident positive relationship between pattern diversity, measured as the dissimilarity

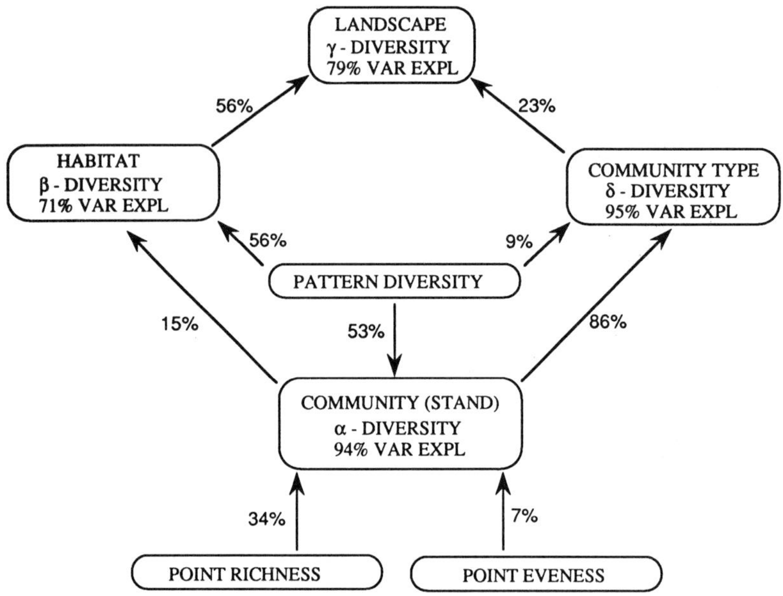

Fig. 4.3 Results of partitioning the variance of Serengeti grassland communities at different hierarchical levels. The percentage within a box is the proportion of the total variance explained and the arrows leading to a box partition that total explained variance among potential contributors.

between point samples, and sward diversity (Fig. 4.4). Although the initial slope is steeper, the positive relation between pattern diversity and α-diversity continued over the entire range recorded. At the level of point diversity, species richness is almost five times as important as evenness. Thus, as a wildebeeste forages through a Serengeti sward, about half the biodiversity it encounters is due to variation from place to place, i.e. vegetation patchiness, about a third is due to the variety of species that surrounds it, and about a tenth due to how similar abundances are distributed among species.

2. *β-Diversity*

Along habitat gradients from rocky hills to flatlands, the *β*-diversity of Serengeti grasslands is principally due to the differentiation of communities along that gradient (Fig. 4.3). The α-diversity of individual stands is only about a third as important as the variation between stands. This indicates considerable difference in species composition of grassland communities along the spatial gradient from rocky hills to flatlands. Associated with this spatial divergence of communities are variations in the mineral nutrient

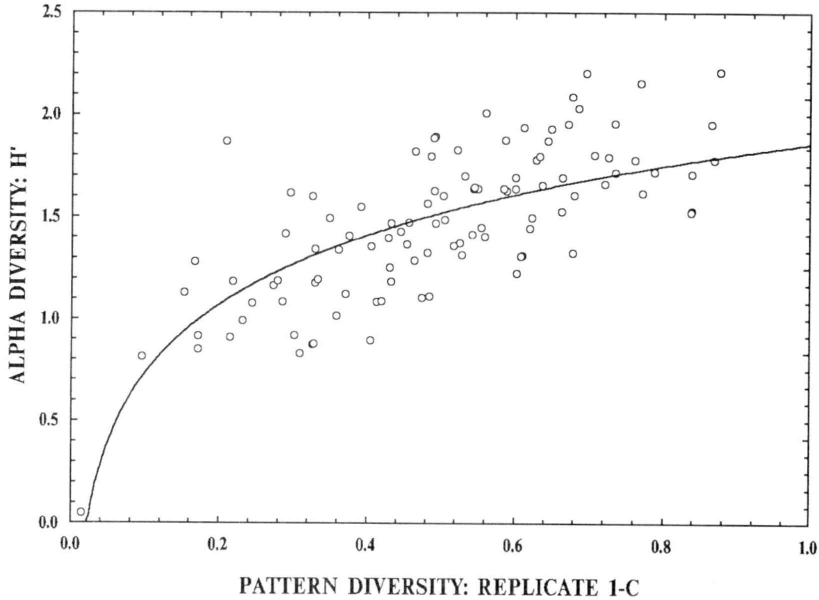

Fig. 4.4 Relationship between pattern diversity, calculated as the average dissimilarity between replicate samples (point diversity) within a grassland stand, and alpha-diversity of that stand. [$y = 1.857 + 0.488 \log X$]; [$R^2 = 0.53$; $F = 117$; $P = < 0.000001$]

properties of forages available to grazers (McNaughton 1988, 1989*a*). The data indicate that a grazer is more likely to be able to balance its diet for nutrients due to this vegetation heterogeneity than would be possible in a homogeneous environment (Seagle and McNaughton 1992). Thus, gradients of grassland biodiversity are associated with patterns of functional properties important to herbivores in the ecosystem. In addition, vegetation stature, productivity, and the grazing that occurs (McNaughton 1985) vary along these gradients of β-diversity.

3. δ-*Diversity*

At the level of δ-diversity, communities occupying similar habitats and classified by the similarity of species compositions of different swards, stand diversity is the overriding determinant (Fig. 4.3). If this was not so, grasslands with different species composition would occupy similar habitats. This integrity of species composition in similar habitats indicates that grassland-types are spread through the ecosystem in similar habitats. These communities are quite similar in species composition and diversity, although they occur in different places. That is, similar communities occur in similar habitats scattered through a landscape or even in different landscape regions. Short grasslands occur on heavily grazed top catena positions in the Western Corridor (Bell 1970) that are similar in species composition to some grasslands on the Serengeti Plains (McNaughton 1983).

Ordination of those communities (McNaughton 1983) indicates that both a biotic factor, grazing, and a physical factor, soil texture, are closely associated with the type of grassland community that occurs in a particular location (Fig. 4.5). The first ordination axis, which recognized a continuum from short (<50 cm canopy height) to tall (>150 cm height) grasses, was principally associated with the sustained grazing intensity that a habitat experiences. Short grasses (communities 1–6) are associated with levels of consumption that maintained standing crop more than 60 per cent below that of fenced stands in the same vegetation. Medium to tall grasslands experience grazing intensities generally below 0.5, ranging close to zero. The second ordination axis, which separated outlier floodplain, tall grasslands, and segregated the mid-height to tall grasslands, was associated with soil

Table 4.1 Dominant species of grasslands in ordination of Fig. 4.5 (McNaughton 1983, modified to conform to Clayton and Renvoize 1982).

Community number	Dominant species
1	*Sporobolus kentophyllus, S. iocaldos, Kyllinga nervosa*
2	*S. ioclados, Digitaria macrblephara*
3	*D. macroblephara, Pennisetum mezianum*
4	*Andropogon greenwayi*
5	*Chrysochloa orientalis, S. ioclados*
6	*Chloris pycnothrix*
7	*S. fimbriatus, Harpachne schimperi*
8	*Heteropogon contortus*
9	*Themeda triandra, Eustachys paspaloides*
10	*T. triandra, Panicum coloratum*
11	*T. triandra, Eragrostis tenuifolia*
12	*T. triandra, S. pyramidalis*
13	*T. triandra, P. mezianum*
14	*T. triandra, Bothriochloa insculpta*
15	*Hyparrhenia filipendula*
16	*Cymbopogon caesius*
17	*Echinochloa haploclada*

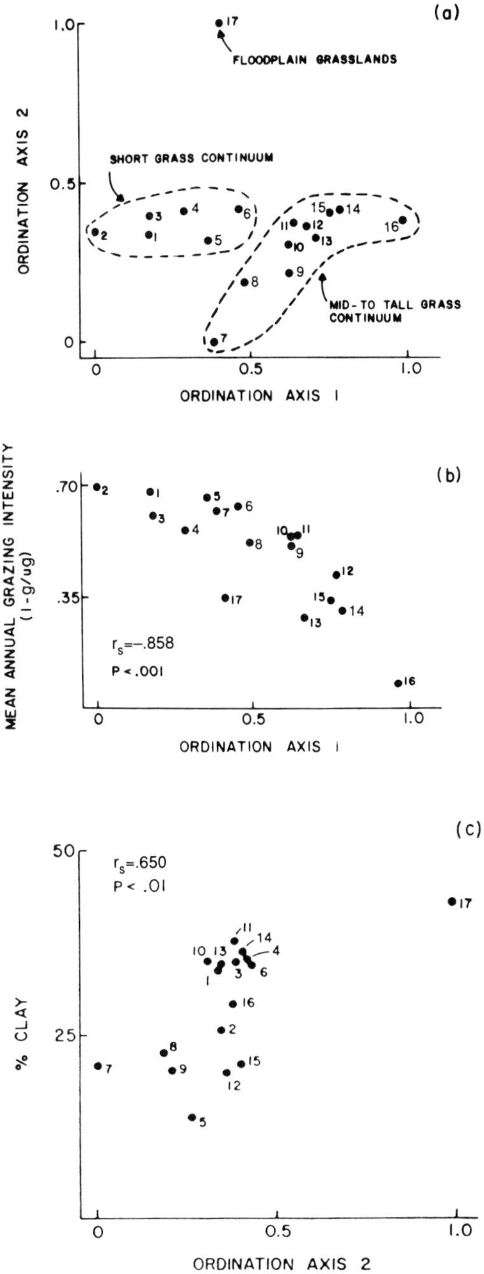

Fig. 4.5 (a) ordination of Serengeti grassland communities in relation to grass stature, with (b) mean annual grazing intensity as the principal correlate of the first ordination axis, and (c) soil texture, here indexed as % clay, as the principal correlate of the second ordination axis. After McNaughton (1983).

texture, here indexed by clay content. Low positions on the axis were associated with low clay, well-drained soils, while high axis positions were associated with tall and mid-grasses on heavy clay soils.

4. *γ-Diversity*

Based on vegetation (Herlocker 1974; McNaughton 1983) and geomorphological (Gerresheim 1970; Epp 1978) criteria, I recognize nine landscape regions in Serengeti National Park (Fig. 4.6). At the scale of γ-

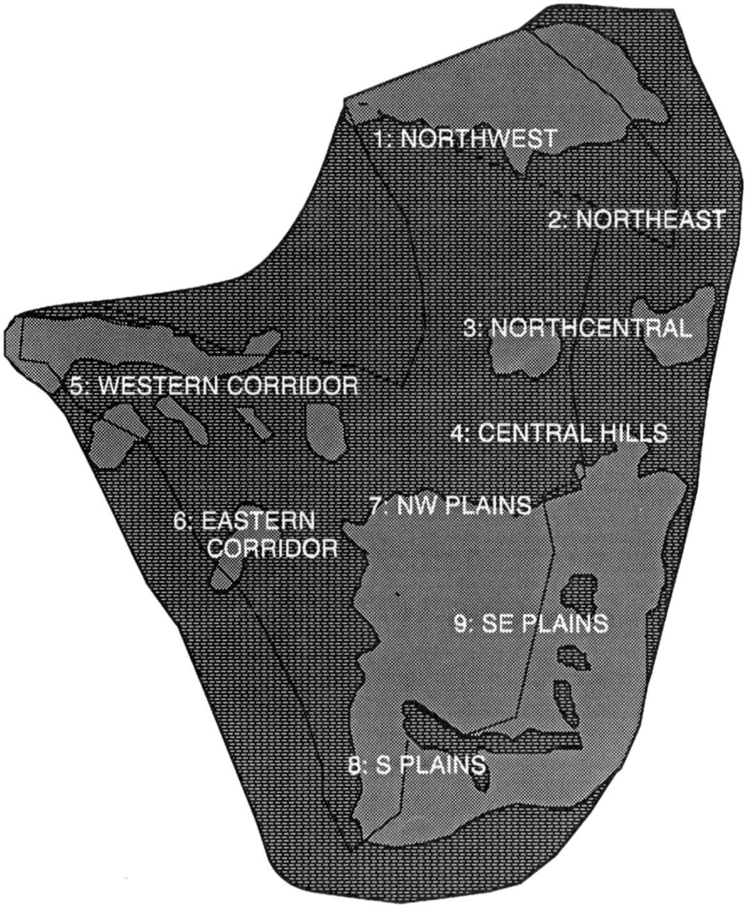

Fig. 4.6 Map of the Serengeti ecosystem indicating nine landscape regions based on grassland species composition and plant and soil nutritional properties (data from McNaughton 1983, 1988, 1990). As in Fig. 4.1, dark areas are savannah woodlands and light areas are open grasslands. Numbers of landscape regions correspond to graphical axes of landscapes in subsequent Figs. 4.7–4.9.

diversity in those landscape regions, the vistas we and herbivores observe at large spatial scales, pattern diversity again is an important element of vegetation diversity (Fig. 4.3). At the landscape level, β-diversity is about twice as important as γ-diversity. Therefore, much of the biodiversity at larger spatial scales is associated with different communities along habitat-gradients in the landscape.

5. *Diversity hierarchy*

Overall, then, there is an evident regularity to the contribution of pattern diversity to grassland biodiversity in the Serengeti (Fig. 4.3). At the level of a stand, the operational unit that we and grazers experience on an immediate basis, α-diversity is about half due to spatial differentiation and half due to localized, point diversity. As we proceed to larger spatial scales along the left hand side of Fig. 4.3, dissimilarity of vegetation composition makes the major contribution to both β-diversity and γ-diversity. Therefore, pattern diversity is a major contributor to vegetation diversity at all spatial scales in the Serengeti grasslands. However, δ-diversity, the biodiversity of vegetation-types in similar habitats, largely associated with grazing intensity and soil texture, is principally due to α-diversity.

Spatially explicit biodiversity

The covariation of α-diversity with landscape region and topographic position provides readily interpretable geographic distributions of community diversity within the Serengeti ecosystem (Fig. 4.7). The highest α-diversity occurs on hilltops in the far NW landscape region. These are tall grasslands with *Hyperthelia dissoluta* as a major species, and with greater species richness than any other grasslands. The soils are heavily leached loamy sands with low fertility and among the lowest pH levels (near 5.5) in the Serengeti (McNaughton 1983, 1985). They commonly occur as an understory in a savanna of *Terminalia mollis* and *Combretum molle* (Herlocker 1974).

The lowest diversity grasslands occur on flatlands in the far Western Corridor landscape; the floodplain grasslands of this landscape region approach monocultures of *Echinochloa haploclada* (McNaughton 1983). Those grasslands can be flooded for several weeks during the rainy season and are often associated with high clay soils, colloquially known as 'black cotton', and a widely dispersed woody overstory of *Acacia seyal* (Herlocker 1974).

Throughout the ecosystem, there is a distinct tendency for α-diversity to be highest on rocky hills and to decline consistently toward lower topographic positions (Fig. 4.7-inset). The high microtopographic heterogeneity on rocky hills, which creates pockets of high moisture due to drainage from exposed rock faces and very shallow soils over large rocks, is associated with their high α-diversity (McNaughton 1983). Therefore, as a grazer

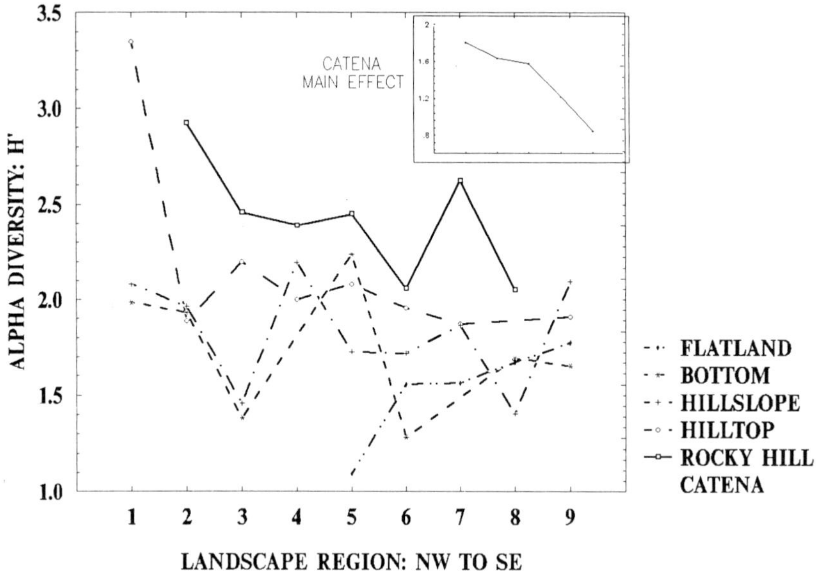

Fig. 4.7 Distribution of alpha-diversity across Serengeti landscape regions (coded as in Fig. 4.6) and topographic (catena) position from rocky hills to flatlands. The catena main effect appears as the inset. [2-way interaction $F (32, 69) = 28.16$; $P < 0.0001$]

proceeds downward from the highest topographic positions in the ecosystem, it progressively encounters, on average, grasslands of lower α-diversity.

An examination of β-diversity as the difference between species compositions of grasslands at adjacent topographic positions reveals a peak at intermediate topographic positions and in landscapes in the middle of the ecosystem (Fig. 4.8). This pattern of spatial variation reveals a tendency for topographic heterogeneity to increase in the northern landscapes (numbered 1–4) at intermediate topographic positions. Highest levels of β-diversity occur in the landscapes (numbered 5–6) of the western Serengeti (McNaughton 1983), and community separation there tends to be highest at intermediate locations along the catena from rocky hills to lowlands. Finally, in the landscape regions of the Serengeti Plains (numbered 7–9), β-diversity tends to increase progressively from upper to lower topographic locations.

This pattern of variation in β-diversity with landscape regions leads to a concordant pattern of variation in γ-diversity across the landscape regions (Fig. 4.9). Because of high spatial differentiation, γ-diversity reaches a decided maximum in the landscape regions of the western Serengeti. It is lower in the northern Serengeti than in the west, but is lowest on the Serengeti Plains. This reflects variation in both topographic variation and

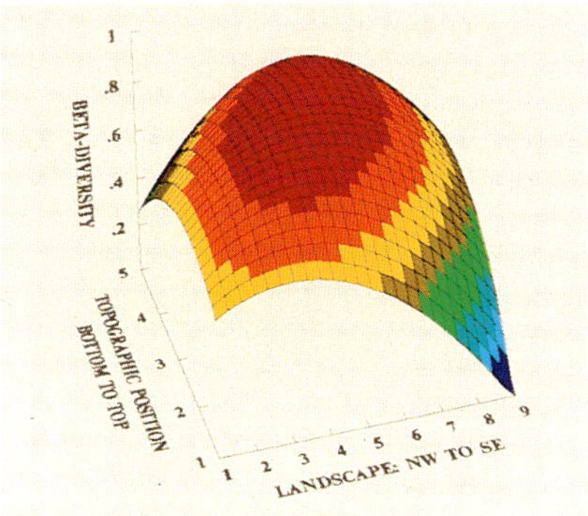

Fig. 4.8 Pattern of beta-diversity in Serengeti grasslands in relation to topographic (catena) position and landscape region (coded as in Fig. 4.6). The surface is a distance-weighted least squares fit derived from 105 grassland stands sampled throughout the Serengeti.

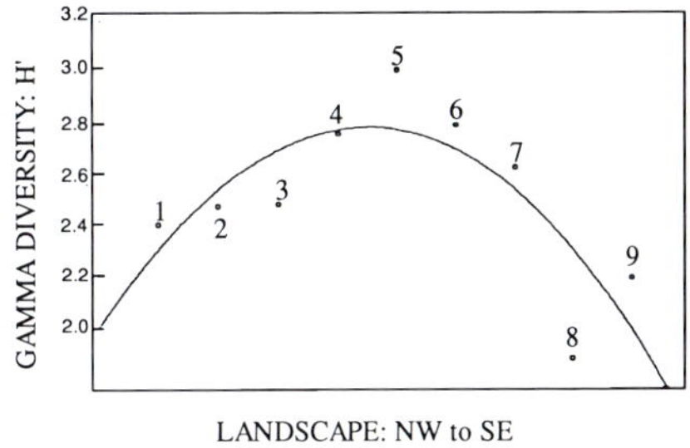

Fig. 4.9 Gamma (landscape)-diversity of Serengeti grasslands in the landscape regions coded in Fig. 4.6. [$y = 2.002 + 0.338x - 0.037x^2$; $t(x) = 2.49$, $P = 0.047$; $t(x)^2 = 2.82$; $P = 0.03$]

(a) 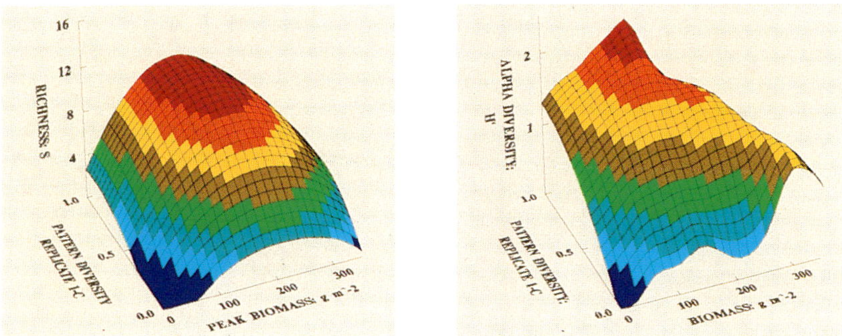 (b)

Fig. 4.10 Pattern of (a) species richness and (b) alpha-diversity of Serengeti grassland stands in relation to peak above-ground biomass and pattern diversity within stands. The surface is the distance-weighted least squares fit to data from 105 grasslands throughout the ecosystem.

(a) 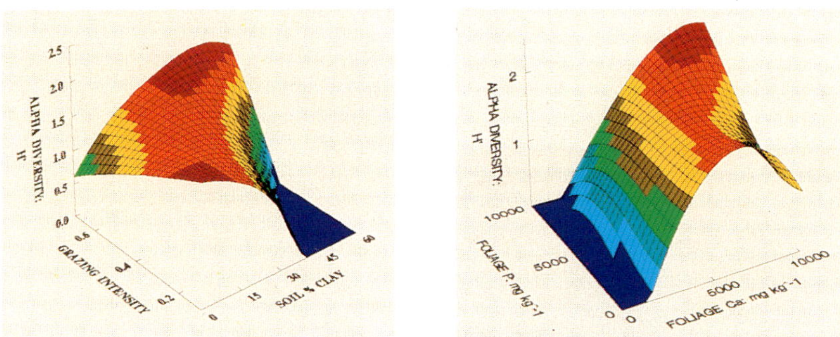 (b)

Fig. 4.11 Grassland alpha-diversity in relation to (a) the two ordination variables of Fig. 4.5, grazing intensity and soil clay content, and (b) two nutrient levels in young leaf blades of the dominant forage grasses in a stand. The surfaces are distance-weighted least squares fit to data from 24 stands for both (a) and (b).

underlying geological heterogeneity (McNaughton 1983). The west is a region of interspersed rocky hills and broad open plains, with a considerable variation in geological substrates from acid dykes, to lake sediments, to greenstones, and ironstones. The Serengeti Plains, in contrast, are gently undulating topographies underlain by deep volcanic deposits from the Crater Highlands.

Vegetation diversity and herbivore coexistence

The configuration of Serengeti grassland biodiversity has several implications for grazers (McNaughton 1983) and conservation evaluation. Patchiness in the local sward will allow a grazer with preferred food sources to walk from patch to patch to feed. If that pattern diversity was missing, forage species would be spread equally across the stand in proportion to their abundance in the sward as a whole. A grazer with preferred food sources would then have to search the entire vegetation for those foods. In addition, different herbivores with different preferred foods can coexist with minimal competition by exploiting different vegetation patches. Vegetation pattern at this hierarchical scale can contribute to the high local diversity of grazers in the Serengeti.

The existence of plant communities, spatially separated stands with very similar species compositions, also can promote the foraging efficiencies of herbivores. By moving from grassland to grassland in similar habitats within a landscape, a grazer need not search the entire landscape for a certain type of preferred vegetation, but merely travel to similar habitats. This also can contribute to biodiversity by allowing the partitioning of resources between herbivore species based on their spatial location.

Grassland biodiversity: a synthetic perspective

In the herbaceous vegetation surrounding Sheffield, England, a peak of species richness was documented at a total above-ground standing crop (biomass+litter) of 600 g m^{-2}, declining above and below this with a long plateau of low richness extending from 900–2600 g m^{-2} (Al-Mufti *et al.* 1977). Grime (1973, 1979) hypothesized that this optimum diversity level is due to an interaction between disturbance and stress, which limits richness at low standing crops, and competition, which limits richness at high standing crops.

Litter and standing dead material never accumulate in such massive quantities in the Serengeti (McNaughton 1985), or other natural African grasslands (Kinyamario and Imbamba 1992), for three reasons. First, large mammals consume much of the material before it dies. Second, they also trample much of the material into the soil where it can be decomposed. And third, fires are a pervasive feature of those grasslands during the dry season, burning much of the dead material residual from grazing and trampling (Stronach and McNaughton 1989).

Nevertheless, species richness of Serengeti grasslands is also related to peak biomass, when pattern diversity is incorporated as an additional variant (Fig. 4.10a). This leads to a dome-shaped relationship between the three variables. At low levels of pattern diversity, S varies only slightly, but is slightly higher at intermediate levels of biomass. As pattern diversity increases, however, the peaked relationship between biomass and richness becomes increasingly evident, reaching a maximum at a standing crop of 150–250 g m^{-2}, and declining at both lower and higher biomass.

The pattern is somewhat more complex when evenness is incorporated and comprehensive α-diversity is considered (Fig. 4.10b). Then, the surface is more convoluted at low levels of pattern diversity and there is a very sharp peak at high levels of pattern diversity near a biomass of 200 g m^{-2}. The contrast between these two surfaces indicate that capturing grassland biodiversity is best achieved at intermediate levels of peak biomass and a high degree of pattern diversity, but the region of capture is much narrower for α-diversity than for species richness.

1. *Biodiversity and ordination variables*

Next, consider two variables, grazing intensity and soil texture, related to species composition as disclosed by community ordination (Fig. 4.5). When α-diversity is plotted against those ordination variables, another complex, twisted surface is revealed (Fig. 4.11a). At low levels of grazing and high clay, the environment occupied by floodplain grasslands near Lake Victoria (community 17 in Fig. 4.5), a very low diversity grassland is present. As the proportion of clay declines at low grazing intensities, there is a progressive increase in diversity. At very high grazing intensities, in contrast, characteristic of short grasslands (communities 1–6 in Fig. 4.5), there is a tendency for α-diversity to increase with soil clay content. Therefore, this twisted surface suggests an interaction between grazing intensity and soil clay content that influences grassland sward diversity. Highest α-diversities span a continuum from tall grasslands on low clay soils at low grazing intensities, to medium grasslands on medium clay soils at medium grazing intensities, to heavily grazed short grasslands on high clay soils.

2. *Biodiversity and mineral nutrients*

Calcium and phosphorus, two important mineral nutrients for herbivores in Serengeti grasslands (McNaughton 1988, 1990), also vary consistently with α-diversity (Fig. 4.11b). Consider a question: if a grazer was searching for high Ca and P because of, for example, the demands of lactation, how diverse a sward should it seek? There is no evident intuitive reason to expect any particular pattern to relate to diversity, but both high Ca and high P are associated with high α-diversity. Therefore, a lactating female or a rapidly growing calf is most likely to obtain high levels of these mineral nutrients by seeking high diversity swards. However, the most favorable balance, particularly considering milk fever in lactating cows, lies within a range of Ca:P between 1 and 2. This pushes the satisfactory nutritional envelope to the right of the diagonal and off the high diversity peak. Assume further a P requirement around 2000 mg kg^{-1}, and maintaining the Ca:P ratio between 1–2; in these circumstances, grasslands from low to relatively high diversities would meet the animals's nutritional requirements, but climbing a surface from near lower left to mid-right, with α-diversity increasing to a peak and then decreasing again.

Biodiversity: what we seek and what we get

1. *Matters of scale*

If a major explicit goal of conservation policy is preservation of biodiversity in natural environments, we are forced to ask: at what scale? The permeating influence of pattern diversity on Serengeti grassland diversity suggests there is no effortless solution to this question.

A second question that is relevant is: when we preserve biodiversity, what else is preserved? Clearly, no thorough answer to this question is feasible. But, using the Serengeti data as a reference point, we can at least conclude that preservation of biodiversity of different levels is accompanied by preservation of different levels of other properties of biological importance.

2. *Community-based strategy*

Only for communities of a specific composition is pattern a minor factor to be considered. Suppose we select, for instance, a *Themeda triandra/Pennisetum mezianum* vegetation-type as a desirable category to be preserved (#13 in Fig. 4.5 from Table 4.1). That community is consistently found in habitats combining a moderate level of grazing intensities and soil textures. Within that range, we can expect more or less the same combinations of species with similar relative abundances. Pattern, divergence from location to location within that habitat combination, will have little effect upon δ-diversity. Instead, δ-diversity is replicated in the α-diversities of the individual swards dominated by *Themeda* and *Pennisetum*. Therefore, preserving that community requires preserving a specific habitat-type. That is simple to conceive.

Since, however, grazing intensity is the first major ordination axis, the *Themeda-Pennisetum* community is biologically coupled, through the grazers, to other communities in the ordination that provide forage to those grazers. It is conceivable to imagine a system of paddocks, structured by the underlying soil textures, with animals rotated among the paddocks to maintain a certain, specified grazing intensity within each paddock. That is formally imaginable, but operationally fragile as a conservation policy.

3. *System-level strategy*

I am not sanguine about the ability of humans to manipulate ecosystems for conservation purposes, unless that manipulation is a carefully thought out means of achieving the goal stated by Elton in the quote beginning this chapter. To achieve that end, we must have more complete understanding of natural ecosystems. Still, time is too short to require complete understanding before we do something, anything perhaps, in most cases. Do we have the ability to manipulate natural ecosystems, specifically those supporting abundant large mammal populations, with feasible procedures that will maintain their essential character and, particularly, ensure the conservation of biodiversity at all trophic levels, from the soil trophic web to the large predators (McNaughton 1989*b*, 1993)? Human intervention, even inadvertent, in grassland and savanna ecosystems has often had catastrophic effects, extending to ecological processes and ecosystem components far beyond specific sites of intervention (McNaughton 1989*b*, 1992, 1993).

To be sure, it is quite feasible to reconstruct a facsimile of the Serengeti's mammalian herbivore and carnivore communities by introducing them to, say, southern Texas or California. There are, after all, tourist parks housing African mammals throughout the developed world, from London's Whipsnade to American safari parks. But the mammals do not exist as elements of a self-perpetuating and self-regulating trophic web. Like cattle and dogs, they are supported by the industrial societies that surround them. No African grasses, trees, shrubs, or microbes will be found. Rather than recreating self-sustaining ecosystems, we create more expansive menageries.

4. *How pattern matters*

Serengeti grassland findings reveal that spatial variation is important at all ecological levels considered in conservation policy. Locally, within swards, α-diversity increases with spatial patchiness of the vegetation. At this and all larger spatial scales, pattern diversity is pervasive. This is manifested in the differentiation of vegetation along topographic gradients, and the expression of that differentiation in landscape diversity.

Diverse landscapes, which, for aesthetic reasons, can be considered a deserving conservation goal in themselves, are characterized by topographic, edaphic, and biological heterogeneity. Mosaics of different grasslands over short distances, less than 5 km, are associated with differences in the abundance of resident animals due to associated nutritional heterogeneity (McNaughton 1988). The high γ-diversity of the western Serengeti National Park supports a diverse fauna of resident animals, including resident herds of the principal migratory species, wildebeeste and zebra (McNaughton 1983, 1985).

The wet season distribution centers of these herds are crucially related to vegetation that is nutritionally sufficient to the needs of reproductive ungulates (McNaughton 1988). Similarly, the wet season concentration areas of the migratory herbivores on the Serengeti Plains also are nutritionally crucial (McNaughton 1990). And, as the data presented here document, variation in nutritional sufficiency is strikingly, and unexpectedly, related to grassland diversity.

5. *Outside reserves*

Private lands, and public lands outside nature reserves, house much of Earth's biodiversity, even in such areas as East Africa (Western 1989). These cultural landscapes, where conservation is a mere by-product, or tolerated ancillary consequence of, other land-use practices are of utmost importance to conservation (Green 1989; Western 1989). Without them, at least one species of mammal once believed extinct and a bird extinct elsewhere in the wild would, in fact, be so.

The rediscovery by naturalists of the believed-to-be extinct black-footed ferret (*Mustela nigripes*) in September 1981 at Meeteese, Wyoming, USA (May 1986), is a signal example of the importance of land outside reserves. A few tolerant ranchers had allowed these animals, and the prairie dogs they depend upon, to continue to exist on the land they owned. Without those open-minded, even eccentric, ranchers this extraordinary predator, a once-abundant animal characteristic of the North American Great Plains, would not now exist. Similarly, the existence of a pocket of the Hawaiian crow or 'alala (*Corvus hawaiiensis*) on private land provides further convincing evidence for the importance of land outside reserves (Culliton 1992).

But, as Elton (1958) sagely observed, a major consequence of much human land-use is a homogenization of the landscape as agriculture, forestry, and other practices convert the variation of natural ecosystems into systems that are homogeneous from the level of point diversity to that of the landscape. In the landscapes of America's grain belt herbicides and fertilizers create a plant point species richness of 1, extending from horizon to horizon in a featureless monotony of zero diversity: no α-diversity: no β-diversity, no δ-diversity, and no γ-diversity.

Landscape homogenization is, perhaps, economically desirable. But it is a conservation cataclysm. The change in farming practices in Illinois after 1960, converting a mixed agriculture with a patchy landscape of pastures, croplands, hayfields, and woodlots, into sustained, unending monocultures of corn and soybeans was accompanied by a decline in grassland bird populations of 80–90 per cent (Illinois Natural History Survey 1983).

Survival of the black-footed ferret and the 'alala depended upon landowners who allowed departures from this zero diversity. We should be grateful to them, encourage them, and convince society to provide them with incentives for land-use policies that maintain the pattern diversity revealed in the Serengeti grasslands. That pattern diversity, I believe, is essential to conservation.

Since mammalian diversity is concentrated in the dry tropics, in savanna ecosystems that mix grassland, shrubland, and woodland (Mares 1992), many of the properties of the Serengeti ecosystem may be widespread properties associated with mammalian biodiversity. A heterogeneous spatial structure may be foremost.

6. *Conserving the unknown, but important*

Uncovering the importance of forage nutritional sufficiency to herbivore abundances and distributions in natural grazing ecosystems indicates that much is still to be learned about the vast landscapes preserved in major reserves (McNaughton 1988, 1990). Similarly, the relationship between Ca and P in forages and grassland α-diversity was unknown until now. Research in the Serengeti ecosystem indisputably indicates that much hidden information also is conserved within that ecosystem (McNaughton 1989*b*). However, the likelihood that such vast areas will be set aside in the future is remote.

Therefore, it is important to recognize that maintaining such expansive wilderness areas will be significant to attaining increased understanding of the essential properties of comparatively undisturbed ecosystems, including those capable of supporting major large ungulate concentrations. Similarly, landscape variety outside reserves is to be treasured.

Among the important properties of large undisturbed ecosystems are a variety of habitats, and associated vegetation-types, at a range of spatial scales. While it is possible to support African ungulates far beyond their natural habitats, and in very uniform environments, the habitats that support them naturally, without supplementation of minerals, water, or veterinary intervention, are characterized by ubiquitous heterogeneity (McNaughton and Georgiadis 1986). The functional properties of that heterogeneity ranges from the minerals accumulated by grass leaves (McNaughton 1988, 1990), to the variation of primary productivity associated with gradients of β-diversity (McNaughton 1985), to the patchy distribution of carnivores (Longland 1991; Deutsch and Nefdt 1992). Without heterogeneity, there is not Nature.

References

Al-Mufti, M. M., Sydes, C. L., Furness, S. B., Grime, J. P., and Band, S. R. (1977). A quantitative analysis of shoot phenology and dominance in herbaceous vegetation. *Journal of Ecology*, **65**, 759–91.

Beals, E. W. (1969). Vegetational change along altitudinal gradients. *Science*, **165**, 981–5.

Bell, R. H. V. (1970). The use of the herb layer by grazing ungulates in the Serengeti. In *Animal populations in relation to their food sources* (ed. A. Watson), pp. 111–24. Blackwells, Oxford.

Campbell, K. L. I. (1989). *Serengeti ecological monitoring programme, SEMP–89–4*. Serengeti Wildlife Research Centre, PO Box 3134, Arusha, Tanzania.

Clayton, W. D. and Renvoize, S. A. (1982). *Flora of tropical East Africa. Gramineae (Part 3)*. A.A. Balkema, Rotterdam.

Culliton, B.J. (1992). 'alala live on in Hawaiian forest. *Nature*, **357**, 277.

Deutsch, J. C. and Nefdt, R. J. C. (1992). Olfactory cues influence female choice in two lek-breeding antelopes. *Nature*, **356**, 596–8.

Elton, C. S. (1958). *The ecology of invasions by animals and plants*, (reprint 1972). Chapman and Hall, London.

Epp, H. (1978). *A natural resource survey of Serengeti National Park, Tanzania*. Serengeti Research Institute Publication (237). Serengeti Research Institute, PO Box 661, Arusha, Tanzania.

Frank, D. A. and McNaughton, S. J. (1991). Stability increases with diversity in plant communities: empirical evidence form the 1988 Yellowstone drought. *Oikos*, **62**, 360–2.

Frank, D. A. and McNaughton, S. J. (1992). The ecology of plants, large mammalian herbivores, and drought in Yellowstone National Park, *Ecology*. **73**, 2043–58.

Gerresheim, K. (1970). Photo-ecological project. *Annual Report of the Serengeti Research Institute, Arusha, Tanzania*, **1970**, 6–10.

Green, B. H. (1989). Conservation in cultural landscapes. In *Conservation for the twenty-first century* (ed. D. Western and M. Pearl), pp.182–98, Oxford University Press, New York.

Grime, J. P. (1973). Competitive exclusion in herbaceous vegetation. *Nature*, **242**, 344–7.

Grime, J. P. (1979). *Plant strategies and vegetation processes*. John Wiley, Chichester.

Herlocker, D. J. (1974). *Woody vegetation of the Serengeti National Park*, Kleberg Studies in Natural Resources, Texas A and M University, College Station TX, USA.

Illinois Natural History Survey (1983). The declining grassland birds. *Illinois Natural History Survey Report*, (**227**), 1–2.

Kinyamario, J. I. and Imbamba, S. K. (1992). Savanna at Nairobi National Park. In *Primary productivity of grass ecosystems of the tropics and sub-tropics* (ed. S. P. Long, M. B. Jones, and M. J. Roberts), pp. 25–69. Chapman and Hall, London.

Longland, W. S. (1991). Risk of predation and food consumption by black-tailed jackrabbits. *Journal of Range Management*, **44**, 447–50.

McNaughton, S. J. (1983). Serengeti grassland ecology: the role of composite environmental factors and contingency in community organization. *Ecological Monographs*, **53**, 291–320.

McNaughton, S. J. (1984). Grazing lawns: animals in herds, plant form, and coevolution. *American Midland Naturalist*, **124**, 863–86.

McNaughton, S. J. (1985). Ecology of a grazing ecosystem: the Serengeti. *Ecological Monographs*, **55**, 259–94.

McNaughton, S. J. (1988). Mineral nutrition and spatial concentrations of African ungulates. *Nature*, **334**, 343–5.
McNaughton, S. J. (1989*a*). Interactions of plants of the field layer with large herbivores. In *The biology of large African mammals in their environment* (ed. P. A. Jewell and G. M. O. Maloiy), pp.15–29. Zoological Society of London Symposia (61). Clarendon Press, Oxford.
McNaughton, S. J. (1989*b*). Ecosystems and conservation in the twenty-first century. In *Conservation for the twenty-first century* (ed. D. Western and M. Pearl) pp. 109–20. Oxford University Press, New York.
McNaughton, S. J. (1990). Mineral nutrition and seasonal movements of African migratory grazers. *Nature*, **345**, 613–15.
McNaughton, S. J. (1992). The propagation of disturbance in savannas through food webs. *Journal of vegetation science*, **3**, 301–14.
McNaughton, S. J. (1993). Biodiversity and stability of grazing ecosystems. In *Ecosystem function of biodiversity* (ed. E.-D. Shulze and H. A. Mooney), pp. 361–83. Springer-Verlag, Berlin.
McNaughton, S. J. and Georgiadis, N. J. (1986). Ecology of African grazing and browsing mammals. *Annual review of ecology and systematics*, **17**, 39–65.
Mares, M. A. (1992). Neotropical mammals and the myth of Amazonian diversity. *Science*, **255**, 976–9.
May, R. M. (1986). The cautionary tale of the black-footed ferret. *Nature*, **320**, 13–4.
May, R. M. (1988). How many species are there on Earth? *Science*, **241**, 1441–9.
Norton-Griffiths, M., Herlocker, D., and Pennycuick, L. (1975). The patterns of rainfall in the Serengeti ecosystem, Tanzania. *East African Wildlife Journal*, **13**, 347–74.
Seagle, S. W. and McNaughton, S. J. (1992). Spatial variation in forage nutrient concentrations and the distribution of Serengeti grazing ungulates. *Landscape Ecology*, **7**, 229–41.
Senft, R. L., Coughenour, M. B., Bailey, D. W., Rittenhouse, L. R., Sala, O. E., and Swift, D. W. (1987). Large herbivore feeding and ecological hierarchies. *BioScience*, **37**, 789–99.
Sinclair, A. R. E. (1974). The resource limitation of trophic levels in tropical grassland ecosystems. *Journal of animal ecology*, **44**, 497–520.
Sinclair, A. R. E. (1977). *The African buffalo*. Uinversity of Chicago Press.
Stronach, N. R. H. and McNaughton, S. J. (1989). Grassland fire dynamics in the Serengeti ecosystem, and a potential method of retrospectively estimating fire energy. *Journal of applied ecology*, **26**, 1025–34.
Western, D. (1989). Conservation without parks: wildlife in the rural landscape. In *Conservation for the twenty-first century* (ed. D. Western and M. Pearl), pp. 158–65. Oxford University Press.
Whittaker, R. H. (1965). Dominance and diversity in land plant communities. *Science*, **147**, 250–60.
Whittaker, R. H. (1972). Evolution and measurement of species diversity. *Taxon*, **21**, 213–51.

5. Ecological measures for conserving terrestrial biodiversity: a New Zealand perspective

I. A. E. ATKINSON

Landcare Research New Zealand, Private Bag 31902, Lower Hutt, New Zealand

Abstract

Efforts by New Zealand to conserve its land-based biodiversity are reviewed, and successes and shortcomings are identified. Several approaches attempting to combine ecological and taxonomic criteria are described to cover the wide range of conservation problems in a biotically and geologically diverse country. The single-species approach has had some notable successes, but has emphasized birds while neglecting other animals and plants. Parallel to this approach has been work at the habitat and ecosystem level. This also has done much to protect biodiversity, but has emphasized forests to the detriment of non-forest communities. To illustrate lessons that can be derived from New Zealand an imaginary scenario is examined: an attempt to protect biodiversity in the country as it was in the 1840s by taking steps that avoid some of the past mistakes. These steps include giving greater priority to lowland areas of dry climate, identifying regions of high species diversity by locating areas of high geological and landform diversity, identifying centres of endemism by searching for areas of long isolation history and/or unusual habitat, and focusing strongly on areas for protection that are most likely to have long-term sustainability by reason of their inherent characteristics. This analysis is summarized as a 4-part general strategy for reducing future losses of terrestrial biodiversity, elements of which may have wider application to the biodiversity crisis now affecting so many parts of the world.

Introduction

This contribution outlines New Zealand's efforts to conserve its land-based biodiversity. It examines the value of geology and landforms for predicting areas of high endemism, species diversity and ecological diversity, and from this develops a framework for minimizing further losses of biodiversity.

Systematics and Conservation Evaluation (ed. P. L. Forey, C. J. Humphries, and R. I. Vane-Wright), Systematics Association Special Volume No. 50, pp. 63–79, Clarendon Press, Oxford, 1994.

New Zealand is mountainous with a strongly regionalized climate. Prior to arrival of humans, 800–1000 years ago, more than 80 per cent was forest-covered. High endemism and low species diversity is found in most plant and animal groups reflecting long isolation. Efforts to protect biodiversity have been pursued at both species and community levels. More recently, biochemical studies have demonstrated how genetic information can change conservation priorities that were based on inadequate taxonomies (Daugherty *et al.* 1990).

The species approach

Several island bird sanctuaries were established late last century in response to the massive decline of the avifauna that followed the spread of alien mammals such as cats, rats, and mustelids. The re-discovery in 1948 of the takahe (*Porphyrio mantelli*), a large flightless gallinule, resulted in the first serious attempt at captive breeding of a threatened species in New Zealand. Protection of this bird is now achieved through management of the surviving wild population in Fiordland, captive breeding, and translocation of the species to islands.

Recovery programmes are in action for many other threatened species involving both mainland management (e.g. Saunders 1990) and translocation to island refuges. The trend is towards greater intervention, exemplified particularly by the success of the Chatham Island black robin (*Petroica traversi*) programme (Merton 1990). Bats, some reptiles, frogs, and large invertebrates (including landsnails), have active recovery programmes. Plants at risk have been documented (Wilson and Given 1989) and remedial action begun.

The species approach has saved or reduced the risk from extinction for several species, while raising public awareness. It has resulted in highly motivated, highly skilled recovery teams and has increased protection for significant areas of habitat used by other species. Nevertheless, an emphasis on birds has led to partial neglect of other groups, particularly plants and invertebrates. Currently, some 300 species of plants and animals are now considered threatened (Molloy and Davis 1992).

The habitat/community approach

The first National Park, Tongariro, was gifted to the country in 1887, only the fourth such park in the world. By 1920 more than 500 scenic reserves totalling 250 000 ha had been established (Dingwall 1981). Crown-owned reserves continued to grow in number and the formation in 1978 of the Queen Elizabeth II National Trust has promoted protection of many thousands of hectares of privately owned land.

The Protected Natural Areas Programme (PNAP) was initiated in the early 1980s to combat the problem of under-representation of non-forest communities. The country was divided into 268 Ecological Districts using differences in landform, vegetation, biota, and climate (McEwen 1987); so far about 57 districts have been surveyed for areas requiring protection.

The quality of many reserves has been compromised by the impact of introduced animals and plants, and huge efforts have been expended since the 1930s to control these effects. Only partial success has been achieved on the mainland, but on islands many introduced mammals have been eradicated.

Altogether more than 30 per cent of the country has some kind of protective status. On the mainland, control of alien mammals is now more focused, while eradication campaigns on islands are gaining momentum. Two failures of the habitat/community approach deserve comment. Reserves remain unrepresentative of some kinds of system: lowland forest below 300 m, tussocklands, wetlands including salt marshes, and dune systems. The PNA Programme has never received the resources necessary for a strong momentum (Ogle 1989), and mechanisms for implementation need to be more vigorously pursued (Dickinson and Mark 1989). A second failure is that attention has sometimes focused too much on species to be protected with insufficient attention given to the long term viability of the protected area itself.

Using geological history and landforms to predict levels of biodiversity

1. *Relationship of endemism to geological history*

(a) Regions of elevated endemism. Regions of elevated plant endemism (Fig. 5.1) are paralleled to some extent by endemism in certain animal groups, particularly geckos and skinks (Pickard and Towns 1988), large landsnails (Powell 1979; Meads *et al.* 1984), some large insects such as giant wetas (Meads 1990), and earthworms (Lee 1959). However, the original parallels between endemic plant and animal distributions have often been obscured by the impact of introduced predators and habitat-loss.

These regions coincide broadly with areas that are geologically old and which have shown greater tectonic stability than elsewhere. This relationship, particularly the influence of the Pliocene distribution of land, was recognized by Climo (1975), and in developing the idea further McGlone (1985) summarized the tectonic history of the country to show how endemic-rich floras were associated with relatively stable areas.

All regions of greater tectonic stability show elevated numbers of plant endemics (Fig. 5.1). The Northland-Auckland region of northern North Island, with 90 endemics, has had a high degree of tectonic stability during

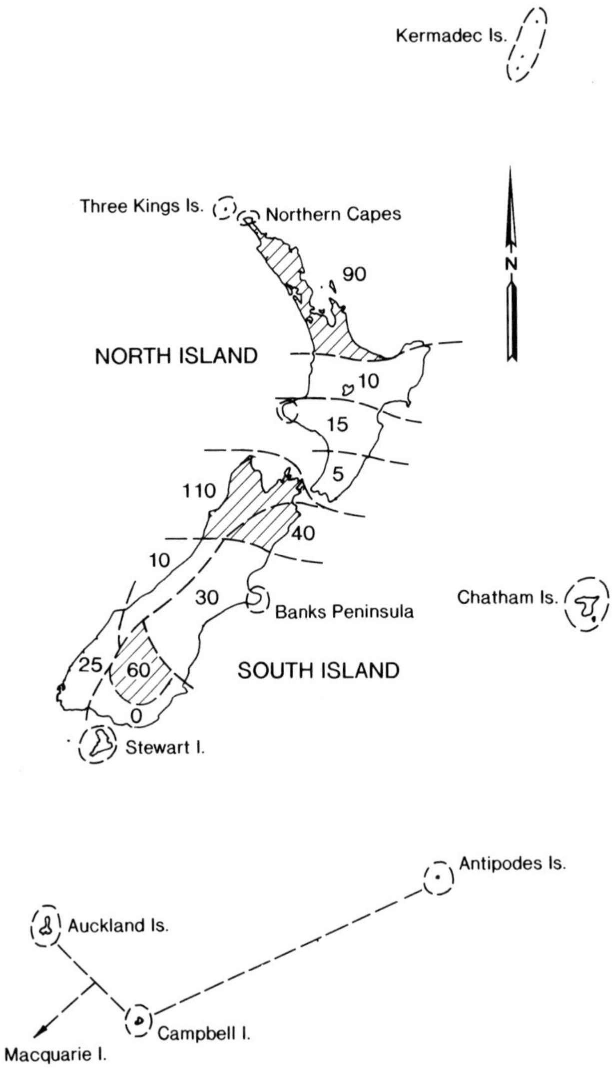

Fig. 5.1 Distribution of endemic species of vascular plants in New Zealand according to regions; numbers rounded to the nearest five species per region. Shaded regions have elevated levels of endemism. Dotted circles indicate localized centres of endemism. Regional boundaries and numbers of plant endemics from A. P. Druce (pers. comm.).

the last 10 to 15 million years (Ballance and Williams 1982). North-west South Island (110 endemics), although subject to uplift and glaciation during the Quaternary, has retained substantial remnants of Tertiary rock cover (Campbell and Johnston 1982). In Inland Otago (60 endemics), a tectonically stable surface covered the area in mid-Tertiary times (Beanland and Barrow-Hurlbert 1988) and remnants of this cover remain.

North-east South Island (40 endemics) is intermediate in number of plant endemics. Much of its landscape is a rugged topography of young ranges with strongly degradational landforms, but the region escaped extensive glaciation (Campbell and Johnston 1982) and many Tertiary cover beds have been preserved (Lensen 1962).

It is apparent that endemism is not related directly to current tectonic stability so much as to past stability, particularly if remnants of the Tertiary cover have survived. Additional factors, such as the history of northern North Island as a warm-climate refuge and the association of specific kinds of rock with certain endemics (Wardle 1991; Lee 1992) have contributed to enhanced endemism.

Remaining regions are geologically more youthful and/or have been subjected to more intense erosion or other disturbance. The central region of the North Island has been subject to repeated volcanic eruptions (Vucetich and Pullar 1973). To the south, in the region centred on latitude 40° S, what is now land became a wide strait in the Pliocene (Fleming 1975). The relatively high number of endemics in the youthful Canterbury region is attributable, in large part, to the survival of Tertiary remnants and older sediments (A. P. Druce, pers. comm.). Fiordland, in south-western South Island, with a comparatively large number of endemics, was severely glaciated. Nevertheless *Nothofagus* forests are extensive there and, given the slow spread of *Nothofagus* spp. from refugia (Rogers 1989), many ice-free sites must have remained between valley glaciers. This has allowed a comparatively large number of endemic species to survive. The south-east part of the South Island, lacking endemic plants, experienced a very severe periglacial climate during the Pleistocene (Fitzharris *et al.* 1982).

(b) Centres of endemism. Within regions of elevated endemism, distributions of endemic species are varied and do not always overlap each other. Centres of endemism (Fig. 5.1) are much smaller, and contain fewer endemics (5–35 plant species) with distributions that overlap to give spatial concentration of plant and animal endemics. All centres are either islands, or have been parts of islands, and are thus likely to be associated with allopatric speciation.

2. *Relationship of species diversity to geological and landform diversity*

(a) Trends in the distribution of species diversity. Features of the distribution of species diversity are shown in Fig. 5.2. The north–south decrease in number

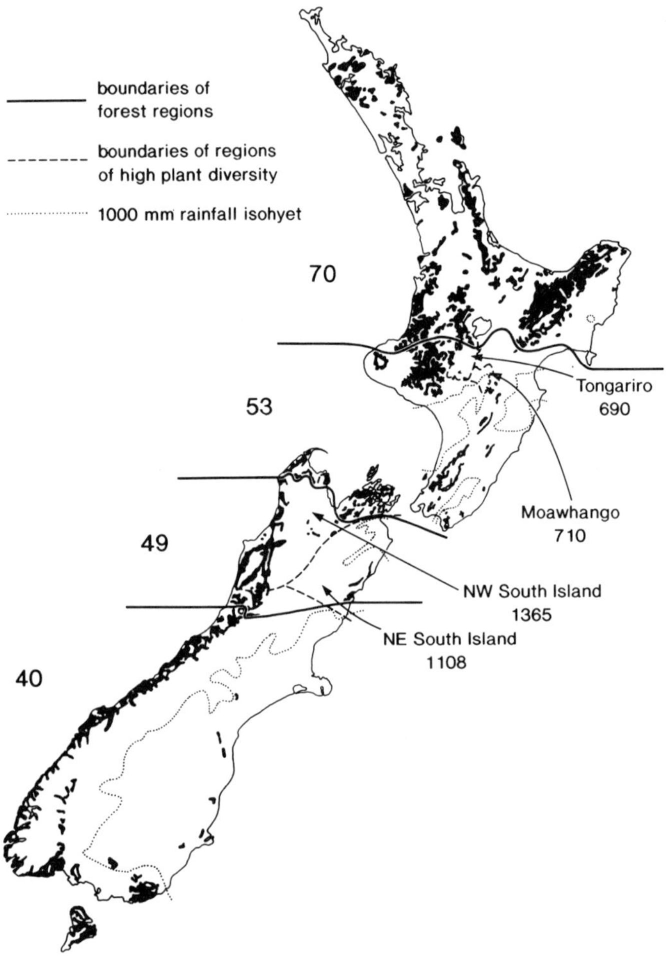

Fig. 5.2 Features of the distribution of species diversity in New Zealand. Shaded areas show extent of remaining lowland forest within each of four regions (after Nicholls 1982). Figures to left of map show species numbers of important canopy trees in each region (Atkinson, unpub.). Figures to right of map are totals of indigenous species of vascular plants recorded in each of the indicated regions and rounded to the nearest five species (data from A. P. Druce, pers. comm.).

of canopy tree species conforms with north to south decrease in species number among woody plants and some animal groups.

Statistical comparisons of regional differences in species diversity are not yet possible because species have been recorded in areas of very different size. It is doubtful whether all regions of relatively high plant diversity are shown in Fig. 5.2, but it is clear that diversity among vascular plants is much more irregularly distributed than the north–south trend described above might suggest.

Some areas of relatively low plant diversity contain a few species that are not found in high-diversity regions. To this extent their biodiversity value is complementary to that of high-diversity regions, regardless of whether or not these sites have other conservation values.

(b) Factors underlying regional differences in plant species diversity. Two factors in addition to tectonic stability appear to underlie the uneven distribution of species diversity. First is the presence of extensive non-forest habitats, as for example those above treeline. The second factor is geological and landform diversity. A variety of landforms implies a variety of habitats. There is no agreed classification for mapping landforms in New Zealand, but the regions of high species diversity shown in Fig. 5.2 are all associated with very topographically diverse landscapes. All contain substantial unforested areas as well.

The relationship between landform diversity and plant species diversity can be illustrated by comparing northern with southern ends of the Ruahine Range, immediately south-east of the Moawhango Ecological Region (Table 5.1). Although the greatest elevation and widest altitudinal range is reached in the central part of the range, it is the northern sector that has the highest diversity of plant species. This appears to be related primarily to the large area of extensive plateaux, mantled with tephra, and the high frequency of Tertiary rock landforms.

North-west South Island is probably the most diverse part of New Zealand in a geological and topographic sense. In some parts, using a map scale of 1:250 000, it is possible to distinguish more than 20 major rock types within an area of 500 km^2 (see Grindley 1961). The figure of 1365 species of vascular plants recorded for this region is approximately 57 per cent of the total New Zealand vascular flora. Some 1230 of these species have been recorded from within the most north-westerly part of the region—north-west Nelson, west of the Motueka R. (A. P. Druce, pers. comm.).

North-east South Island is also topographically varied with localized areas of great geological diversity. In contrast to the north-west sector, however, the greater part is formed from Mesozoic greywacke (Lensen 1962), with a more restricted range of landforms and soils. The limited species diversity of greywacke terrain, compared with terrains of Tertiary rocks, can be seen by comparing the floras of mainly Tertiary rocks and diverse landforms with those where greywacke predominates (Table 5.2). The Ben More–Chalk Range flora includes 31 strictly coastal species from a habitat not represented in the other three areas. Excluding these species still leaves the flora with 81 more species than that of Inland Kaikoura, even though the latter is larger in area and covers more than twice the altitudinal range. Excluding the

Table 5.1 Influence of landform diversity on diversity of vascular plants in the Ruahine Range, North Island

	Northern Ruahine	Central Ruahine	Southern Ruahine
No. vascular plant species[1]	655[3]	601	486
Differentiating factors[2]			
Altitudinal range (m)	200–1437	280–1733	80–1257
Cloud cover and rainfall	Increases from north to south		
Plateaux	Extensive plateaux of Tertiary and greywacke rocks	Extensive greywacke plateaux	Absent
High-altitude limestone landforms (> 1000 m alt.)	Extensive	Absent	Absent
Screes	Unimportant	Extensive	Unimportant
Faulting	Numerous major faults	Numerous major faults	Several major faults
Airfall tephras		Decrease in thickness from north to south	
Non-forest communities	Extensive	Extensive	Minor

[1] Plant species data from surveys by A. P. Druce (unpub.).
[2] Sources of information are Kingma 1962; Elder 1965; A. P. Druce (pers. comm.); and author's observations.
[3] The northern Ruahine mountain range includes part of the Moawhango Ecological Region and this is reflected in the relatively high number of vascular plant species.

Table 5.2 Floras of Tertiary rocks/landforms contrasted with those of greywacke rocks/landforms in Marlborough, NE South Island, New Zealand

	Ben More–Chalk Range	Inland Kaikoura Range	Black Birch Range	Boulder Stream catchment, Wairau Mountains
Indigenous vascular flora	562[1]	450[2]	290[3]	409[3]
Size of study area (km^2)	225	*c*.325	*c*.23	*c*.21
Altitudinal range (m)	sl–1245	240–2885	210–1700	500–2120
Most extensive rock types[4]	Tertiary limestones and Cretaceous sandstones	Mesozoic greywackes	Mesozoic greywackes	Mesozoic greywackes
No. of rock types at map scale of 1:250 000[4]	16	6	1	1
Characteristic landforms	Cliffs, gorges, bedrock steeplands, narrow jagged ridges, limestone screes, colluvial slopes, basins	Long broad slopes, narrow valleys cirque basins, bedrock steeplands	Rounded ridges, bedrock and colluvial steeplands cirque basins	Rounded ridges, bedrock and colluvial steeplands, gorges,
Forest cover	Partial cover	Partial cover	Partial cover	Small areas only

[1] Floristic and environmental data for this range taken from Druce and Williams (1989).
[2] Floristic and environmental data for this range taken from Williams (1989).
[3] A. P. Druce (pers. comm.).
[4] Geological information from Lensen (1962).

coastal species again, the Ben More–Chalk Range flora includes 201 species not recorded for Inland Kaikoura. The remaining areas in the comparison (Table 5.2) are much smaller, but include subalpine and alpine habitats which are absent from the Ben More–Chalk Range. Nevertheless the floras of these two greywacke catchments are much smaller than those recorded for the Tertiary rocks and landforms.

3. *Relationship of ecological diversity to landforms*

Many ecosystem processes are conditioned by landform properties such as slope angle, slope shape, position on the slope, depth and texture of regolith, etc. Differences between ecological systems are often related to landforms, thus landform diversity can be used to identify key areas for protecting a representative range of communities and ecosystem variation.

(a) Examples of the relationship between landforms and ecological diversity. Three cases of very different scale are used to illustrate this relationship. In the indigenous forest catchment of Taita Experimental Station (17 ha), near Wellington (Fig. 5.3), each landform element is required if all types of stand are to be represented. On Little Barrier Island (3080 ha), off the north-east coast of North Island (Fig. 5.4), vegetation types can be related directly to landforms. As before, a full range of landforms must be included if the full range of ecological diversity is to be represented, although this would not require the whole island.

A third example comes from Tongariro National Park (85 215 ha), three orders of magnitude larger than the first example (Fig. 5.5). As with Little

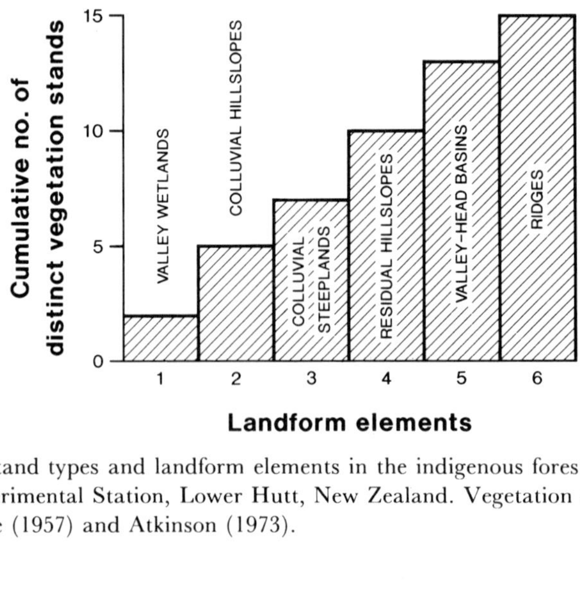

Fig. 5.3 Stand types and landform elements in the indigenous forest catchment at Taita Experimental Station, Lower Hutt, New Zealand. Vegetation and landforms from Druce (1957) and Atkinson (1973).

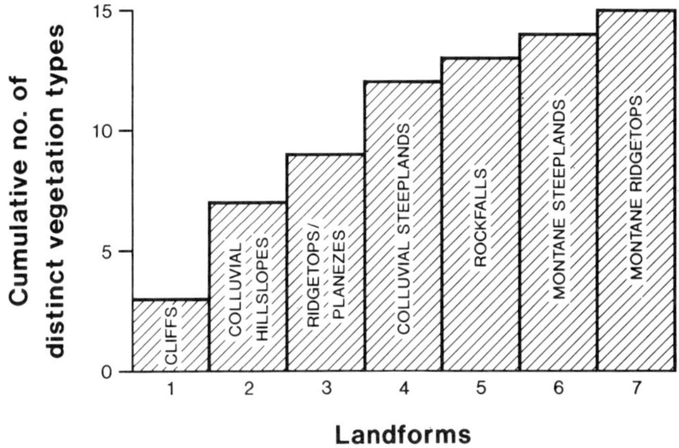

Fig. 5.4 Vegetation types and landforms on Little Barrier Island, Hauraki Gulf, New Zealand. Vegetation and landforms from Hamilton (1961).

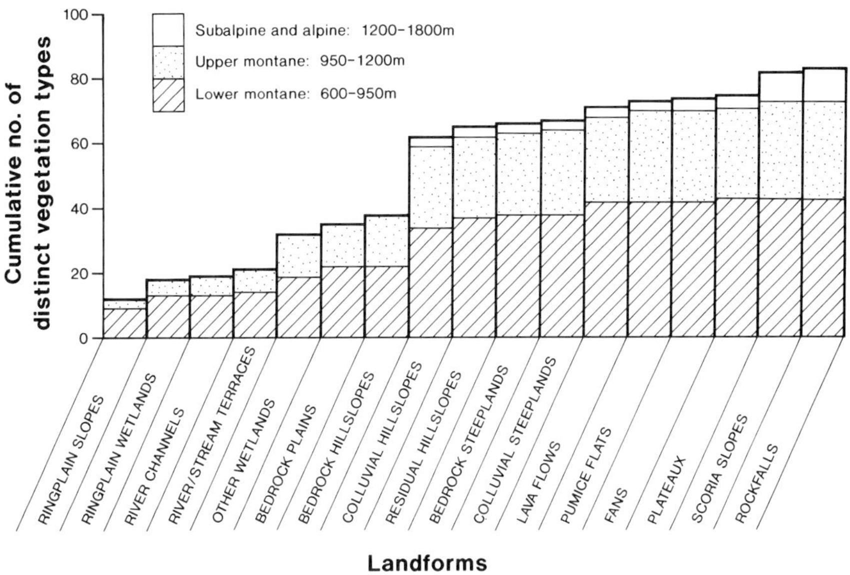

Fig. 5.5 Vegetation types and landforms in Tongariro National Park, North Island, New Zealand. Vegetation and landforms from Atkinson (1981).

Barrier Island, the possibility of having to select part of the area for protection is not an issue. If it were, however, it would not be possible to protect the full complement of 75 vegetation types mapped (Atkinson 1981) without including all 17 types of landform present. Again, this would not require the total area of the Park.

Ecological diversity is not focused directly on species but on spatial differences in communities and ecosystems. An area dominated by a single kind of landform (and community type) may still be of high conservation value if this landform is unrepresented in topographically more diverse parts of the country.

4. *Geological history and landforms as pathfinders to areas of high biodiversity*

Exploration of New Zealand had reached an advanced state by the 1870s, with a quite detailed 1:2 000 000 scale geological map published in 1873 (Burton 1965). Neither the relationship between endemism and geological history, nor the influence of landforms on species and ecological diversity, were then appreciated. Had this been known, it might well have been possible to identify areas requiring protection without waiting for detailed knowledge of species taxonomy and distribution.

These relationships could now be exploited in countries where taxonomic knowledge is still inadequate. Geological history and landform differences could be used as pathfinders to areas of high biodiversity.

A scientific framework for minimizing losses of terrestrial biodiversity

Minimizing further losses of terrestrial biodiversity in New Zealand will require increased effort. Four major priorities are identified, some of which may be relevant to other countries.

1. *Species recovery programmes*

An expanded and comprehensive species recovery programme is needed, and has already been implemented. Targeting threatened species of public concern, or because they are of key significance in the functioning of particular systems, is a necessary and achievable component of protecting communities.

2. *Protection of representative areas of indigenous or semi-indigenous communities*

In New Zealand this priority requires a better resourced Protected Natural Areas Programme that focuses on:

(a) Regions of dry climate. Extensively modified semi-indigenous communities still exist in areas with rainfall below 750 mm. These require protection; even in their damaged state they are often the only sources for dryland genotypes of certain species. Other regions where dry-climate indigenous systems have been destroyed include Madagascar, Hawaii, Australia, countries of the Sahel, and South America.

(b) Regions of relative tectonic stability and/or long isolation. Understanding geological history should be pursued as a way to locate endemic species of plants and animals that have hitherto been missed. The approach is relevant to topographically diverse countries for which detailed faunistic and floristic knowledge is lacking.

(c) Regions of high geological/landform diversity. This can be followed as a method of identifying areas of both high species diversity and high community/ecosystem diversity. As with (b), it should be examined at more than one scale; too broad an approach may miss areas of conservation importance. In countries where taxonomic knowledge of the biota is poor, the use of remote sensing techniques will provide a powerful tool for identifying areas of high biodiversity.

3. *Design of protected areas for maximum sustainability*

This can be achieved within each protected area by:

(a) Maximising the range of landforms and their altitudinal coverage. A greater range of landforms provides greater insurance against the effects of various types of catastrophic event, as well as longer term climatic change. It provides sites for replacement processes that require disturbed surfaces of variable age, and ensures the widest possible availability of habitats for a variety of species.

Some biogeographic principles of reserve design, particularly large size, have been shown to be applicable in New Zealand. Ogle (1987), however, gives examples demonstrating that small patch size can also be necessary for survival of some animal species.

(b) Minimizing edge/area ratios. Designing the shape of a protected area in a way that minimizes edge/area ratios will reduce vulnerability of the area to invasion by alien plants (Timmins and Williams 1991).

(c) Designing protected areas for long-term sustainability. Priority should be given to topographic configurations most likely to contribute to the self-maintenance of the system. It is easier to maintain hydrological and nutrient regimes within whole catchments than with fragmented parts of catchments. The positioning of buffer zones, when these are needed, must be determined by topographic features. Design objectives for a protected area should include landforms or elements of landforms that give additional protection to particular species or systems, e.g. islands, rivers, and cliffs. Cliffs have sometimes provided a second chance to save a species, though often genetically attenuated, from extinction both in New Zealand and elsewhere

(e.g. Cronk 1986). Principles of island biogeography that influence reserve shape should not be allowed to override the principle of landform representation. Conserving the greatest amount of biodiversity in the smallest area will not always be compatible with ensuring long-term sustainability of that area.

4. *Ecological restoration*

This priority stresses re-establishment of both animal and plant species formerly present in an area in order to:

(i) restore the functioning of a damaged indigenous community to a state less influenced by alien factors such as human disturbance, alien plants, and animals or
(ii) restore a particular kind of indigenous community to part of the country from which it has been lost.

The potential of ecological restoration as a method for harnessing human and financial resources in protecting terrestrial biodiversity has been well demonstrated by the Guanacaste National Park programme in Costa Rica (Janzen 1986). Several restoration programmes are being pursued actively in New Zealand (e.g. Towns and Atkinson 1991). Replacing essential trophic links that have been lost from systems, or reinstating more complete feeding guilds, are objectives worthy of attention (Atkinson 1988; Atkinson and Millener 1991).

Conclusion

No single approach for reducing losses of biodiversity will be appropriate for all countries. New Zealand's biodiversity problems differ from those of Australia, as those of both countries differ from the problems facing Europe or tropical America. The framework advocated here is based on studies by many people working in a country that is geologically and topographically very diverse; it is likely to be less applicable to countries of low relief. The challenge is to see how the various taxonomically-based approaches can be integrated with ecological approaches to meet the range of biodiversity problems now facing the world.

Acknowledgements

A substantial part of the species diversity data used in this paper is derived from over 40 years of field and experimental-garden observations made by Tony Druce. I am indebted to him for much helpful discussion and criticism during preparation of this paper and to Colin Ogle, Paul Blaschke, Janice Molloy, Alan Saunders, and Chris Margules for critical comments on the draft. I thank Tessa Roach for typing and

Marcus Smith for preparation of the diagrams. Presentation of the paper in London was made possible by grants from The British Council in Wellington, New Zealand, and The Royal Society of London.

References

Atkinson, I. A. E. (1973). Soils of Taita Experimental Station. *New Zealand Soil Bureau Bulletin*, **32**. Department of Scientific and Industrial Research, Wellington.

Atkinson, I. A. E. (1981). *Vegetation map of Tongariro National Park, North Island, New Zealand. Scale 1:50 000.* Botany Division, Department of Scientific and Industrial Research, Wellington.

Atkinson, I. A. E. (1988). Opportunities for ecological restoration. *New Zealand Journal of Ecology*, **11**, 1–12.

Atkinson, I. A. E. and Millener, P. R. (1991). An ornithological glimpse into New Zealand's pre-human past. *Acta XX Congressus Internationalis Ornithologici*, **1**, 129–92.

Ballance, P. F. and Williams, P. W. (1982). The geomorphology of Auckland and Northland. In *Landforms of New Zealand* (ed. J. M. Soons and M. J. Selby), pp. 127–46. Longman Paul, Auckland.

Beanland, S. and Barrow-Hurlbert, S. A. (1988). The Nevis-Cardrona fault system, Central Otago, New Zealand: late Quaternary tectonics and structural development. *New Zealand Journal of Geology and Geophysics*, **31**, 337–52.

Burton, P. (1965). The New Zealand Geological Survey 1865–1965. *New Zealand Department of Scientific and Industrial Research Information Series*, **52**.

Campbell, I. B. and Johnston, M. R. (1982). Nelson and Marlborough. In *Landforms of New Zealand* (ed. J. M. Soons and M. J. Selby), pp. 285–98. Longman Paul, Auckland.

Climo, F. M. (1975). The landsnail fauna. In *Biogeography and ecology in New Zealand* (ed. G. Kuschel), pp. 459–89. W. Junk, The Hague.

Cronk, Q. C. B. (1986). The decline of the St Helena ebony *Trochetiopsis melanoxylon*. *Biological Conservation*, **35**, 159–72.

Daugherty, C. H., Cree, A., Hay, J. M., and Thompson, M. B. (1990). Neglected taxonomy and continuing extinctions of tuatura *Sphenodon*. *Nature*, **347**, 177–9.

Dickinson, K. J. M. and Mark, A. F. 1989. Tussock grassland reserve design: some practical considerations. In *Management of New Zealand's Natural Estate* (ed. D. A. Norton), pp. 26–30. New Zealand Ecological Society Occasional Publication No. 1.

Dingwall, P. R. (1981). Harry Ell's vision in nature conservation. *Landscape*, **10**, 23–7.

Druce, A. P. (1957). Botanical survey of an experimental catchment, Taita, New Zealand. *New Zealand Department of Scientific and Industrial Research Bulletin*, **124**.

Druce, A. P. and Williams, P. A. (1989). Vegetation and flora of the Ben More–Chalk Range area of southern Marlborough, South Island. *New Zealand Journal of Botany*, **27**, 167–99.

Elder, N. L. (1965). Vegetation of the Ruahine Range: an introduction. *Transactions of the Royal Society of New Zealand Botany*, **3**, 13–66.

Fitzharris, B. B., Mansergh, G. D., and Soons, J. M. (1982). Basins and lowlands of the South Island. In *Landforms of New Zealand* (ed. J. M. Soons and M. J. Selby), pp. 343–66. Longman Paul, Auckland.

Fleming, C. A. (1975). I. The geological history of New Zealand and its biota. In

Biogeography and ecology in New Zealand (ed. G. Kuschel), pp. 1–86. W. Junk, The Hague.

Grindley, G. W. (1961). Sheet 13, Golden Bay (1st ed.). *Geological map of New Zealand 1:250 000.* Department of Scientific and Industrial Research, Wellington.

Hamilton, W. M. (ed). (1961). Little Barrier Island (Hauturu). *New Zealand Department of Scientific and Industrial Research Bulletin*, **137.**

Janzen, D. H. (1986). *Guanacaste National Park: tropical ecological and cultural restoration.* Editorial Universidad Estatal a Distancia. San José, Costa Rica.

Lee, K. E. (1959). The earthworm fauna of New Zealand. *New Zealand Department of Scientific and Industrial Research Bulletin*, **130.**

Lee, W. G. (1992). New Zealand ultramafics. In *The ecology of areas with serpentinized rocks. A world view* (ed. B. A. Roberts and J. Proctor), pp. 375–418. Kluwer Academic Publishers, Amsterdam.

Lensen, G. J. (1962). Sheet 16 Kaikoura (1st ed.). *Geological map of New Zealand 1:250,000.* Department of Scientific and Industrial Research, Wellington.

McEwen, W. M. (ed.) (1987). Ecological regions and districts of New Zealand. Third revised edition in four 1:500 000 maps. *New Zealand Biological Resources Centre Publication No. 5.* Department of Conservation, Wellington.

McGlone, M. S. (1985). Plant biogeography and the late Cenozoic history of New Zealand. *New Zealand Journal of Botany*, **23**, 723–49.

Meads, M. J. (1990). *The Weta Book. A guide to the identification of wetas.* Land Resources Division, New Zealand Department of Scientific and Industrial Research, Lower Hutt, New Zealand.

Meads, M. J., Walker, K. J., and Elliott, G. P. (1984). Status, conservation, and management of the land snails of the genus *Powelliphanta* (Mollusca:Pulmonata). *New Zealand Journal of Zoology*, **11**, 277–306.

Merton, D. V. (1990). The Chatham Island black robin. *Forest and Bird*, **21**, 14–19.

Molloy, J. and Davis, A. (1992). *Setting priorities for the conservation of New Zealand's threatened plants and animals.* Department of Conservation, Wellington.

Nicholls, J. L. (1982). The extent and variability of the native lowland forests. In *Lowland forests in New Zealand. Proceedings of a symposium* (ed. K. Thompson, A. P. Hodder, and A. S. Edmonds), pp. 79–92. University of Waikato, Hamilton.

Ogle, C. C. (1987). The incidence and conservation of animal and plant species in remnants of native vegetation within New Zealand. In *Nature conservation: the role of remnants of native vegetation* (ed. D. A. Saunders, G. W. Arnold, A. A. Burbidge, and A. J. M. Hopkins), pp. 79–87. Surrey Beatty, Chipping Norton, NSW, Australia.

Ogle, C. C. (1989). An overview of reserve design and location in New Zealand. In *Management of New Zealand's Natural Estate* (ed. D. A. Norton), pp. 11–18. New Zealand Ecological Society Occasional Publication No. 1.

Pickard, C. R. and Towns, D. R. (1988). *Atlas of the amphibians and reptiles of New Zealand.* Conservation Sciences Publication No. 1. Department of Conservation, Wellington.

Powell, A. W. B. (1979). *New Zealand Mollusca.* Collins, Auckland.

Rogers, G. M. (1989). Beech and conifer community interactions in Moawhango Ecological Region, North Island, New Zealand. *New Zealand Journal of Ecology*, **12**, 47–61.

Saunders, A. (1990). Mapara: island management 'mainland' style. In *Ecological restoration of New Zealand islands* (ed. D. R. Towns, C. H. Daugherty, and I. A. E. Atkinson), pp. 147–49. Conservation Sciences Publication No. 2. Department of Conservation, Wellington.

Timmins, S. M., and Williams, P. A. (1991). Weed numbers in New Zealand's forest and scrub reserves. *New Zealand Journal of Ecology*, **15**, 153–62.

Towns, D. R. and Atkinson, I. A. E. (1991). New Zealand's restoration ecology. *New Scientist*, **1765**, 36–9.

Vucetich, C. G. and Pullar, W. A. (1973). Holocene tephra formations erupted in the Taupo area, and interbedded tephras from other volcanic sources. *New Zealand Journal of Science and Technology*, **16**, 745–80.

Wardle, P. (1991). *Vegetation of New Zealand.* Cambridge University Press.

Williams, P. A. (1989). Vegetation of the Inland Kaikoura Range, Marlborough. *New Zealand Journal of Botany*, **27**, 201–20.

Wilson, C. M. and Given, D. R. (1989). *Threatened Plants of New Zealand.* Department of Scientific and Industrial Research Publishing, Wellington.

6. Inventories of biodiversity: more than a question of numbers

N. E. STORK

Biodiversity Division, Department of Entomology, The Natural History Museum, London SW7 5BD, UK

Abstract

Estimates for the number of described species in the world of 1.4–1.8 million, and for the total number of species of around 5–15 million, would suggest that we know only a fraction of the Earth's fauna and flora. Other data would suggest that we know little even of the majority of described species. To comprehend the implications of environmental change on biodiversity we need a better understanding of the distribution of species and of the role they play in ecosystems. Increased efforts are required towards compiling inventories of the world's fauna and flora, and monitoring changes. The importance is stressed of focussing such efforts on a range of sites, with the biota of some being intensively studied and fully documented. For wider biodiversity inventories it is necessary to select a range of indicator groups, and criteria for their selection are suggested. The importance is stressed of using 'sampling packages' for study of hyperdiverse groups such as insects. The organisation of international efforts to provide inventories is discussed and a list of recommendations is proposed on these and other relevant issues.

Introduction

In different ways a number of organisations and individuals have recently called for renewed and focused efforts to make inventories and to monitor changes in the Earth's fauna and flora (di Castri and Younès 1990; Margules and Austin 1991; Solbrig 1991*a*, 1991*b*; di Castri *et al.* 1992; Groombridge 1992). These calls have been stimulated by the growing concern about the so-called 'Biodiversity Crisis'. Biodiversity, a term coined in the early '80s,

Systematics and Conservation Evaluation (ed. P. L. Forey, C. J. Humphries, and R. I. Vane-Wright), Systematics Association Special Volume No. 50, pp. 81–100, Clarendon Press, Oxford, 1994.

represents the diversity of life expressed at various levels including genetic, population, species, community, landscape, and ecosystem (IUCN, UNEP, WRI, 1990; Holloway and Stork 1991). However, it is more than this, encompassing concern for the safe management and sustainable utilization of these resources (Reid and Miller 1989).

Several issues, in particular, have raised the importance of biodiversity on the biological and political agenda. First, the FAO (1981) report and others (Myers 1980, 1989) on tropical forest destruction have increased awareness of the escalating rate of deforestation. All indicate a dramatic reduction in the extent of tropical forests over the last 20 years. Second, changes in atmospheric carbon dioxide levels threaten to cause global warming through 'the greenhouse effect'. Coupled with estimates of forest destruction and habitat change are dramatic predictions of loss of species (see Groombridge 1992). Many have used data on the present distributions of groups such as birds and mammals to predict the extinction of many of their species. In doing this they have used species–area theory suggested by MacArthur and Wilson (1967) and extended by many others. This predicts that there will be approximately twice as many species in an area ten times larger than another area; conversely, it is suggested that a ten-fold reduction in area results in halving of the species count. Using this theory, Wilson (1988) extrapolated that 17 500 species are being lost each year (based on a conservative estimate of 5 million species of organisms on earth). In practice, we still know little about how large-scale reductions in area affects species extinction for most groups of organisms.

Finally, our understanding of the true extent of biodiversity has been turned upside down by recent studies which have increased estimates of the world's total species count by ten-fold or more. It is this problem, the fact that we know so little of these species, many of which are more likely to become extinct than to be discovered and described, and the question of how to inventory and monitor the earth's living resources, that I address here. I first examine what we know of life on earth. I then examine the problem of inventorying at the species level. In particular, I discuss the need for both intensive and extensive studies, and for standard sampling methods and indicator groups. I do not consider monitoring in detail but examine some of the issues involved. My comments largely concern terrestrial environments and, in particular, arthropods, since that has been my research focus, although many of these ideas apply equally to other groups of organisms such as those in marine ecosystems. As defined above, biodiversity represents a complex set of interactions at different levels from genes to ecosystems (see Solbrig 1991*a*,*b*), but to many field biologists species are the most recognizable and practical units of biodiversity. I therefore consider inventories of biodiversity largely at the species level rather than at the genetic, ecosystem, or any other level.

What do we know of life on earth?

1. *Species inventory*

The number of recognized species is currently considered to be around 1.4–1.8 million (see Fig. 6.1; and Hammond 1992). These totals are based on various counts made for different groups from the literature (e.g. Zoological Record), and since we lack a single computerized list of all described species the precise total is unknown. For some groups (e.g. flowering plants) computerized lists of species have been or are being prepared. For others such as the insects, which comprise more than 50 per cent of all described organisms, comparable lists are surprisingly absent. If we consider that the mean rate of description since Linnaeus established the binominal system for naming species as 6–8000 species/year, then present rates of about 13 000/year (not including embryophyte plants, see Hammond 1992) are disappointing. Even with modern technology it seems that the rate of species description is depressingly slow. Several authors have drawn attention to the imbalance in the ratio of numbers of specialists to the number of described species for speciose and non-speciose groups of organisms. For example, May (1988) showed that the mean numbers of papers/described species/year (for the years 1978–87, papers listed in Zoological Record) on mammals, birds, insects, and nematodes were 1.8, 1.0, 0.02, 0.002, respectively (but note that figures for mammals and birds were only for taxonomic papers, whereas those for insects and nematodes included papers on all aspects of their biology). Gaston and May (1992) showed also that the geographical distribution of taxonomists, as measured by the distribution of 1200 taxonomists borrowing specimens from The Natural History Museum in London, is heavily biased towards the north temperate regions (Fig. 6.2). Similar results were obtained from analysis of the loans from the Royal Botanic Gardens, Kew (Gaston and May 1992).

Given that there are about 1500–2000 entomological taxonomists in North America alone (Gaston and May 1992), why is it that so few species of organisms are described each year? There are a number of possible reasons and some are listed here. First, in most countries, particularly those that are host to most of the world's species, there is an acute shortage of taxonomists (Gaston and May 1992). Second, is the low priority given to describing species for those best-placed to carry out this task. This is often reflected in poor funding for basic species descriptions. Third, there is a lack of opportunity for existing and potential taxonomists in developing countries to study historical collections and, in particular, the type material of previously described species, which is almost entirely housed in the major museums in Europe and North America. Fourth, most species-rich groups are poorly sampled, with many individual species being known from only single sites.

Lack of availability of new species to describe is not usually a problem,

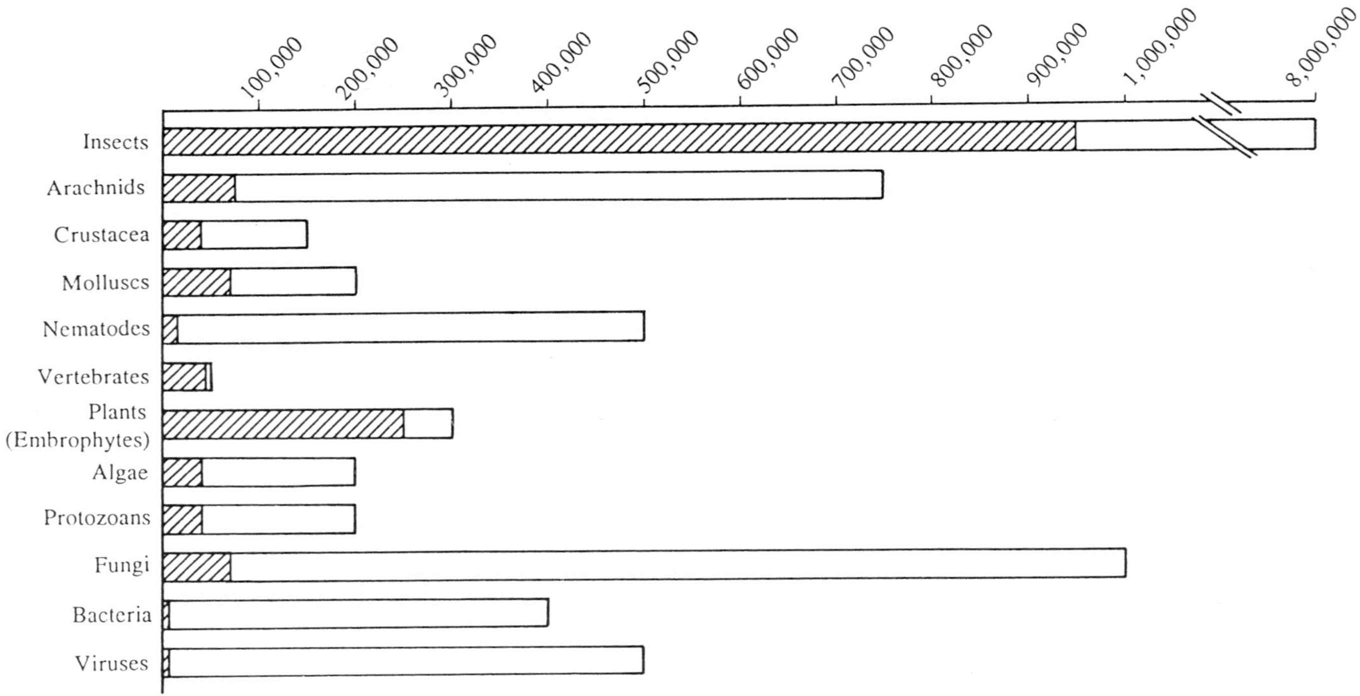

Fig. 6.1 Histograms showing estimated number of species on earth of the major taxa including described (shaded) and undescribed (unshaded) species. After Hammond (1992); note that Hammond was not prepared to extrapolate to more than 100 times the known number of species for groups such as nematodes, fungi, protozoa, bacteria, and viruses.

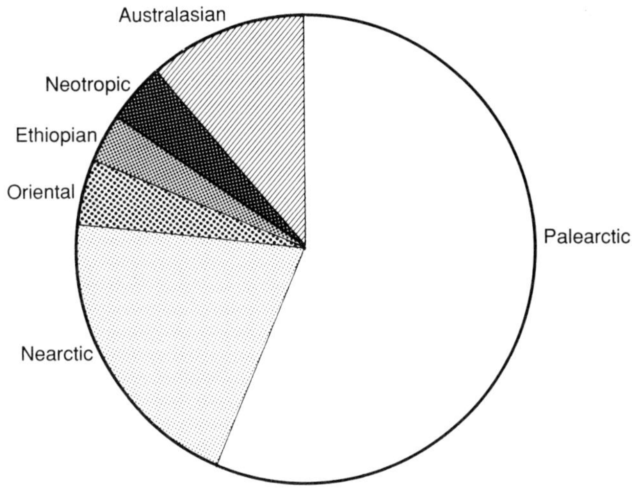

Fig. 6.2 Distribution of about 1200 taxonomists borrowing insect specimens from The Natural History Museum, London (1986–1991) (after Gaston and May 1992).

with sampling programmes in the tropics producing thousands of new species for groups such as insects. The sole world expert on Corylophidae (Coleoptera), J. Pakaluk, found that from a sample of 65 species collected in knockdown insecticide samples from trees in Sulawesi (Stork and Brendell 1990), only two or three were described. Just what proportion of the 2000 species of all beetle groups in these samples are new is not known, but we should be aware that this corylophid example may be unusual. Hodkinson and Casson (1991) found that on average 62.5 per cent of Hemiptera species collected in Sulawesi were undescribed. In a survey of taxonomist's views of the proportion of undescribed species in the world, or different regions of the world, very few taxonomists were prepared to predict more than five times the number of described species for their specialist group (Gaston 1991). So how long will it take to describe all species? Wilson (1988) suggests that the task could be completed by 25 000 professional full-time taxonomists.

2. *How many species are there?*

Having discussed the number of species known and the rate of description, the next question to ask must be 'How many species are there in total?' (May 1988, 1990, 1992; Stork 1988, 1993; Stork and Gaston 1990; Gaston 1991). Most studies indicate that the total number of species on earth is probably around 5–15 million (Barnes 1989; Gaston 1991; Hammond 1992; Holloway and Stork 1991; Stork 1993). Since insects comprise more than half of all described species it is not surprising that the greatest margin of error concerns the number of arthropod species, with estimates ranging from

2 million species (Hodkinson and Casson 1991) to 30 million or more (Erwin 1982, 1988). Fungi and nematodes are also possible super-diverse groups. Hawksworth (1991) suggested, on grounds similar to those proposed by Erwin (1982) for insects, that the number of species of fungi could be as high as 1.5 million. Nematodes are a super-abundant group occurring in almost all terrestrial, marine, and freshwater ecosystems. Small samples of marine sediments are known to contain hundreds of species (Grassle and Maciolek 1992). Some have argued that many nematodes parasitic on plants and animals are host-specific, which would suggest that there may be at least as many species of nematodes as there are species of, say, insects (J. Lambshead, pers. comm.). However, evidence for such a conclusion is, as yet, not forthcoming. Furthermore, species definitions are generally poor for most nematodes and until more is known about their genotypic and phenotypic variation it is very difficult, if not impossible, to estimate the total number of nematode species.

3. *Species abundance and biology*

How much do we know of each of the 1.4–1.8 million described species? For many species of birds and large mammals, but not all, we have a precise knowledge not only of the geographical distribution, reproductive biology, and ecology of individual species, but also their abundance. However, there are still many of the world's birds, bats, and small mammals about which essentially nothing is known. What then of most plants, 'lower' vertebrates, and invertebrates? Cursory examination of larger museum collections and of the literature associated with them suggests that many species are known only from their original descriptions, often from single specimens and single localities. Indeed, one study of a random sample of papers on any aspect of beetle systematics (as included in Zoological Record) indicated that 45 per cent of beetle species (n=186spp) are known from single localities and 13 per cent from single specimens (Stork, unpublished).

Although we can be fairly confident that there are many species with restricted ranges (e.g. species of small isolated mountains and islands), we must assume that many species are considerably more widespread than is presently known (Gaston 1991). For groups such as insects the majority of described species are, and have been, collected by techniques that give little indication as to the feeding or breeding biology of the species concerned. Much is made therefore, of even poorly documented anecdotes. For example, most tiger beetles (Cicindelidae) have sedentary larvae of fairly conservative morphology living in holes in the ground. However, for one pan-tropical lineage of this group, numbering perhaps 600 or more species, and whose adults are unusual in being strictly arboreal, the remarkable twig-dwelling biology of their larvae is known only from one anecdote for one species (Shelford 1905).

4. *The closer you look the more you find*

There is growing evidence that biologists have failed to comprehend other parameters of natural diversity as well as the number of species on earth. Although we know a great deal about biomass in agricultural and fishery systems, and estimates of global primary productivity have been made on a number of occasions, we really know little of the true abundance of life on earth and hence consequently of its biomass. Several recent findings support such conclusions. Much has been made of possibly new, but probably rare, groups of organisms in the deep sea. However, apparently extraordinarily widespread and common marine organisms have also been overlooked. The bacterial genus *Synechococcus*, discovered only some fifteen years ago, is now believed to occur at densities of 1 000–10 000 cells per ml throughout the surface of the oceans. Furthermore, molecular studies indicate that so-called 'strains' of *Synechococcus* are in fact as distantly related to each other as are orders and classes of eukaryote organisms. Considering that more than two-thirds of the Earth is covered by oceans, this taxon must rank as the commonest on Earth (Solbrig 1991*b*).

Our estimates of numbers of terrestrial arthropods are questionable. In one extremely simple examination of arthropod abundance in one hectare of Indonesian lowland rain forest, samples were collected from each of five main habitats; soil, leaf litter, herb layer, tree trunks, and tree canopy (Stork 1988; Stork and Brendell 1993). All sampling methods produced quantitative samples of the arthropods in a given volume of the sampled habitat. From the resulting data a minimum estimate of 45 million individuals for total arthropod abundance in a hectare of forest was obtained. Tree trunks and the herb layer apparently contributed less than 5 per cent to this total, with soil (55 per cent) and leaf litter (14 per cent) combining to produce twice as much as the canopy (27 per cent). However, several studies indicate that the results of this hectare analysis are only a first step, and that more sophisticated sampling methods are required for more accurate estimates. Adis (1987, 1988), using a modified Kempson extraction apparatus, found that arthropod abundance in Amazonian soils was more than 10 times that recorded by Tullgren funnels in the Indonesian study. This difference is probably due to the greater efficiency of the Kempson apparatus at extracting micro-arthropods, particularly Collembola and Acarina (although there may be true differences in arthropod abundance between these sites). Evidence from another extraction method, where arthropods are removed from soil by washing in carbon tetrachloride, indicate that even Kempson-based estimates could be low. André *et al.* (1992) quote figures of 175 000 and 190 000 individuals/m^2 (mostly Collembola and mites) for arthropod densities in French and Belgian sites. If the same densities were true for Indonesian lowland forests then the 45 million estimate discussed above would be too low by more than a factor of 10. Unfortunately, the efficiency of carbon tetrachloride extraction is very low for soils that are rich in organic matter. Recent studies of lowland forest in Cameroon indicate an abundance and density of termites equivalent to 90 million individuals per ha in some forest types (P. Eggleton pers. comm.).

These examples demonstrate the difficulties in comparing data derived from different study sites and continents, different seasons, and from different sampling methods. Furthermore, comparative studies of the efficiencies of these methods in temperate and tropical ecosystems are urgently required (see *Sampling packages* below).

The biodiversity inventory

1. *Why make inventories?*

The sections above indicate just how little we know of the abundance of different groups of organisms, the variation in that abundance through time and space, the number of species for different groups, and their geographical distributions. Only for a very few groups, such as birds, large mammals, and some plants are these patterns clear. Changes in the abundance of these groups as a result of environmental change have been more or less monitored over the last 100 years. From such data, Simberloff (1984) has suggested that, at present rates of forest destruction, 12 per cent of bird species will be extinct in the next 100 years. Similarly, Myers' (1988) examination of ten areas of the world with high endemism, his so-called 'hot spots', led to his suggestion that 7 per cent of plant species would become extinct through loss of habitat in these areas. Intensive studies of some groups have provided clear warnings to biologists and policy makers, and have played a major part in stimulating the protection of biodiversity through the Rio Convention. However, we are really unsure as to how the conclusions of studies such as those for birds and flowering plants can be interpreted for other groups of organisms, particularly the super-diverse groups such as insects, fungi, and microorganisms. Are similar or even higher proportions of species in these groups likely to become extinct in the future, or are the population dynamics and distributions of the species of the super-diverse groups such that the majority are not as threatened as better known groups of organisms? Through compiling inventories we are more likely to determine the extent of species ranges and to make better predictions of the consequences of climate change. Such information also will be invaluable to those making decisions regarding areas to conserve or manage for biodiversity (see Vane-Wright *et al.* 1991).

Recent attention has focussed on the role of biodiversity in the functioning of ecosystems (di Castri and Younès 1990; Lawton and Brown 1993). Are all species essential for the stable existence of an ecosystem or can many be lost before the system deteriorates? This and similar questions are being examined at a time when we know that an unknown number of species are being lost. We know that some species have a critical role in ecosystems and that their loss results in the breakdown or change of that ecosystem. Such species vary from large plants (e.g oak trees in an oak woodland) to small insects (e.g euglossine bees or fig wasps that pollinate brazil nut and fig trees, respectively). Some species may play a more subtle role in generating and maintaining the ecosystem that others utilize. For example, such 'ecosystem generators' or 'ecosystem engineers' as earthworms and termites help to produce soil in many parts of the world (Stork and Eggleton 1992; see also Hawksworth 1992 for the role of microorganisms). Knowing what organisms are present and in what numbers, through intensive study of sites, will help to determine the roles played by groups and by individual species, and the ecological principles underlying the maintenance of species assemblages.

To those concerned with the sustainable utilization of land resources it is important to know how biodiversity is affected by different land use strategies. At present it is impossible to make realistic predictions or to put a biodiversity cost on particular land uses. Many now believe that site inventories are a priority for conservation and sustainable utilisation, and for determining that cost. Janzen (1993) has suggested that conservation can only be achieved through education and commitment of local peoples, such that they become convinced it is in their own interests to conserve their biodiversity. Education and involvement of local peoples in the natural history of their lands is the way forward. Janzen believes that the first step towards sustainability is through compiling an inventory of biodiversity on an unprecedented scale. Such an endeavour is being carried out in Costa Rica by specialist taxonomists throughout the world and by locals and nationals at INBio (Instituto Nacional de Biodiversidad), the new non-governmental institution. One of INBio's functions is to make possible the inventory of Costa Rica's estimated 500 000 species of animals and plants in 10 years. An important additional function of INBio is to expand the benefits to the endemic population, which come from chemical companies prospecting plants and animals for useful chemicals. The agreement between INBio and Merck & Co. is a major landmark in this process. INBio has also led the way with a new concept: 'parataxonomists' (Gamez 1991). These are local people with no formal biological training, who are being taught by skilled taxonomists to collect, prepare, and sort specimens of plants and animals. This expansion of the taxonomic workforce has been welcomed by many, although possible dangers lie in considering parataxonomists as substitutes for trained graduate taxonomists. Cranston and Hillman (1992) for example, discuss the limitations of using technicians rather than taxonomists for assessments of biodiversity from freshwater habitats (see also Oliver and Beattie 1993, and this volume).

2. *Indicator groups and their selection: choices?*

It is clear that unless there are radical changes in the way we describe species, and unless there is massive increase in taxonomic man-power, the vision of completing an inventory of the Earth's fauna and flora in the next 20–50 years seems out of the question (Solbrig 1991*b*; di Castri *et al.* 1992). Therefore, in order to have a clearer picture of the parameters of biodiversity, how species are threatened and of their role in ecosystem function, we need to study selected groups of organisms such that these should act as indicators for others. No one group is typical of all organisms, and therefore a range of groups needs to be selected with criteria for selection depending on the questions being asked (Sutton and Collins 1991; di Castri *et al.* 1992). Important factors in choosing a range of groups are listed in Table 6.1.

Invariably, taxonomists give excellent reasons for using their own specialist groups as an indicator group for change (e.g. Pearson and Cassola 1992 for Cicindelidae), and therefore selecting a list of, say, less than 100 taxa is not easy. However, in practice, both funding for and availability of taxonomic expertise are often limiting (hence, a suitable definition for an indicator group could be 'that for which a taxonomist can be found'!). Well-organized attempts at compiling inventories will invariably include birds, mammals, and plants. For the insects, groups such as butterflies, dung beetles, or tiger beetles are often selected. However, although taxonomically well-known, these groups are not usually biologically representative of insects as a whole. In many instances a 'shopping basket' of insect groups of varied biologies might be more suitable. The trade-off for using such a 'shopping basket' of groups is that almost certainly little will be known of the biologies of individual species. The advantage is that a broader view of the effects of change on large numbers of species of insects may be obtained. An additional advantage of the 'shopping basket' approach is that a large number of groups can be sampled using standard 'sampling packages' (see below).

3. *Sampling packages*

For a few sites around the world we have a great deal of information about the diversity of their faunas and floras, and for some we also have a reasonable historical record. Using this information to help determine where and what to conserve, or how to monitor change is often made extremely difficult, largely through lack of comparability of the sampling methods and effort used to collect specimens. Even where such data are available, lack of continuity renders comparative analysis extremely difficult (see earlier discussions on abundance of soil arthropods). Historical collections of organisms in museums and herbaria can provide valuable information on past and present distributions. However, because they have been compiled to

Table 6.1 Important factors for choosing indicator groups (see text for explanation).

(a) Logistic factors	
	(1) Ease of sampling
	(2) Ease of sorting to species
	(3) Availability of relevant taxonomic expertise
(b) Ecological factors	
	(4) Representatives of all trophic levels and major functional guilds
	(5) Full ranges of body sizes and growth forms
	(6) Representatives from low, medium and high diversity groups
	(7) Wide range of host-specificities
	(8) Wide range of known geographic distributions
(c) Environmental factors	
	(9) Recognized importance to agriculture, environment etc.

maximise the biological representation of faunas and floras of different regions—and usually the most accessible parts of those regions—their use is generally very limited.

Standardization and calibration of sampling methods are extremely important for comparisons of sites in time and space (Hammond 1990, 1992; Hammond and Harding 1991; Coddington *et al.* 1991). For a few groups of organisms, such as birds and amphibians, detailed sampling packages for inventorying and monitoring have been produced. The 'shopping basket' approach of sampling representative ranges of invertebrate indicator groups, as discussed above, also requires use of standardized sampling package. Many different methods are used by entomologists to collect insects, but until recently little research had been undertaken to calibrate the resulting samples against known standards. Hammond (1990) has attempted to do this for tropical forest beetles in an intensive study of the beetle fauna of an area of forest in N. Sulawesi. The need for low cost standard sampling packages is greatest in tropical regions. One such package has been devised and tested by Gadagkar *et al.* (1990) and serves well as a case study of this technology.

4. *Indian case study*

Gadagkar *et al.* (1990) tested a sampling package for insects suitable for use by those with modest research budgets. This included a small light trap run on dry cell batteries, sweep net sampling, pitfall- and baited-trapping. These methods were operated under standard conditions and for standard periods of time, in 3 ha plots at each of 12 sites representing different types of forest reserves in the Uttara Kannada district of the state of Karnataka, southern India. Samples of insects were sorted to orders and then to Recognisable Taxonomic Units (RTUs, i.e. sorted to apparent species without naming them). A total of 16 852 individuals were sorted to 1789 RTUs, cross-referenced between sites. The mean number of RTUs/ha plot was 101.1 (max. 239, min. 43, standard deviation 45.2). Amongst other things, Gadagkar *et al.* (1990) used these data to show that the highest and lowest numbers of species were found in *Eucalyptus* and teak plantations, respectively. Natural forest plots were between these two extremes of monoculture plantations. They made no attempt to compare the species composition of the sites.

This exercise shows that sampling a wide range of insect groups across a number of sites is feasible, rapid, and relatively cheap. However, the sampling methods are all activity based (i.e. sample moving insects) and hence can be influenced by environmental conditions. Light traps, for example, are notoriously variable in their catch depending on the lunar cycle and exact location. Many of the methods used are open to considerable human bias (e.g. collecting and sorting of over 1700 RTUs). For example, were trained taxonomists involved in the sorting or checking RTUs (e.g. see Cranston and Hillman 1992)?

Interpreting the results of a sampling programme such as that of Gadagkar *et al.* (1990) is difficult. The main result of their study, that monoculture forests have more diverse insect faunas than those of mixed forests, may be spurious and indicate that a larger proportion of these faunas were actively seeking scarcer resources. Another question to ask is what the samples actually represent. In other words, what parts of the local fauna have been sampled and, more importantly, what portion has been omitted. In the Indian case study we do not know whether 1, 50, or 100 per cent of the fauna has been sampled and how this percentage varies from site to site. This is where intensive and long-term sampling of at least one site would have been useful. Hammond (1990) and Stork and Brendell (1990) tested this for an area of forest in N. Sulawesi. They sampled at least 5000 beetle species from the Toraut region in the Dumoga-Bone National Park, using a wide range of intensive sampling methods which allowed the calibration of these methods against a 'known' total.

One factor rarely considered in ecological papers is the standard of taxonomic sorting. This can be extremely variable but, unlike statistical tests, no formal standards are possible. In my experience, editors of ecological journals often take authors to task over minor statistical problems, but rarely question their taxonomic standards even when a paper deals with a group of organisms that is species-rich or taxonomically difficult.

Organizing future efforts to inventory

1. *Intensive versus extensive sampling*

The IUBS-SCOPE-UNESCO initiative to understand how ecosystems function (di Castri and Younès 1990) proposes to determine the role of organisms in ecosystems, and posed the question as to whether all species are essential for sustainable maintenance of ecosystems. This implies that we know all the species, their population dynamics, and their food web structure in ecosystems. In reality it would be fair to say that we do not know this for any site in the world, not even for the least species-rich temperate or tundra ecosystems. A detailed understanding of the total biodiversity of a few sites around the world, and the way that organisms at these sites interact, would provide the necessary relational context to understand less intensive sampling programmes at other sites. For example, an extraordinary observation such as the discovery of 739 species of chalcidoid wasps represented by a total of 1455 individuals in a sample of 25 000 individuals of all arthropods from ten trees in Borneo (Noyes 1984; Stork 1991), has little value unless we know what total chalcidoid diversity is at similar SE Asian sites. Second, we need to be able to compare this sample with those from other sites. It is therefore essential that at least a few sites in the world are intensively sampled and their levels of biodiversity calibrated against scales of time, season, ecosystem, habitat, and sampling effort. For these reasons intensive sampling of a few sites is seen as an essential part of an international programme to compile inventories and to monitor changes in the world's fauna and flora (Solbrig 1991*b*; di Castri *et al.*. 1992). Intensive studies would more clearly identify key species and indicator groups for the extensive inventorying process. Less intensive studies of indicator groups or key species at a wide range of sites (so-called extensive studies) will aid in more rapid listing and mapping of these taxa and monitoring changes (Fig. 6.3).

2. *How many sites and how many ecosystems?*

Compiling species lists has always been an *ad hoc* process, with intensity of collecting effort having more to do with logistics and politics than biology. Thus, for example, the faunas and floras of Java and Panama are better known than New Guinea or Guatemala because of the historical interests of The Netherlands and USA, respectively, in the former countries. The *ad hoc* nature of past (and present) efforts to make inventories is responsible, to a great extent, for our poor understanding of biodiversity. A more organized programme, such as DIVERSITAS (di Castri *et al.* 1992), is seen by some as a way of overcoming this problem. Unfortunately, the distribution of land masses and of oceans is not regular, neither are political boundaries, and

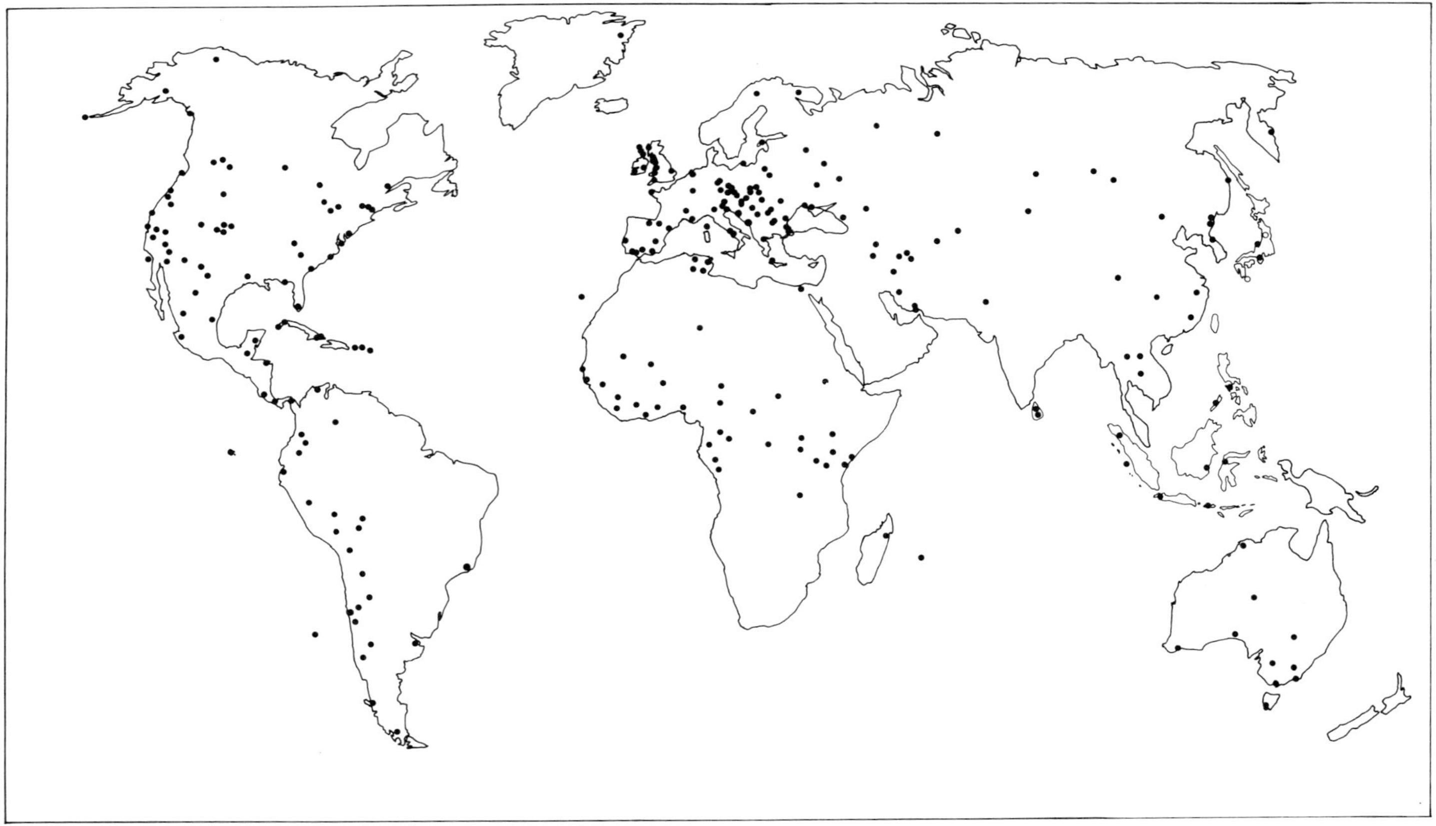

Fig. 6.3 Map of the world showing the distribution of all Man and the Biosphere Reserves.

therefore any such programme is faced with problems of which ecosystems and which sites should be given highest priority, selection once again having to be based more on non-biological than biological criteria. DIVERSITAS aims to inventory and monitor selected taxa of, initially, 25 sites distributed through five ecosystems (tropical and temperate forests, savannas, grasslands, and tundra). These sites are all Man and the Biosphere Reserves, part of the MAB UNESCO programme, and have some protected status. DIVERSITAS will succeed only if those managing sites coordinate activities and exchange information through a well structured network. Another initiative, BIONET, promoted by CAB International, is seeking to increase taxonomic skills and resources in developing countries through the establishment of a network of regional centres.

As signatories of the Biodiversity Convention, many countries are developing biodiversity action plans. The international initiatives, which provide a top-down approach, will have to be flexible if they are to integrate with sometimes widely differing existing and rapidly developing country programmes.

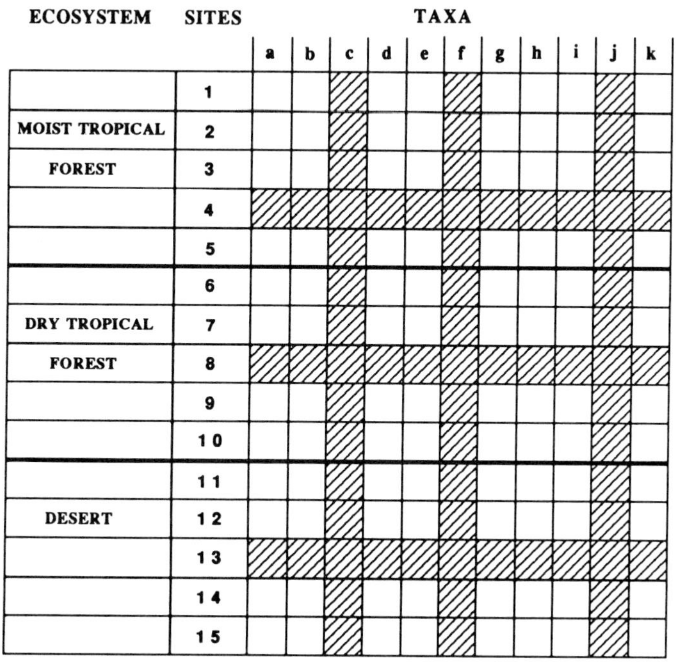

Fig. 6.4 Diagrammatic representation of an intensive and extensive sampling programme (after di Castri *et al.* 1992). Inventories would be made of all groups of organisms (a–k) at a few selected sites (4, 8, 13). At a much wider range of sites (1–15) inventories would be made of only selected focal taxa (c, f, j).

Man and the Biosphere (MAB) reserves have been proposed as suitable first sites for DIVERSITAS. However, the distribution of these is unrepresentative of biodiversity as can be seen in Fig. 6.3, and additional sites are necessary. What criteria, therefore, should be considered for site selection? On the practical side, sites should be secure, both politically and in terms of long term stability of land use, and should be readily accessible. They should have an existing infrastucture capable of maintaining an inventory programme. Sites should be of a large size, incorporating 'used' and 'unused' areas which can be utilized for examination of the effects of exploitation. Previous scientific studies and local availability of human taxonomic and ecological resources would strengthen the usefulness of a particular site. Finally, the distribution of sites should be representative of a range of ecosystems, of biodiversity, and of major continents.

Monitoring of biodiversity

Even more difficult than organizing efforts to provide inventories of biodiversity is organising efforts to monitor changes in biodiversity. I do not propose to discuss this in detail here. Monitoring involves regular checks on changes in populations of species (or even simply presence/absence), or of changes in communities. Interpretation of these changes is extremely complex as an understanding of natural change, often over long or short periods of time, is required. Isolated populations of organisms are known to die out and establish as a natural process (e.g. see Hanski and Gilpin 1991). Identifying which changes in diversity or population structure are due to natural phenomena and which are due to man-induced environmental changes is therefore not an easy process. Establishment of a network of sites where the same sampling procedures and indicator groups are being used will help monitoring processes. Use of geographical information systems (GIS) and development of computer software such as BIOCLIM (Busby 1991), where geographical information and species distribution data can be used to predict future distributions under certain conditions of environmental change, will help enormously.

Recommendations

This paper has discussed a number of issues related directly to the processes of compiling inventories and monitoring for change, and the following is a summary of recommendations requiring urgent action:

1. Integration of international programmes for making inventories.
2. Monitoring change with national programmes.
3. Input of funding from national and international agencies.

4. Improvement of taxonomic and ecological resources in developing countries through training programmes and establishment of country and regional centres.
5. Selection of intensive and extensive study sites.
6. Selection of indicator groups and key species based on sound scientific rather than historical criteria.
7. Improve availibility of information through networking and sharing of databases.
8. Preparation of manuals/handbooks and 'sampling packages' for sampling indicator groups.

Acknowledgements

Some of the ideas discussed in this paper were raised by the author and others during meetings (Harvard Forest, USA; Paris; La Selva, Costa Rica) organized principally by IUBS and UNESCO. I thank the organisers of these meetings for inviting me to participate and the other participants for stimulating discussions. I thank M. Brendell, K. Gaston, I. Gauld, P. Hammond, J. Lawton, C. Lyal, and R. Vane-Wright for commenting on earlier versions of this paper.

References

André, H. M., Lebrun, P., and Noti, M. (1992). Editorial. Biodiversity in Africa: a plea for more data. *Journal of African Zoology*, **106**, 3–15.

Adis, J. (1987). Extraction of arthropods from neotropical soils with a modified Kempson apparatus. *Journal of Tropical Ecology*, **3**, 131–8.

Adis, J. (1988). On the abundance and density of terrestrial arthropods in central Amazonian dryland forests. *Journal of Tropical Ecology*, **4**, 19–24.

Barnes, R. D. (1989). Diversity of organisms: how much do we know? *American Zoologist*, **29**, 1075–84.

Busby, J. R. (1991). BIOCLIM—a climate analysis and prediction system. In *Nature conservation: cost effective biological surveys and data analysis* (ed. C. R. Margules and M. P. Austin) pp. 64–73. CSIRO, Melbourne.

Coddington, J. A., Griswold, C. E., Silva Da'vila, D., Penaranda, E., and Larcher, S. F. (1991). Designing and testing sampling protocols to estimate biodiversity in tropical ecosystems. In *The unity of evolutionary biology* (ed. E. C. Dudley), pp. 44–60. *Proceedings of the fourth International Congress of Systematic and Evolutionary Biology*. 2 vols. Dioscorides Press, Portland, Oregon.

Cranston, P. and Hillman, T. (1992). Rapid assessment of biodiversity using 'Biological Diversity Technicians'. *Australian Biologist*, **5**, 144–54.

di Castri, F., and Younès, T. (1990). Ecosystem function of biological diversity. *Biology International*, Special Issue **22**, 20 pp.

di Castri, F., Robertson Vernhes, J., and Younès, T. (1992). Inventorying and monitoring biodiversity. *Biology International*, Special Issue **27**, 28 pages.

Erwin, T. L. (1982). Tropical forests: their richness in Coleoptera and other arthropod species. *Coleopterists Bulletin*, **36**, 74–5.

Erwin, T. L. (1988). The tropical forest canopy: the heart of biotic diversity. In *Biodiversity* (ed. E. O. Wilson and F. M. Peter), pp. 123–9. National Academy Press, Washington DC.

FAO (1981). *Tropical forest resources*. Food and Agriculture Organisation, Rome.

Gadagkar, R., Chandrashekara, K., and Nair, P. (1990). Insect species diversity in the tropics: sampling methods and a case study. *Journal of the Bombay Natural History Society*, **87**, 337–53.

Gamez, R. (1991). Biodiversity conservation through facilitation of its sustainable use: Costa Rica's National Biodiversity Institute. *Trends in Ecology and Evolution*, **6**, 377–8.

Gaston, K. J. (1991). The magnitude of global insect species richness. *Conservation Biology*, **5**, 283–96.

Gaston, K. J. and May, R. M. (1992). Taxonomy of taxonomists. *Nature*, **356**, 281–2.

Grassle, J. F. and Maciolek, N. J. (1992). Deep-sea species richness: regional and local diversity estimates from quantitative bottom samples. *American Naturalist*, **139**, 313–41.

Groombridge, B. (ed.) (1992). *Global biodiversity. status of the Earth's living resources*. Chapman and Hall, London.

Hammond, P. M. (1990). Insect abundance and diversity in the Dumoga-Bone National Park, N. Sulawesi, with special reference to the beetle fauna of lowland rain forest in the Toraut region. In *Insects and the rain forests of South East Asia (Wallacea)* (ed. W. J. Knight and J. D. Holloway), pp. 197–254. Royal Entomological Society of London.

Hammond, P. M. (1992). Species inventory. In *Global diversity. status of the Earth's living resources* (ed. B. Groombridge), pp. 17–39. World Conservation Monitoring Centre, Chapman and Hall, London.

Hammond, P. M. and Harding, P. T. (1991). Saproxylic invertebrate assemblages in British woodlands: their conservation significance and its evaluation. In *Pollard and veteran tree management* (ed. H.J. Read), pp. 29–37. Richmond Publications, Slough, UK.

Hanski, I. and Gilpin, M. E. (1991). Metapopulation dynamics—brief history and conceptual domain. *Biological Journal of the Linnean Society*, **42**, 3–16.

Hawksworth, D. L. (1991). The fungal dimension of biodiversity: magnitude, significance and conservation. *Mycological Research*, **95**, 641–55.

Hawksworth, D. L. (1992). Biodiversity in microorganisms and its role in ecosystem function. In *Biodiversity and global change* (ed. O. T. Solbrig, H. M. van Emden, and P. G. W. J. van Oordt), pp. 83–93. IUBS, Paris.

Hodkinson, I. D. and Casson, D. (1991). A lesser predilection for bugs: Hemiptera (Insecta) diversity in tropical forests. *Biological Journal of the Linnean Society*, **43**, 101–9.

Holloway, J. D. and Stork, N. E. (1991). The dimensions of biodiversity: the use of invertebrates as indicators of human impact. In *The biodiversity of microorganisms and invertebrates: its role in sustainable agriculture* (ed. D. L. Hawksworth), pp. 37–62. CAB International, Wallingford, UK.

International Union for Conservation of Nature and Natural Resources, United

Nations Environment Programme, and World Resources Institute (1990) *The biodiversity conservation strategy programme, mobilizing worldwide action to sustain the living resources of our planet.* IUCN, Gland.

Janzen, D. H. (1993). What does tropical society want from the taxonomists? In: *Hymenoptera and biodiversity* (ed. J. LaSalle and I. D. Gauld), pp. 295–307. CAB International, Wallingford, UK.

Lawton, J. H. and Brown, V. K. (1993). Redundancy in ecosystems. In: *Biodiversity and ecosystem function* (ed. E.-D. Schulze and H. A. Mooney), pp. 255–70. Springer-Verlag, Frankfurt.

MacArthur, R. and Wilson, E. O. (1967). *The theory of island biogeography.* Princeton University Press.

Margules, C. R. and Austin, M. P. (ed.) (1991). *Nature conservation: cost effective biological surveys and data analysis.* CSIRO, Canberra.

May, R. M. (1988). How many species are there on earth? *Science*, **241**, 1441–9.

May, R. M. (1990). How many species? *Philosophical Transactions of the Royal Society, B*, **330**, 293–304.

May, R. M. (1992). How many species inhabit the earth? *Scientific American*, October, 18–24.

Myers, N. (1980). *Conversion of tropical moist forests* (Report to the National Academy of Sciences). National Research Council, Washington, DC.

Myers, N. (1988). Threatened biotas: 'hot spots' in tropical forests. *The Environmentalist*, **8**, 187–208.

Myers, N. (1989). *Deforestation rates in tropical forests and their climatic implications.* Friends of the Earth Report, London.

Noyes, J. S. (1984). In a fog. *Chalcid Forum*, **3**, 4–5.

Reid, W. V. and Miller, K. R. (1989). *Keeping options alive: the scientific basis for conserving biodiversity.* World Resources Institute, Washington DC.

Oliver, I. and Beattie, A. J. (in press). A possible method for the rapid assessment of biodiversity. *Conservation Biology.*

Pearson, D. L. and Cassola, F. (1992). World-wide species richness patterns of tiger beetles (Coleoptera: Cicindelidae): indicator taxon for biodiversity and conservation studies. *Conservation Biology*, **6**, 376–91.

Shelford, R. (1905). Report on exhibit of the larvae of *Collyris emarginatus*, Dej., and of *Mormolyce. Proceedings of the Entomological Society of London*, **1905**, lxxii–iii.

Simberloff, D. (1984). Mass extinctions and the destruction of moist tropical forest. *Zhurnal Obshchei Biologii* (Journal of General Biology) USSR, **45**, 767–78.

Solbrig, O. T. (1991*a*). *Biodiversity, scientific issues and collaborative research proposals.* MAB Digest 9, UNESCO, Paris.

Solbrig, O. T. (1991b). *From genes to ecosystems: a research agenda for biodiversity.* IUBS Monograph Series, Paris.

Stork, N. E. (1988). Insect diversity: fact, fiction and speculation. *Biological Journal of the Linnean Society*, **35**, 321–37.

Stork, N. E. (1991). The composition of the arthropod fauna of Bornean lowland forest trees. *Journal of Tropical Ecology*, **7**, 161–80.

Stork, N. E. (1993) How many species are there? *Biodiversity and Conservation*, **2**, 215–32.

Stork, N. E. and Brendell, M. J. D. (1990). Variation in the insect fauna of Sulawesi trees with season, altitude and forest type. In *Insects and the rain forests of South East Asia (Wallacea)* (ed. W. J. Knight and J. D. Holloway), pp. 173–90, Royal Entomological Society of London.

Stork, N. E. and Brendell, M. J. D. (1993). Arthropod diversity studies in lowland rain forest of Seram, Indonesia. In *The Natural History of Seram* (ed. J. Proctor and I. Edwards), pp. 115–30. Intercept, Andover, UK.

Stork, N. E. and Eggleton, P. (1992). Invertebrates as determinants and indicators of soil quality. *American Journal of Alternative Agriculture*, **7**, 23–32.

Stork, N. E. and Gaston, K. J. (1990). Counting species one by one. *New Scientist*, **1729**, 43–7.

Sutton, S. L. and Collins, N. M. (1991). Insects and tropical forest conservation. In *The conservation of insects and their habitats* (ed. N. M. Collins and J. A. Thomas), pp.405–24. Academic Press, London.

Vane-Wright, R. I., Humphries, C. J., and Williams, P. H. (1991). What to protect?—systematics and the agony of choice. *Biological Conservation*, **55**, 235–54.

Wilson, E. O. (1988). The current state of biological diversity. In *Biodiversity* (ed. E. O. Wilson and M. P. Peters), pp. 3–18, National Academy Press, Washington, DC.

7. Biodiversity research and biological collections: transfer of information

EBBE S. NIELSEN* and JUDY G. WEST[†]
*Australian National Insect Collection, CSIRO Division of Entomology, GPO Box 1700, Canberra, ACT 2601, Australia
[†]Australian National Herbarium, CSIRO Division of Plant Industry, GPO Box 1600, Canberra, ACT 2601, Australia

Abstract

The role of collections in biodiversity research is reviewed. It is shown that there are major problems that need to be addressed: the priorities for future collection-based research with respect to biodiversity issues. It is shown that the vast amounts of data within biological collections already exist that, with appropriate planning, could be mobilized into useful published lists and incorporated into electronic databases. On a more strategic level the information required for decision-taking on biodiversity issues is reviewed. Statistics are provided on the size of the task, and the scope of the task is assessed using Australian models and examples, in relation to life lists, monographs and checklists, and future inventory projects.

Introduction

Biodiversity research is based on knowledge of taxa, their interrelationships, classification, evolution, past and present distributions, and an understanding of their role in the ecosystems in which they occur. For most organisms this information is available only from well-maintained biological collections and research databases. In order to generate the required knowledge about taxic diversity we must coordinate and plan the development of collections, inventory work, and improve our overall ability to process specimens and data from field-based inventory work.

The vast amount of information already contained in collections must be mobilized and made more readily available. Computer technology permits

Systematics and Conservation Evaluation (ed. P. L. Forey, C. J. Humphries, and R. I. Vane-Wright), Systematics Association Special Volume No. 50, pp. 101–21, Clarendon Press, Oxford, 1994.

the development of high-quality inventories of all named organisms so that our current knowledge of taxonomy, nomenclature, and classification is reflected in the form of databased checklists and catalogues. Likewise, computer databasing technology makes it possible to capture, examine, and extract information from label data of specimens held in biological collections. Computerized descriptive databases (taxonomic databases) make the taxonomic descriptive process more effective and concise, and will permit on-line interactive identification supported by user-friendly graphics. Computer technology enables the storage and effective use of large taxonomic databases; such databases can also be shared on electronic networks.

Information required for biodiversity decisions

Biological collections and the staff associated with them can provide much of the critical baseline data needed for biodiversity decision-making and management. Biosystematic data available from these collections and from systematists associated with them can be summarised under four headings: (1) taxa and their identification, (2) the distribution of taxa (both past and present), (3) the phylogeny (interrelationships) of taxa, and (4) the general biology of taxa. Environmental biologists and planners seek the involvement of biosystematists with inventories; this may include design of sampling protocols (ideally with an appropriate ecological input), field sampling, sorting, processing, identification, storage of voucher material, and capture of label data in electronic form.

While the larger and more prominent organisms, such as vertebrates and vascular plants, are relatively well-known, the amount of immediately available knowledge for many groups of invertebrates, lower plants, and fungi often exclude them from consideration for inventory work. However, since these groups clearly represent the majority of the world's taxic diversity it is necessary to expend some of the available resources on them in a targeted and well planned manner. This should include coordination of inventory work, selection of study groups, determination of the apppropriate level of research effort, and an increase in the efficiency with which taxonomic work is undertaken.

Assessment of level of knowledge of total biological diversity

In order to establish priorities for biodiversity research, it is important to have as complete an understanding as possible of the magnitude of the task. For the purpose of planning taxonomic research, it is critical to have good estimates of the total number of species with which we are dealing. These estimates have varied considerably in the past, depending on the organisms and ecosystems used for making the predictions, and on the methods of extrapolation. Hawksworth (1992) estimated that there are 5 450 000 species of bacteria, fungi, protozoa, and viruses worldwide, of which 159 000 are described. The number of estimated species is 14 000 bryophytes, 17 000 lichens, approximately 6000 algae, 12 000 ferns, and 250 000 flowering plants. Nematode figures vary from 500 000 to 1 million, vertebrates comprise 43 000, of which half are fishes. However, the greatest uncertainty is with regard to the number of species of insects. Over the past 30 years the total insect fauna had been estimated to be within the range of 3–10 million species, until estimates by Erwin (1982) projected that figure beyond 30 million. Most taxa were suggested to be confined to rainforests and assumed to have restricted geographical ranges. The insect diversity issue was reviewed by Gaston (1991), and using a range of examples and projections he concluded that 'a figure of less than 10 million seems more tenable, and one of around five million feasible'. It is important to note that these figures are estimates made to the best of our ability at any one time.

While the high proportion of rainforest insects reported caused near sensational interest in numbers of species, the estimates appear insufficiently justified (Monteith 1990; Gaston 1991) and probably did little to support the notion that it is indeed possible to survey Life on Earth and make practical use of the resulting information. May (1992) reviewed the different methods of estimating species numbers, including both direct and empirical estimates, and indirect and theoretical estimates. Wilson (1985) estimated the level of resources required to handle 10 million species based on traditional description and illustration of all species. 'Systematics Agenda 2000' (1993), which is the American systematics' community response to the biodiversity crisis, also addresses the cost of describing all living organisms on Earth. Estimates of species richness indicate the need to identify priority areas for allocation of resources to address the most pressing gaps in our knowledge.

Figure 7.1 summarizes the known and estimated total biota on the Australian continent, categorized into species that are named, species known in biological collections but not named, and the estimated number of undocumented taxa (single celled organisms are not included). Undoubtedly these data will be further refined, but they are unlikely to change radically unless our species concepts change significantly. Nielsen and Halliday (in prep.) have dissected this dataset further and, using estimates from biological collections and recent surveys, predict that there are about 140 000 insects in Australia. This stands in stark contrast to much larger figures discussed recently in the Australian media. While these inflated figures have been refuted (Monteith 1990), their release adds little to the credibility of biosystematics and the belief that it is possible to establish meaningful priorities.

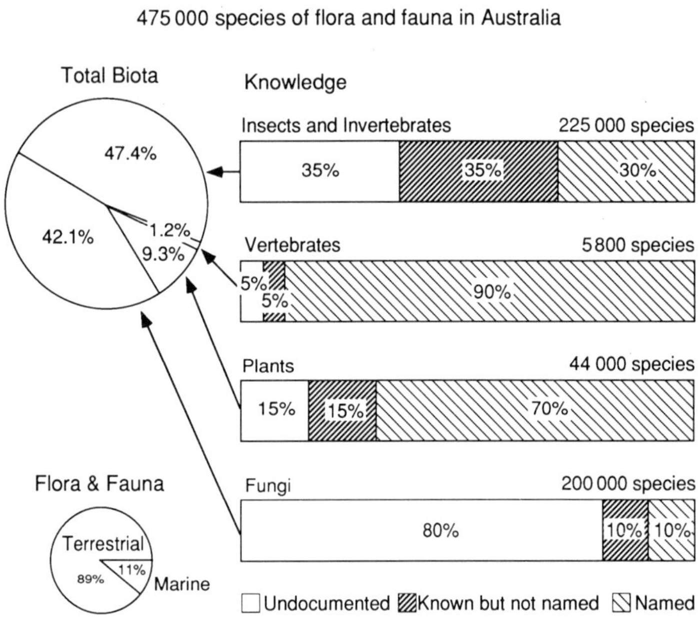

Fig. 7.1 Species diversity in Australia. The total species diversity is estimated at 475 000 species of plants, animals (invertebrates and vertebrates), and fungi (single celled organisms have been excluded from this estimate). For each major group the size of the named, known but unnamed, and undocumented has been estimated. The figure is modified from Walker *et al.* (1992) by the authors assisted by the Australian Biological Resources Study (ABRS).

Information access and transfer

Priority setting for conservation, and for land and ecosystem management, requires access to information about the world's living organisms. While less than half of all living organisms are formally named, this is not always indicative of the level of information available: there is no simple relationship between an organism being named and being known. At the simplest level, the collections held in a museum or herbarium will collectively hold much information about an undescribed species.

An important constraint on access to information is the worldwide general decline and underfunding of biosystematics, biological collection maintenance, and taxonomy. If critical baseline data are to be collated for biodiversity research and assist in decision-making for sustainable development, these problems will need to be addressed. The recent rapid increase in demand for information has placed further pressure on institutions responsible for collection maintenance, and their staff, to respond to these needs in the face of diminishing available resources.

Even with the above constraints, we do have the knowledge and capacity to set priorities for particular groups to be studied and target research to appropriate levels. A simplified cost-benefit diagram is presented in Fig. 7.2. The two examples given are a comprehensive taxonomic revision, the traditional taxonomic output, and a checklist. The taxonomic revision permits identification, includes thorough synonymy and bibliography, and has full descriptions or redescriptions of all taxa, and furthermore contains a complete summary of all available information about the species treated. The revision is relatively expensive research, the time used per taxon is high, and from a user perspective the overall frequency of use is low; the published account is generally only useful to a small user-group which seeks identification or information about the group treated. We do not argue against taxonomic revisions; they are fundamental to all biosystematics since they represent the research product which usually names new species and provides comprehensive accounts of all species of a natural group. However, taxonomic revisions require significant resources and should be ranked according to priority for the important groups, such as those containing major pests or good indicator species, or groups that are particularly environmentally and biologically informative. At the other end of the scale, the checklist contains minimal information only: usually the correct binomial name, author and year of publication set out in systematic order. Accordingly, it does not permit identification, but the availability of the

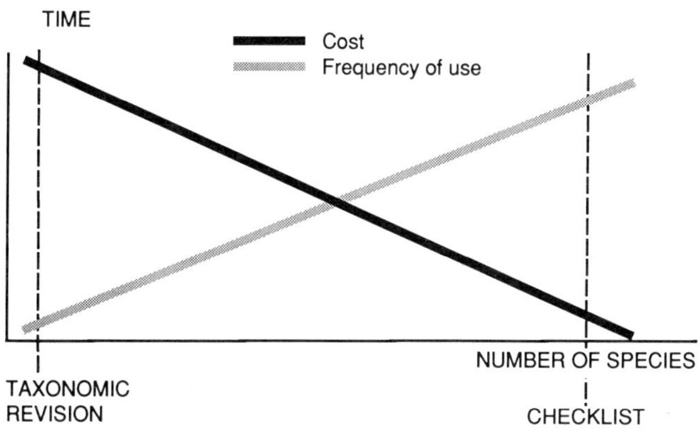

Fig. 7.2 Schematic cost-benefit diagram of biosystematics research. Detailed and high quality taxonomic species revisions are very expensive and have a relatively small frequency of usage, but are often prerequisites for resolving a range of specific problems. Checklists are relatively inexpensive to produce and attract a high level of usage. The information content is usually closely correlated with the cost curve.

correct taxon name allows access to the literature, databases, and communication. For checklists the frequency of use is high and the user community is broad.

Biosystematic information and outputs can be broadly categorized as shown in Fig. 7.3, in which the different kinds of taxonomic treatments are listed and ranked from the relatively least expensive (the checklist) to the relatively more expensive (the taxonomic revision). In addition to the categories listed, it is important to record that formats such as family or generic overviews can compartmentalize large groups into more manageable subsets for further study and provide taxonomic reviews which summarize the most important taxonomic information.

We need also to improve the efficiency of collection management and development without loss of curatorial standards. It is important to adopt and develop new methods and make use of new tested materials as they become available. But the problem facing curation and collection maintenance is that it appears to be an open-ended task, and it can be difficult to establish meaningful objectives and assess achievements (West and Nielsen 1992). This can cause managerial difficulties, both with regard to external funding bodies and internal institutional assessment. McGinley (1989) has suggested one tool for setting priorities and demonstrating real progress that has been successfully applied in several large biological collections. In terms of investment in gaining access to information, collection maintenance is the most cost-effective of all: it is possible to extract information for any species from a well managed collection and capture specimen data in electronic form. The information from collections is usually available only to those with

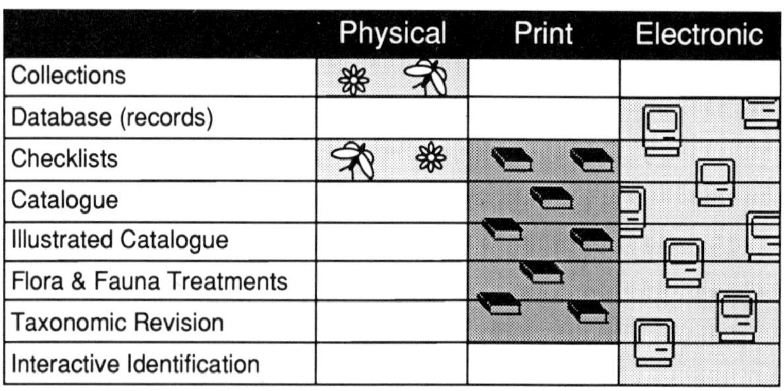

Fig. 7.3 Schematic outline showing methods for capturing and relaying information about taxa. The three media are physical (i.e. the actual use of curated biological collections), print (journals, monographs, flora, and faunas), and electronic (disk, CD-ROM, electronic networks).

physical access to the collection since only a very small proportion of the holdings in collections have been entered on a computer database.

To make the most effective progress in collection management, taxonomic research and databasing, including checklists and catalogues, the keywords are an 'established protocol' and 'quality control'. These critical aspects will ensure that all resources invested are effective and that any additional resources will add extra value to the data sets; nothing should need to be revisited. With reference to descriptions of new taxa, Gaston and Mound (1993) commented on the level of redundant descriptions; a significant percentage of these new names are later found to apply to species described previously (they are synonyms). It would be interesting to learn how much quality control was implemented by the institutions and journals that published the papers with the highest level of synonymy.

The distribution of the biosystematic workforce also needs to be addressed in major institutions, nationally, and globally. There is frequently no positive relationship between size of groups, level of knowledge, and the distribution of the employed human resources. Figure 7.4 gives an example from Australia. We have estimated that there are about 20 000 vascular plants

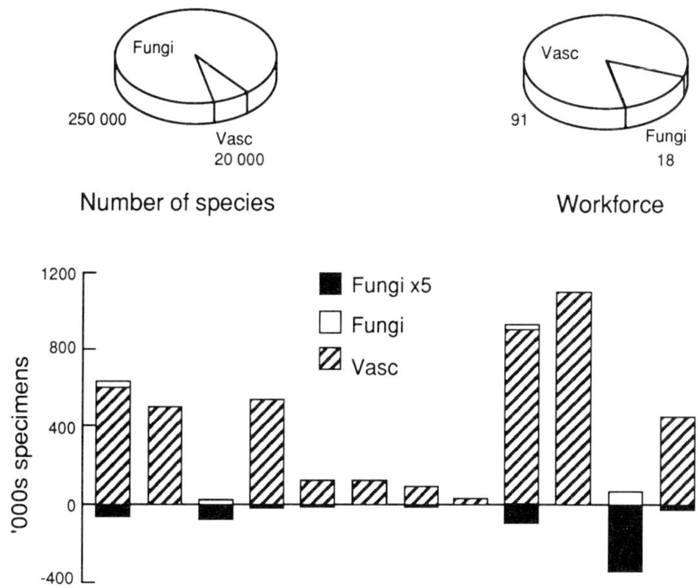

Fig. 7.4 Comparison of holdings of vascular plants and fungi in Australian institutions and the systematics workforce with estimated diversity of species. For example, a list of valid names only will be useful, but has a lower level of comprehensiveness than a list providing valid names with synonyms and original citations. Redrawn from West and Nielsen (1992).

and possibly 250 000 fungi in Australia, and whilst there are 91 people working on vascular plants there are only 18 mycologists. The histogram shows holdings of vascular plants and fungi in Australian herbaria, which are similarly not proportionally representative of the flora.

Predictive classification

An important objective of biosystematics is to construct classifications that reflect the grouping of closely related species (or taxa) on the basis of shared unique characters. Such phylogenetic classification not only reflects genealogical relationships but also increases the level of stability and utility of the classifications. It also provides a hierarchical system in which taxa and the information associated with them is stored, and permits making predictions about the characters and other properties of taxa.

With the available information about the geographic distribution of taxa and the classification, it is now possible to make precise measurement of biodiversity and apply this information to conservation management. The significance of this information in nature conservation and in assigning priorities to species or groups of species that require special conservation attention is discussed by May (1990), Vane-Wright *et al.* (1991), Faith (1992) and in detail elsewhere in this volume (Williams and Humphries; Oosterbroek; Vane-Wright *et al.*). The availability of powerful software, such as WORLDMAP (Williams *et al.* 1993), makes it possible to utilize taxonomic biodiversity information in priority-area analysis and land use decision-making to a level not previously possible.

Lists of recorded species: life lists

We are able to communicate unambiguously about living organisms by using the binominal name. There are internationally agreed rules regarding the use and application of names for all living organisms. However, it is of major concern that the available knowledge of the nomenclature of all named taxa and their classification is generally incomplete nationally, by continent, or globally, although there are notable exceptions. With currently available computer technology there are no technical constraints on producing high quality checklists and catalogues of all names ever introduced into the literature irrespective of status. Such lists, often referred to as 'lists of recorded species' (May 1992), can be designed to be comprehensive to various degrees (Fig. 7.5) and must truly be a priority. While there is frequent concern regarding our level of knowledge of biological resources, using the number of named species as an indicator, it is a major weakness—particularly amongst zoologists—that to keep an informative global record is not held as a widely accepted priority. It seems incongruous that we have

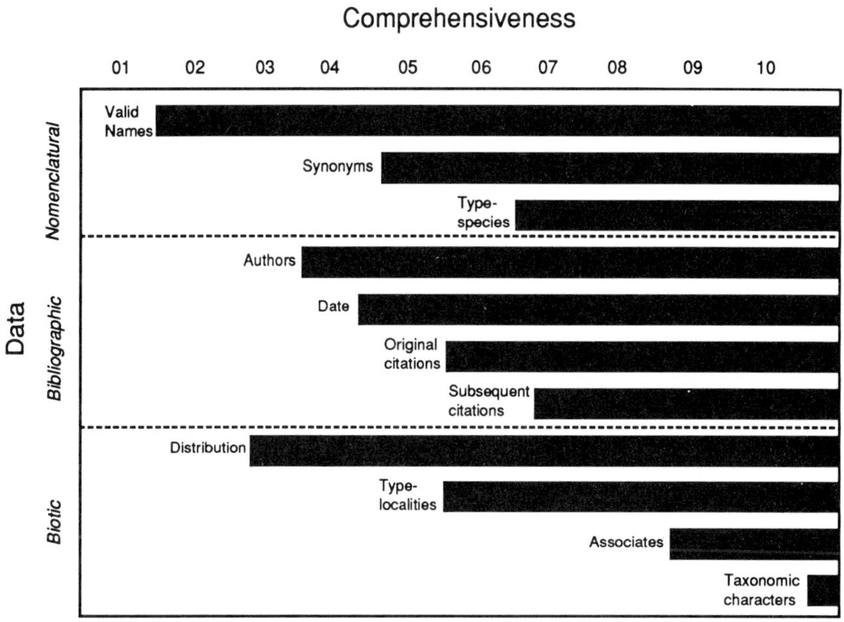

Fig. 7.5 Schematic outline of levels of comprehensiveness in biosystematic publications. Data in publications have been grouped into nomenclatural, bibliographic, and biotic. For example, a list of valid names only will be useful, but has a lower level of comprehensiveness compared with a list providing valid names with synonyms and original citations. Modified from Thompson and Knutson (1987).

comprehensive records of books in the world's libraries and that we can access these records electronically, whereas there is no such access to even the most basic information about Life on Earth. Clearly, it must be a managerial responsibility of the world's larger biosystematics institutions to ensure that such work has priority and is undertaken according to a well established protocol, and with a high level of quality control. Such work is a basic prerequisite for the most effective resource use in biodiversity research and management, and is essential to avoid confusion and duplication of effort.

A range of such checklist/catalogue projects has been undertaken for the names of some nomenclatural groups or systematic groups of the species-rich groups of fauna. Three examples are given here. Based on the virtually complete card index of the approximate 280 000 genus, species, and infraspecific names of Lepidoptera held at The Natural History Museum in London, Nye and Fletcher (1991) have compiled a catalogue, *The Generic Names of Moths of the World*, as complete as possible within the available time and resources, of the approximately 26 400 genus-group names of Lepidoptera. This six-volume work provides full cross-referenced access to all available generic names with listings of homonyms, objective synonyms, misspellings and other nomenclaturally unavailable names, full reference to dates and places of publication, type species and their designations, and (in some volumes only) types and location of type specimens. This work is of immense value to anyone working anywhere on Lepidoptera, one of the largest orders of insects. It is an outstanding example of how nomenclature can be made available to users: names and nomenclature are tools for communication and should not be an impediment that only few workers understand and can apply correctly.

The only large order of invertebrates for which a virtually complete world catalogue is available is Diptera. This is not a single document but a range of regional catalogues. The most recent include Evenhuis (1989) *Catalog of the Diptera of the Australasian and Oceanian Regions* and Soós and Papp (ed.) (1992) series *Catalogue of Palaearctic Diptera* published in 14 parts. These regional catalogues have been published since 1965 under various editorships, using a range of formats and protocols and various degrees of computer technology. However, the important observation is that there is in fact a world catalogue of Diptera which can be further developed, corrected, and maintained as an up to date source of information.

While the impracticality of undertaking complete accounts of all names of large groups for whole continents is frequently stressed by nomenclaturists it is demonstrably possible. An example is the near complete *Checklist of the Lepidoptera of Australia* which is a computerized reference to the taxonomy, nomenclature, and classification of the entire Australian Lepidopteran fauna. It contains all names ever associated with the Australian Lepidopteran fauna including names used because of misidentification or misspelling. This checklist is based on (1) a complete alphabetic index of all original descriptions of Lepidoptera named from Australia with label information from primary types, and (2) an archive, as complete as possible, of colour transparencies of the corresponding primary types. The index of all original descriptions permits ready checking of original spellings without searching the often rare primary literature, and the availability of the colour slides makes it possible to confirm the identity of the name-bearing specimen to species or species group without troubling the institution that houses the specimen in question. When the first version is finished in late 1993, the checklist database will be fully maintained and updated.

In Australia, one of the objectives for the Australian Biological Resources Study (ABRS) is to produce the *Zoological Catalogue of Australia*. This comprehensive catalogue is planned to be completed in 87 volumes each containing about 1800–2000 named species. Currently the ABRS plans to publish three volumes per year. To date, 10 volumes have been published; all in printed form and typeset from databased files. Currently, ways of maintaining updated databases for completed groups, electronic access via internet, and subscription to possible annual CD-ROM updates of all completed and available groups are being investigated. To our knowledge the Australian *Zoological Catalogue* is the only national planned 'list of recorded species' of all taxa of the animal kingdom.

Further projects of ABRS have been the development of major national botanical data sets. The *Census of Australian Vascular Plants* (CAVP) is a list of currently accepted names for Australia's vascular plants (ca. 17 000 names) with distributional information. This was compiled largely from existing state censuses and checklists, and published in hardcopy (Hnatiuk 1990). It has since been linked to a database system, the *Australian Plant Name Index* (APNI); a list of all vascular plant names (more than 62 000) applicable to the Australian flora with details of typification and publication (Chapman 1991).

Despite the widespread utilization and exploitation of plant species, the growing concern about loss of biodiversity, and the urgent need for conservation of genetic resources, there is still no modern unified conspectus of plant species of the world, their distribution and attributes. Recently, this problem has received attention and will be addressed partly through the formation of the International Organization for Plant Information (IOPI). IOPI is an organization formed with the principal aim of promoting and preparing a series of integrated, dispersed, computerized databases summarizing the basic taxonomic information, and biological and other attributes of all kinds of plants in the world. As its first project, IOPI intends to compile a world species checklist for vascular plants. This is a major task involving many experts from around the world. Once a basic checklist is compiled, it will be possible to cross-link and integrate further databases concerned with other attributes, e.g. taxonomic, biological, and ecological, and then to use the results for example, in agriculture, forestry, horticulture, conservation, medicine, biotechnology, or in other ways.

Taxonomic databases and interactive identification

Taxonomic database programs are being developed by many organizations around the world. Some of these include special modules for interactive identification: for example, the packages under development at the Royal Botanic Garden, Edinburgh (Pankhurst and Aitchison 1975; Pankhurst 1986); CAB International, Wallingford; the Expert Centre for Taxonomic Identification (ETI), Amsterdam; and the CSIRO Division of Entomology, Canberra (Dallwitz 1993). Some of these programs do not include features such as description writing and key generation, and there are other systems for interactive identification alone. Increasingly, some of the software available for taxonomic databases will permit the inclusion of text, illustrations, sound, and video.

In the following, we briefly describe the DELTA/INTKEY software package of the CSIRO Division of Entomology. The DELTA data format (Dallwitz 1980; Dallwitz and Paine 1986) is based on ordinary text files which may be created and modified with any text editor. The data are in free format, i.e. they do not have to be positioned in fixed fields in the records. Descriptions are coded with reference to a character list and five types of characters are available: text characters, ordered or unordered multi-state characters, continuously variable or integer numeric characters. There are no restrictions on the numbers of characters, character states for each character, or the amount of text, and comments can be placed both in the character list and the coded descriptions. DELTA was designated as a standard format for taxonomic descriptions at the 1988 meeting of TDWG (Taxonomic Databases Working Group for Plant Sciences) in St Louis, Missouri.

The central program of the DELTA system is CONFOR, which translates the data into natural language, or into the formats required by various other taxonomic programs for cladistic and phenetic analysis, key construction (Dallwitz 1974) and interactive identification and information retrieval. If the character list is translated into another language, all of the products of the system, such as descriptions, keys, and interactive identification, can be produced in that language. So far, the system has been used in English, Chinese (Xu Zhu *et al.*, in press), French (Watson *et al.* 1986), Greek (Watson *et al.* 1988), and Portuguese and Spanish (Webster, Kirkbride, and Reyna 1989). Descriptions and keys can be typeset automatically from the program output (Dallwitz 1984, Watson and Dallwitz 1992).

The latest addition to the DELTA package is INTKEY, an interactive program for identification and information retrieval (Watson and Dallwitz 1989; Watson *et al.* 1989; Bruhl *et al.* 1992). Its facilities include the following: the ability to allow for errors in observation or interpretation; direct handling of numeric values, which therefore do not have to be arbitrarily ranged; the ability to alter the treatments of unknown, inapplicable, and overlapping values, as required for different applications (flexibility in this respect being particularly significant in relation to identification versus information retrieval); retrieving text information (such as synonyms, lists of representative taxa, references etc.); restricting operations to subsets of characters or taxa; defining keywords to represent subsets of characters and taxa; locating characters by included words; specifying 'character reliabilities' appropriate for particular purposes (making practical the calculation of 'best' identificatory characters for very large sets of characters and taxa); obtaining lists of taxa possessing or lacking particular attributes or combinations of attributes; preparing lists of taxa uncoded for particular characters or sets of characters; listing similarities or differences among taxa, with the ability to vary the interpretations of 'similarity' and 'difference'; describing, comparing and contrasting taxa in terms of nominated sets of characters; generating diagnostic descriptions for specimens or taxa, to specified degrees of redundancy (permitting automatic generation of descriptions with italicized diagnostic characters); coalescing descriptions (e.g. to generate generic descriptions from species descriptions); screen display of illustrations of characters and taxa; input of complex or lengthy sequences of commands from files; selective output of results to files; complete on-line help; and acceptable response times with large sets of data. All of the program's text-commands, prompts, error messages, and help are in simple text files, so that foreign-language versions can easily be produced by non-programmers. INTKEY is currently being rewritten for Microsoft Windows.

Taxonomic databases not only permit consistent descriptions, and keys with effective emphasis on reliable characters, but they allow user-friendly interactive identification that can be used by students or naturalists with little formal training, as well as professional users with needs for taxonomic identification. Programs such as INTKEY and ETI allow full integration of illustrations which reduce the need for knowledge of technical terms and morphology (see Lawrence *et al.* 1993). Taxonomic database techniques allow storage, maintenance, and retrieval from very large databases; available software will support databases of entire continental floras and faunas when the data are available. Datasets can be transferred via disks, tapes, or networks, or made available on CD-ROMs (e.g. ETI 1992 *Birds of Europe*; Lawrence *et al.* 1993; Watson and Dallwitz 1993).

Most applications of DELTA and INTKEY and similar packages provide taxonomic databases and interactive identification systems for closely related taxa, e.g. species of a genus, or genera of a family. There are some examples of interactive identification systems for the faunas or floras of specific regions. For instance, Hyland and Whiffin (1993) have produced a key to the rain forest trees of northern Australia. The specimen-based database underlying this interactive system covers all major morphological characters for the 1050 tree species of rainforest areas. Such applications obviously supply the user of our taxonomic research with a user-friendly, multiple entry system for a group of plants and an ecosystem in which it is often difficult to collect complete specimens.

Once taxonomic databases are established for the purpose of inventory and monitoring, it is possible to generate hardcopy or electronic keys and descriptions for the taxa known from a smaller area under study. As new morphological information becomes available or populations with new character states are discovered, it is possible to enhance and correct the database rather than generate redescriptions of already described taxa.

Specimen databases

It is important that staff and management associated with biological collections realize that the physical collection database is the major asset for their institution and the community they serve.

The biological collections of the museums and herbaria around the world contain vast amounts of information associated with the specimens. Many biological collections contain over 10 million specimens; the natural history collections of The Natural History Museum in London alone contain over 67 million specimens (*The Natural History Museum Annual Report for Science* 1991). While the data from specimen labels have always been available in the past, extraction has been time consuming and risks specimen damage. These data have not been very accessible.

Until recently, very few institutions had a coordinated effort with regard to specimen databases or any other databases for that matter. Electronic databases were not usually seen as productive areas for long-term development compared with developing and maintaining the physical collections. Without a standardized effort individual databases may serve the immediate short-term needs of individual researchers or research groups, but the investment made in the data capture must contribute to long-term maintained databases.

The realization by institutions housing collections that computerization of specimen data is part of the curatorial process means that these data are now more accessible. With the data stored in electronic databases, information is accessible through a multiple entry system, and those responsible for collections are able to provide information and services more readily. As shown in Fig. 7.6, the cost of capturing the data associated with a

	Herbaria	Museums
Collecting a specimen	$14.2 (12-20)(35)	$49.1 (4-30)(250)
Identifying a specimen	$7.2 (4-10)	$17.1 (5-30)
Lending a specimen	$4.0 (3.8 -4.5)	$10.4 (3-20)
Databasing a specimen	$2.5 (1.4-3)	$4.8 (1-10)
Estimated cost of preparing a specimen	$15	Not Known
Total cost of curating a specimen	$52.5	Not Known

Fig. 7.6 The average costs ($A) of collecting and processing natural history specimens in Australian biological collections. The range of reported costs is indicated in parentheses. The total cost and preparation cost have not been estimated for fauna because of extreme variation between large mammals and small invertebrates. Data from Armstrong (1992).

curated specimen is less than 5 per cent of the cost of maintaining the specimen (based on data from Australian herbaria).

The percentage of collections already computerized varies considerably among the different organisms, and reflects the amount of resources allocated to databasing tasks, and past institutional policies. Since many collections, both plant and animal, hold vast amounts of material for which the specimen data are not databased, collection managers must identify and prioritize the appropriate groups to be worked on first. They should concentrate resources on those tasks (priority criteria can for example include accessibility of specimens, level of curation, completeness of collections, presumed information content of each group, and the complementarity of existing data).

The advantages of extracting data from specimen databases are many. Once the collection information is computerized it can be retrieved in a more efficient manner. Examples of enquiries and information concerning specimen data that are greatly enhanced through specimen databases include:

geographic distribution of a species, including changes in distribution over time;
the flora and fauna of a specific region;
the presence of rare and endangered species in a particular area;
species in a particular ecosystem; and
range increase over time of an introduced pest (weed or insect).

Hence, collections assembled primarily for taxonomic work can now provide comprehensive and readily accessible input to other disciplines and endeavours such as ecology, agriculture, and urban planning, as well as to biodiversity management.

Features of the particular software packages should enable the database manager and developer to include some sophisticated validations and checks for data quality and manipulation. For instance, the concept of linked hierarchies for taxonomic information (family, genus, species, etc.) and geographic location, and lookup tables or related databases for taxonomic names and collector names provide means of controlling quality at the point of data entry.

For those initiating specimen databasing projects there is much to be learnt from others who are already achieving results. Limited resources for such endeavours dictate that we build on the experiences of institutions already encompassing the technology. It is also important that institutions responsible for similar collections aim for close linkage and interchange of data. An example from the efforts made by herbaria in Australia provides some insights of the success of such a collaborative venture. Most Australian herbaria had either established or were developing collection databases by the late 1980s. The initial development was co-ordinated by three major herbaria, but it soon became clear that there was a need to co-ordinate the projects more closely and to establish compatible systems across all the major collections, i.e. the nine larger Australian herbaria.

The establishment of such compatible systems must consider the State, political, financial, historical, and personal preferences and constraints of the various institutions and their staff. Inevitably, this means that agreement will almost certainly never be reached on hardware and software alternatives; and probably does not need to be! Realizing that the structural consistency of the databases was not important, representatives from all major Australian herbaria agreed to establish a set of data interchange standards.

In 1989 general agreement was reached on the type and standard of data to be included in each of the 105 possible data elements already compiled. The result was the first working version of HISPID—Herbarium Information Standards and Protocols for Interchange of Data. Since that time appropriate representatives from each institution have met twice to update and review elements of HISPID.

Although HISPID still needs further refinement, the very existence of such a set of standards has been advantageous. It is clear that it has helped in the consistent development of the various herbarium specimen databases. The more compatible the databases the easier it is to transfer data. So, to some extent HISPID has dictated the development and structure of the more recently established databases, and has played a significant role in refining those established in the mid 1980s. While HISPID provides data interchange standards, it is also flexible enough to accommodate the particular requirements of individual institutions.

HISPID is regarded by the Australian herbaria as a dynamic set of standards. It is regularly reviewed and refined in light of new ways of handling particular data elements, new software developments, changing technology and concepts, and further usage.

The working-group basis of HISPID development has established close links between the institutions responsible for collection management. Even though the Australian botanical community is relatively small and well networked, we now have a better appreciation of institutional requirements and desires.

The success of HISPID in the Australian herbaria context might suggest that it be considered for adoption at the international level. Those responsible for entomological collections in Australia have commenced the development of a similar set of standards.

The development of compatible collection databases and standards such as HISPID yield some fundamental advances for systematics. Probably the most significant is the general agreement to share specimen data via sophisticated collection information management systems. Also, computerizing the specimen data has been totally incorporated into the curatorial process, and has extended our practices well beyond the quill pen. These specimen databases are accepted as part of the herbarium infrastructure for information management and research purposes.

Inventory

Our current knowledge of biological diversity available from biological collections or published accounts is based predominantly on relatively non-quantitative and often opportunistic collecting. While such collecting, termed 'museum collecting' in Coddington *et al.* (1991), does not usually permit estimation of relative abundance of species or total number of yet undiscovered species, it is this collecting effort that has generated the understanding we have today of the diversity of the world's living organisms. The knowledge of Australian biodiversity as illustrated in Fig. 7.1 has been generated through four phases of sampling: (1) *opportunistic sampling* by the early explorers, settlers, and early collectors; (2) *general surveys* where well equipped expeditions to various parts of the continent sampled as much of the total species richness as possible, within a limited amount of time at any one site before moving to the next site; (3) *specialized collecting* where experienced field collectors concentrate on their special group of interest but ignore most or all other groups, and (4) *baseline sampling* where intensive non-quantitative collecting techniques are used for often significant periods of time (for example one to two years for insect base-line studies), sometimes in combination with standardized quantitative sampling methods.

Through baseline sampling it is possible to record a very large proportion of the biodiversity present. However, given constraints on available resources it has probably never been possible to collect, sort and process, and identify to species level all taxa from any one site. While planning is under way for such all encompassing taxonomic inventories (D. Janzen and W. Hallwachs, pers. comm.), no such inventory has as yet been undertaken anywhere in the world. However, when relatively intensive inventories are undertaken it is important to develop a stepwise sampling and processing protocol for the individual site so as to maximize the information acquired about diversity of the site studied, and to permit subsequent comparative sampling and analysis (Coddington *et al.* 1991). General recommendations with regard to comparative inventory and monitoring are made by di Castri *et al.* (1992) who emphasize subsampling of focal groups (termed 'target groups' in Fig. 7.7)—see Stork (this volume).

Figure 7.7 shows the general protocol for the insect survey component of an integrated multi-disciplinary survey of the Cape York Peninsula in tropical Australia. The insect survey is based around a series of permanent sampling sites and the priority groups are those called target groups. The permanent sites (where traps are continuously sampled for one year) are established so that: (1) the total number of trap-sites will maximize the

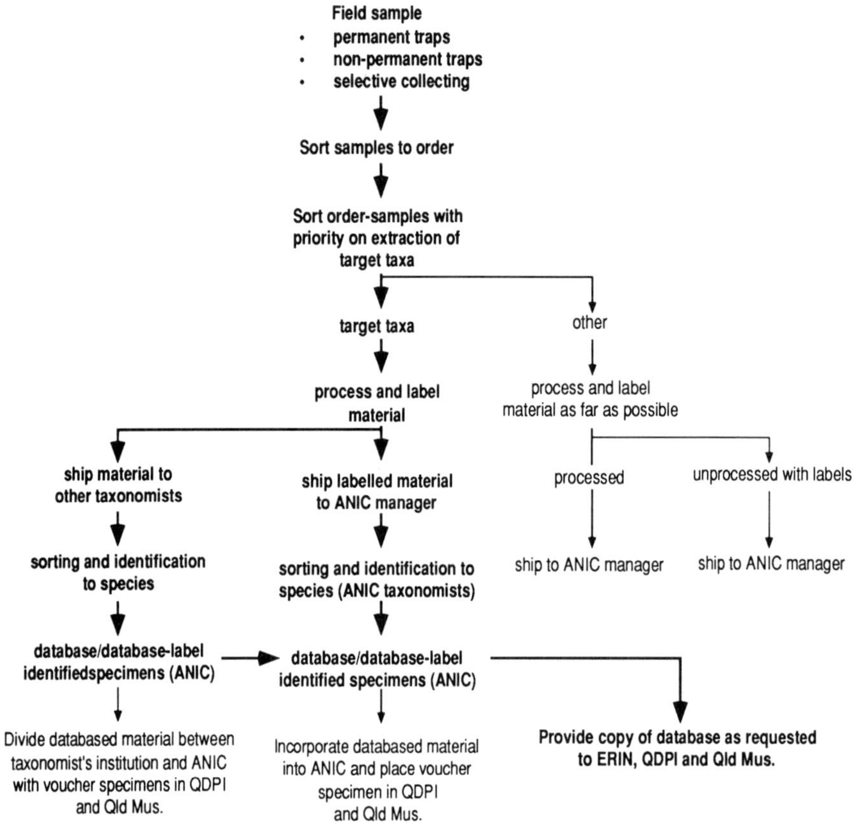

Fig. 7.7 Example of survey protocol for insects. This example is taken from the overall pathway protocol of the insect survey of the Cape York Peninsula Land Use Strategy (CYPLUS) in Australia. The insect survey is jointly undertaken between the Queensland Department of Primary Industry (QDPI), the Queensland Museum (Qld. Mus.), and the Australian National Insect Collection (ANIC). The pathway in bold is that of the target groups; the label data have been captured in a database and specimens are shared between the participating institutions; the database is further processed by the Environmental Resources Information Network (ERIN).

number of ecosystems sampled (based on vegetation types), (2) there is year-round access in this remote part of Australia, and (3) consideration is taken of the activities of other projects in the survey; the number of traps is determined by the resources available. Non-permanent traps and selective collection is intended to maximize the number of sites for which presence of a taxon is recorded, while absence does not necessarily indicate that the taxon is absent from the site collected. The target groups are groups of insects from which sizeable datasets are already available from the area, groups in which the majority of the species are readily identified, or groups for which there are taxonomists who will undertake identification within the timeframe available.

Acknowledgments

We wish to thank colleagues at our institutions and the ABRS for many stimulating discussions and particularly Drs Bryan Barlow, Peter Cranston, Bruce Halliday, and Ms Maggie Sutton of CSIRO for commenting on drafts of this paper.

References

Armstrong, J. A. (1992). The funding base for Australia's biological collections. *Australian Biologist*, **5**, 80–8.

Bruhl, J. J., Watson, L., and Dallwitz, M. J. (1992). Genera of Cyperaceae: interactive identification and information retrieval. *Taxon*, **41**, 225–35.

Chapman, A. D. (1991). Australian plant name index. In 4 vols. *Australian flora and fauna series* **12, 13, 14, 15**. Australian Government Publishing Service, Canberra.

Coddington, J. A., Griswold, C. E., Dávila, D. S., Peñaranda, E., and Larcher, S. F. (1991). Designing and testing sampling protocols to estimate biodiversity in tropical ecosystems, In *The unity of evolutionary biology* (ed. E. C. Dudley), pp. 44–60. *Proceedings of the fourth International Congress of Systematic and Evolutionary Biology*. 2 vols. Dioscorides Press, Portland, Oregon.

Dallwitz, M. J. (1974). A flexible computer program for generating identification keys. *Systematic Zoology*, **23**, 50–7.

Dallwitz, M. J. (1980). A general system for coding taxonomic descriptions. *Taxon*, **29**, 41–6.

Dallwitz, M. J. (1984). Automatic typesetting of computer-generated keys and descriptions. In *Databases in systematics* (ed. R. Allkin and F. A. Bisby.), pp. 279–90. *Systematics Association Special Volume*, **26**, Academic Press, London.

Dallwitz, M. J. (1993). DELTA and INTKEY. In *Advances in computer methods for systematic biology: artificial intelligence, databases, computer vision* (ed. R. Fortuner), pp. 287–96. Johns Hopkins University Press, Baltimore.

Dallwitz, M. J. and Paine, T. A. (1986). User's guide to the DELTA system: a general system for processing taxonomic descriptions, (3rd edn). *CSIRO Division of Entomology, Report 13*, 1–106.

di Castri, F., Vernhes, J. R., and Younès, T. (1992). Inventorying and monitoring biodiversity. A proposal for an international network. *Biology International*, Special Issue 27, 1–28.

Erwin, T. L. (1982). Tropical forests: their richness in Coleoptera and other arthropod species. *Coleopterists Bulletin*, **36**, 74–5.

ETI. (1992). *Birds of Europe CD-ROM, Macintosh Version 1.0.* ETI, University of Amsterdam.

Evenhuis, N. L. (ed.) (1989). *Catalog of the Diptera of the Australasian and Oceanian Regions.* Bishop Museum Press, Honolulu and E. J. Brill, Leiden.

Faith, D. P. (1992). Systematics and conservation: on predicting the feature diversity of subsets of taxa. *Cladistics*, **8**, 361–73.

Gaston, K. J. (1991). The magnitude of global insect species richness. *Conservation Biology*, **5**, 283–96.

Gaston, K. J. and Mound, L. A. (1993). Taxonomy, hypothesis testing and the biodiversity crisis. *Proceedings of the Royal Society of London*, B, **251**, 139–42.

Hawksworth, D. L. (1992). Microorganisms. In *Global biodiversity. Status of the Earth's living resources* (ed. B. Groombridge), pp. XV111 and 585. Chapman and Hall, London.

Hnatiuk, R. J. (1990). *Census of Australian vascular plants*. Australian Government Publishing Service, Canberra.

Hyland, B. P. M. and Whiffin, T. (1993). *Australian tropical rain forest trees*. Volumes 1–3, software. CSIRO Publications, Melbourne.

Lawrence, J. L., Hastings, A., Dallwitz, M. J., and Paine, T. (1993). *Beetle larvae of the world. Interactive identification and information retrieval for families and sub-families*. CD-ROM. Version 1.0 for MS-DOS. CSIRO Publications, Melbourne.

May, R. M. (1990). Taxonomy and destiny. *Nature*, **347**, 129–30.

May, R. M. (1992). Past efforts and future prospects towards understanding how many species there are. In *Biodiversity and global change* (ed. O.T. Solbrig *et al.*), pp. 71–81. IUBS Monograph No. 8.

McGinley, R. (1989). Entomological collections management—are we really managing? *Insect Collection News*, **2**(2), 19–27.

Monteith, G. (1990). Rainforest insects: biodiversity, bioguessmation, or just hand-waving. *Myrmecia*, **26**, 93–5.

Nye, I. W. B. and Fletcher, D. S. (1991). *Microlepidoptera. The generic names of moths of the world*. **6**, British Museum (Natural History), London.

Pankhurst, R. J. (1986). A package of computer programs for handling taxonomic databases. *Computer Applications in the Biosciences*, **2**, 33–9.

Pankhurst, R. J. and Aitchison, R. R. (1975). An on-line identification program. In *Biological identification with computers* (ed. R. J. Pankhurst.), pp. 181–5. Academic Press, London.

Soós, A. and Papp, L. (ed.). (1992). *Catalogue of Palaearctic Diptera. Volume 1. Trichoceridae-Nymphomyiidae*. Hungarian Natural History Museum, Budapest.

Systematics Agenda 2000. *Charting the biosphere*. (Draft 1993). Systematics Agenda 2000 Steering Committee, Department of Ornithology, American Museum of Natural History, New York.

Thompson, F. C. and Knutson, L. (1987). Catalogues, checklists and lists: a need for some definitions, new words and ideas. *Antenna*, **11**, 131–4.

Vane-Wright, R. I., Humphries, C. J., and Williams, P. H. (1991). What to protect—systematics and the agony of choice. *Biological Conservation*, **55**, 235–54.

Walker, B. *et al.* (1992). Australia's biodiversity: what it is, its significance and what is happening to it. In *Scientific aspects of major environmental issues: biodiversity* (ed. Prime Minister's Science Council), pp. 1–7. Australian Government Publishing Service, Canberra.

Watson, L., Aiken, S. G., Dallwitz, M. J., Lefkovitch, L. P., and Dube, M. (1986). Canadian grass genera: keys and descriptions in English and French from an automated data bank. *Canadian Journal of Botany*, **64**, 53–70.

Watson, L. and Dallwitz, M. J. (1989). Grass genera of the world: interactive identification and information retrieval. *Flora Online*, **22**.

Watson, L. and Dallwitz, M. J. (1992). *The grass genera of the world.* CAB International, Wallingford, UK.

Watson, L. and Dallwitz, M. J. (1993). *Families of flowering plants—interactive identification and information retrieval.* CD-ROM, Version 1.0 for MS-DOS. CSIRO Publications, Melbourne.

Watson, L., Damanakis, M., and Dallwitz, M. J. (1988). *The grass genera of Greece: descriptions, classification, keys (in Greek).* University of Crete, Heraklion.

Webster, R. D., Kirkbride, J. H., and Reyna, J. V. (1989). New World genera of the Paniceae (Poaceae: Panicoideae). *SIDA*, **13**, 393–417.

West, J. and Nielsen, E. S. (1992). Management and accessibility of biological collections. *Australian Biologist*, **5**, 68–75.

Williams, P. H., Vane-Wright, R. I., and Humphries, C. J. (1993). Measuring biodiversity for choosing conservation areas. In *Hymenoptera and biodiversity* (ed. J. LaSalle and I. Gauld), pp. 309–28. CAB International, Wallingford, UK.

Wilson, E. O. (1985). The biological diversity crisis: a challenge to science. *Issues in Science and Technology*, **2**, 20–9.

Xu Zhu, Watson, L. and Dallwitz, M. J. (in press). *Grass genera of China (in Chinese).* Chinese Agricultural Press.

8. The industrialization of scientific information

B. J. RICHARDSON
Faculty of Science and Technology, University of Western Sydney, Hawkesbury, Bourke St, Richmond, NSW, Australia

Abstract

Because of the large volume of potential data available, the short time frames set for their integration, and the user-driven nature of modern environmental databases, their development often requires an industrial rather than research orientation. This paper considers some of the issues involved in developing such systems.

The establishment of national and international information systems of biological information requires careful planning (systems analysis) if work is not to be wasted. Failure of such systems can be due to unreal expectations, or to unexpected changes in user requirements before the system is established. It is important that any such information system be designed with the needs of the users clearly in mind. A first step therefore may be to decide who the users are.

In national environmental databases, the data sets are very large and the information to be included will have been collected by different agencies and collectors over many decades using different standards. The data are of varying quality and any attempt to integrate such data requires an effective and automated method of validation of each record and of record sets. The Australian government has established such a system, the Environmental Resources Information Network (ERIN), to draw together and analyse environmental information related to the continent of Australia. The experience gained and the methods used in developing this programme will be used to exemplify the issues.

Introduction

National information systems of environmental data have been established in several parts of the world over the past two decades. These systems have not

Systematics and Conservation Evaluation (ed. P. L. Forey, C. J. Humphries, and R. I. Vane-Wright), Systematics Association Special Volume No. 50, pp. 123–31, Clarendon Press, Oxford, 1994.

been successfully used by decision makers for a variety of reasons, but mostly due to a lack of coordination between the needs of users and the perceptions of system builders: that is, there has often been a significant lack of planning. Many groups responsible for developing such systems have an 'academic' view of the systems they develop and find it difficult to accept the nature and needs of their 'real' as opposed to 'ideal' clientele. However, it is important that any such system be designed with the actual needs of the users clearly in mind. Establishing these needs can be quite difficult given the lack of relevant knowledge on the part of policy officers and environmental managers. As well as these user-based issues, there are a number of matters of practical concern in the design of the database. This paper summarizes a number of issues that arise and draws on the Australian experience (Anon. 1991).

The Environmental Resources Information Network

The problems that result from the absence of adequate environmental information necessary to underpin national or regional decision-making have become evident in the past few years. As a consequence the Australian government established the Environmental Resources Information Network (ERIN): 'to draw together, upgrade and supplement information on the distribution of endangered species, vegetation types and heritage sites' (Hawke 1989).

The programme was to use recently developed computer-based tools to combine, analyse and display such data in the form of a Spatial Information System (SIS).

SIS development is fraught with problems. Many different options are available and a very large number of different data sets can be included in the system. Data sets can be very large (for example, the specimen backed records in the museums and herbaria of Australia include more than 34 million specimens or lots (Richardson 1992). Furthermore, SIS tends to be complex, difficult to use, and the results obtained subject to easy misinterpretation by the inexperienced. For these reasons, it is essential before starting to define the purpose of establishing such systems, and to question closely the potential users as to the specific questions and in what time frame they require answers. In ERIN, professional consultants were used with experience both in SIS and in the sociology of carrying out such user-analyses. This step took 80 per cent of the time of the start-up phase of ERIN, and great self-control was needed not to be 'busy' purchasing hardware, software, and data until these matters are settled. Unlike the use of such systems for research programmes, the establishment of large user-oriented systems is analogous to industrial development of a new manufactured product.

An observation made during the user survey was the complete lack of any concept of how the system might help users. As a consequence, a systematic education programme of potential users was started. This included simple information on what a SIS is, and demonstrations of several systems were set up to carry out the kind of analysis that users might need to undertake. The level of ignorance could be frustrating, and every effort had to be made to avoid telling the users what they needed. It is envisaged that education of users, especially in conceptually new ways to use such powerful new tools in making environmental policy decisions, will be a major task of the ERIN unit for several years.

A third important step was to place the programme firmly under the management of a Steering Committee consisting of line managers in relevant policy areas. This gave them a sense of ownership and responsibility for ensuring that it met their needs. These processes have converted potential users in the portfolio from an attitude of deep suspicion and unhappiness in having to 'waste time' talking to the consultants, to one of some enthusiasm and support for ERIN.

What must be appreciated is that no one system and host organization can handle responsibility for all of the needs for environmental information and analysis. Environmental information systems must be based on established and well co-ordinated interdisciplinary and multiagency collaboration and cooperation. Gone also are the days of mega-centralised computing. Present solutions lie in the direction of open systems and distributed processing combined with responsibility for the quality, management, and availability of the data remaining in the hands of the custodian of each data set. To this must be added extensive training and support for the user community, including senior management.

Standards

In establishing national databases of environmental information it is necessary to integrate data from many sources. For this reason the data collected by different workers and agencies need to be data based in a consistent manner. To reach this consistency there is need to specify and adopt national standards.

Because many people will be collecting data relevant to the national purpose but for their own purposes, it is essential to develop a commitment by the relevant communities to national standards. This can best be done by involving these people in the development of the standards so that they will meet everyone's needs and be compatible with the different goals of different workers. As a consequence of this process, all involved will have a sense of ownership of the standards and commitment to their use. In establishing the Australian system, consultations were carried out nationally with all of those agencies involved in collecting relevant data. Thus, standard development was facilitated rather than imposed.

In establishing national standards of this nature a number of problems need to be dealt with. These can be considered below.

1. *Semantics.* Care must be taken to ensure that the same meaning is applied to words and phrases by each agency. If data are to be integrated from several sources and made available to a range of users, we must be sure that they all mean the same thing when they use the same term.

2. *Taxonomic nomenclature.* The international codes of taxonomic nomenclature constitute proper standards as the names are formally defined. Other commonly used descriptors are less clear, for example vegetation types, where different types may be described using the same word by different agencies.

3. *Diversity of standards.* Ideally we would wish to ensure that all incoming data were collected by using the same standards. This is often not achievable as much useful data have been already collected. As a consequence, we will need to attach to each data set information on the standards used in its collection, preferably in a form that will allow the data to be 'translated' into a form compatible with other data sets. Unfortunately, it may not be always possible to integrate different data sets, as they have been collected using incompatible standards. This most frequently occurs when pre-classified data have been entered into the database rather than the primary information.

4. *Absence of standards.* Occasionally, it is found impossible to establish standards useful under all conceivable situations and then some compromise, preferably related to categories of users, will be needed. An example of such problems can occur in the definition of conservation status where the world standards established by IUCN may not be suitable for national or local use. Problems especially arise with respect to distributional data. For example, a simple policy query asking for information on 'threatened species' at 'Mt Quamby' contains several semantic difficulties; what is meant by 'threatened' (locally, nationally, or internationally?), which 'Mt Quamby', and within what distance from Mt Quamby?

5. *Core data sets.* While complete integration of data sets may never be possible, it might well be possible to agree on some core criteria, for example, taxonomic name, latitude and longitude, time and date of collection, and name of collector. Other attributes of a data set would then be defined in the data dictionary for the data set.

Clearly the use of a diversity of standards, especially those leading to pre-classification of data, cause great difficulty and wherever possible primary data should be entered into the database. Where necessary a data dictionary needs to be attached to each data set to ensure no semantic problems occur. This ensures that conclusions based on those data can be rigorously reviewed, alternative analyses can be performed, and baselines for monitoring established.

Databases

Databases may occur in many forms. They may be, for example, digital or paper records. No matter what form the database takes, it is essential to attach key information to the base before making it available to others. Matters to be taken into account are:

1. *Database identifier*. This should be a short convenient working name to be used by all those interested in the database.

2. *Database name*. This should be a longer and clear description of the database.

3. *Version*. As databases are frequently edited or updated it is important to mark the version on any copy of the data set.

4. *Custodianship identifier*. This briefly identifies the owner of the data.

5. *Custodianship name*. Formal title of the custodial institution or individual.

6. *Custodianship address*. The postal address of the custodian.

7. *Contact officer*. The name or position of the contact officer for information relating to the database.

8. *Recipients*. Anyone receiving a copy of the database should provide information equivalent to that above for the custodian.

9. *Security class*. It is frequently important to the custodian of data to restrict access to the database. Examples of such restrictions might be, exact locations of endangered species, sites sacred to Aboriginal people, or data supplied by companies and of commercial value. It may be necessary to establish three or more categories of users with different restrictions on their access to primary data.

10. *Release instructions*. This allows the custodian of the data to define who recipients may be. Clearly, such restrictions increase the difficulty in managing the database.

11. *Review date.* A cycle of review of restrictions should be put in place with the aim of removing restrictions whenever possible.

12. *Database summary.* Information needs to be supplied that explains what the data base contains. This should cover in textual form the themes of the data base, the spatial domain covered by the database, the data or date range of the observations, the geographic scale used, and standards used, the name of the collator, the lineage of the database (that is the textual description of the origins and process used in generating the database), caveats regarding the quality of the data, and the completeness of the data.

Taxonomic databases

Because clearly recognized and logically consistent international standards are in place for taxonomic nomenclature, the design of a database of taxonomic names is relatively straightforward. It must be remembered, however, that although the nomenclatural codes for botany and zoology are each internally consistent, they do differ and an integrated database would need to take account of these differences. In establishing a taxon database the aim is essentially to develop a register of recognized taxa for a country. However, any such database must be able to take account of alternative and changing taxonomies. The structure of such a database should not only supply taxonomic names, but also form the basis for searching other data sets by hierarchical units (for example, to allow the available databases to be searched for all the species in a particular family). Ideally a taxonomic database should also allow cross-referencing of taxa to other useful categories of interest to users, for example, 'high seas pelagic' or 'endangered'.

As well as the international codes, a number of other standards related to taxonomic nomenclature have been developed, for example, HISPID (Herbaria Information Standards and Protocols For Interchange Of Data) and ABIS (Australian Biogeographic Interchange Standards). In Australia the data base structure of several databases takes account of taxonomic needs, for example, CAP (Census of Australian Plants), CAVS (Census of Australian Vertebrate Species), Zoological Catalogue of Australia, Census of Bryophytes and Lichens. Internationally, standards have been developed for plant data (Hollis and Brummitt, 1992).

The database structure should take into account the following:

1. *Formal categories.* The categories family, genus, species, sub-species, other infra specific categories, and authorities for these names should be included. The family name is needed because of the possibility that the same name may be used for genera of plants and animals. A taxonomy above the family level, probably as a separate database, may be needed for some forms of integrated analysis.

2. *Applied categories.* Users may well need to relate other names to the formal categories and attaching standard common names to the database can be very helpful.

There are a number of problems in establishing an effective taxonomic data base structure and these need to be taken account of in the design. Firstly, there are historical changes in taxonomy with time. These result from improved knowledge of the phylogenetic relationships of the species involved. Unfortunately, databases supplied by others may use old names and the integrated database structure must take account of these synonyms. Furthermore, it is sometimes discovered that taxa previously thought of as one species consist of two or more species, and as a consequence part of the range data for a species appears in the database under another valid species name. A further problem can occur when different individuals or agencies use alternative taxonomies. This is especially common at the generic level where different concepts can lead to the 'splitting' or 'lumping' of genera. These problems require the database to include synonyms and also dated times of change of usage.

Space

One of the key parameters in environmental information describes the exact location where data were collected. There are a number of factors that must be taken into account in describing the exact location.

1. *Geometric description.* It is not uncommon for there to be several different systems in use in any one country. Biologists commonly use latitude and longitude and this can be accurate to tens of metres. Surveyors and engineers, however, need much more accurate systems of location for developing the cadaster. In Australia, the Australian Mapping Grid is used for this purpose. If it is intended to overlay biological information with a cadaster in a SIS, care must be taken to ensure that the same location in the different systems match properly.

2. *Zonal units.* It is common to use pre-classifications in describing collecting localities. These derivative classifications should be avoided or derived as needed after digitizing the primary data. Examples include, data collected by grid squares, by electorate, by census district, by local government area, or by statistical unit.

3. *Locality.* Textual descriptions are often used and these can cause problems in digitizing. For example, '15km SW of Brisbane'. It is very difficult to establish a good geometric location for such a locality. On the other hand, the textual description can be used as an effective check on the accuracy of the given latitude and longitude.

4. *Precision.* Precision means the level of mismatch between the actual and recorded collecting position, and may occur either in the field or at the time of digitizing the data. It is essential that any report of location include a measure of precision.

5. *Scale.* The scale of the map used, or the accuracy of the estimation of the position affect the way the information can be interpreted in making generalizations. Commonly, coarse-scale representations of environmental boundaries are smoothed and small pockets of habitat excluded. If such information is combined with other data sets collected at a much finer scale, incorrect conclusions can be drawn. It is essential that scale information be attached to any geometric description.

Spatial data transfer standards

A national system of transfer standards is essential. For example, the US Spatial Data Transfer Standard has recently been developed and an analogous standard is being developed for Australia. Each country will need to determine its own standards.

Time

Biological distributions and other meaningful environmental information frequently change with time. As a consequence it is essential to 'date stamp' any records in the database. Changes are of two types, *definition changes* and *value changes*. Definition changes relate to attributes of items in the data dictionary. For example, the boundaries of a park may be changed at a certain date so that a species not previously in the park now is reported as present, though there has been no actual change in its distribution. Value changes relate to variations in measures taken in the field. For example, diurnal changes in the position of an animal between feeding and roosting areas, or seasonal changes in the distribution of a species.

Methods of verifying distribution data

When data sets, especially if records are derived from several agencies, are combined it is commonly found that there are inconsistencies or anomalies between records. As a consequence, it is essential to establish a formal process of checking the accuracy of records and sets of records. In Australia for biological records this is done by an automated system that: checks that the required data fields are complete in each record; checks that the scientific name is valid in each record; checks that the point locality is located over land not the sea for each record; runs a model (BIOCLIM) that predicts the distribution of a species on the basis of the bioclimatic profile of known locality points and identifies points with outlier values; maps such data sets and aberrant record locations.

The map and other information is then returned to the originating agency for checking (A. Chapman, pers. comm.). Because of the large size and number of data sets to be included in a national system such validation programmes need to be developed for each situation.

It can be seen from the issues raised above that the design of a national system is not a trivial exercise, and the integrated efforts of many agencies are needed for an effective outcome.

References

Anon. (1991). *ERIN Database design and development: final report.* CSIRO Division of Information Technology, Canberra.

Hawke, R. J. L. (1989). *Our Country our future.* Australian Government Publishing Service, Canberra.

Hollis, S. and Brummitt, R. K. (1992). *World geographic scheme for recording plant distributions. Taxonomic Database Standards No. 2, version 1.0, January 1992.* Published for the International Working group for Plant Sciences (TDWG), by the Hunt Institute for Botanical Documentation, Carnegie Mellon University, Pittsburgh.

Richardson, B. J. (1992). Australia's biological collections and those who use them. *Australian Biologist*, **5**, 19–30.

9. A possible method for the rapid assessment of biodiversity

I. OLIVER and A. J. BEATTIE
School of Biological Sciences, Macquarie University, NSW 2109, Australia

Abstract

A method is described for rapid assessment of biodiversity using a wider variety of taxa than the usual inventories of flowering plants and vertebrates. To explore the value of invertebrates and cryptogamic plants as surrogates of biodiversity, estimates of species richness of spiders, ants, polychaetes, and mosses were undertaken by classifying them as Recognisable Taxonomic Units (RTUs), and comparing them with specialist identifications. RTU estimates turned out to be close to formal taxonomic estimates for the invertebrates but were misleading for mosses. It was concluded that trials of all groups deemed to be useful for RTU analysis needed to be evaluated before reliability could be attained. Taxonomic specialists are considered indispensable to train personnel, to undertake the preliminary trials, and to check the results of RTU estimates before wholesale studies were undertaken by 'biodiversity technicians'.

Rapid assessment

Inventories of vertebrates and flowering plants are frequently used as surrogates for estimates of total biodiversity. This is, in part, because the inclusion of invertebrates and non-flowering plants can be time-consuming, costly, and difficult given the shortage of specialists. In view of these problems, it is important to explore other measures of the variety of taxa that may be useful for the rapid assessment of biodiversity. In this study we explored this objective by classifying invertebrates and mosses into Recognisable Taxonomic Units (RTUs). The goal was to discover whether or not RTUs provided reasonable estimates of species richness, by comparison with formal Latin binominals, in a comparatively short time.

Systematics and Conservation Evaluation (ed. P. L. Forey, C. J. Humphries, and R. I. Vane-Wright), Systematics Association Special Volume No. 50, pp. 133–6, Clarendon Press, Oxford, 1994.

RTUs are taxa that are readily separable by morphological differences obvious to individuals with less training than professional taxonomists. These individuals are called biodiversity technicians in Australia.

Estimates of the species richness of field samples of spiders, ants, polychaetes, and mosses made by a technician and by specialist taxonomists were compared. The technician received a few hours training in the basic taxonomy of each group, including observation and utilization of the most important and easily identified taxonomic characters, and an outline of possible problems such as sexual dimorphism and polymorphism. Specimens were sorted into RTU's using a standard dissecting microscope. No scientific keys were used. A reference collection was compiled as new RTUs were encountered, and illustrations of distinguishing characters made. The specialists examined the same samples and sorted to species.

For all taxa the biodiversity technician recorded 252 RTUs and the specialists recorded 233 species. The closest estimate was for the mosses, being 87 RTUs compared to 86 species, then the ants 33 to 35, spiders 103 to 91, and the polychaetes with 29 RTUs compared to 21 species. These results indicated that some splitting and lumping of species had occurred. As the entire sample of each group had been analysed by specialists, we were able to determine how splitting and lumping affected these estimates. For each group we determined the number of cases in which one RTU corresponded to one species (hereafter referred to as the '1 to 1' category), the number of cases in which one RTU corresponded to more than one species, and the number of cases where a single species was split into different RTUs.

Analysis revealed the true level of accuracy. In the ants and the spiders 88 per cent and 83 per cent of cases respectively were represented by the '1 to 1' category, the RTU estimates being relatively unaffected by splitting and lumping. The greatest level of overestimation of species occurred in the polychaetes where single species were split into 2, 3, or 4 RTUs on several occasions and, consequently, only 43 per cent of cases were represented by the '1 to 1' category. The RTU estimate for mosses was the closest to the formal number of species. However, the frequency of splitting and lumping was high, with only 23 per cent of cases being in the '1 to 1' category. The high concordance between estimates for the mosses was due to the fortuitous symmetrical distribution of splitting and lumping. This problem was reduced when only specimens bearing the character rich sporophytes were included in the analysis. When this was done, the frequency of splitting and lumping was reduced, and the percentage of cases in the '1 to 1' category increased from 23 per cent to 50 per cent.

We also analysed a subset of the data to determine whether or not smaller samples might estimate the degree of overestimation and underestimation of species with reasonable accuracy. This was considered important because, if true, a smaller number of specimens per RTU sent for specialist analysis would greatly increase the speed of the process. As a first approximation we took 10 specimens at random (or all specimens where there were fewer than 10 in the sample) from each RTU. Knowing the numbers of different species in RTUs and the number of specimens within species in the whole collection, we calculated the probability of representing all species in smaller samples. Splitting was detected by comparing RTU subsamples and identifying those species that appeared in more than one. We anticipated that in subsamples of 10 specimens there would be a proportion of cases in which species were not sampled by these procedures. The results showed that of the 147 species listed by the specialists, 145 were represented in the subsamples. The mosses were not included in this analysis because they were, on average, only 2.5 specimens per species.

The RTU estimates for the ants, spiders, and mosses were close to the formal taxonomic estimates. However, the moss data show that RTU estimates can be misleading. This demonstrates that different groups of organisms must be analysed by the procedures described here before they can be included confidently in rapid biodiversity assessment using RTUs.

The data for the mosses also suggest that the use of RTUs requires a minimum level of unambiguous morphological variation that a relatively untrained individual can easily and quickly utilize. Ambiguity may arise from sexual dimorphism and developmental polymorphism. Where taxa exhibit extreme sexual dimorphism, we suggest that only one sex should be used. Although this approach may result in missing some species, the inclusion of all life stages or sexes in rapid biodiversity assessment may lead to large errors. Our experience also strongly suggests that with species-rich samples pre-sorting into major groups may be advantageous. Keys may be useful for this phase.

Taxonomic specialists are vital at three stages of rapid biodiversity assessment: (i) to train the biodiversity technicians, (ii) to provide the initial formal estimates for comparison with the RTU estimates, and (iii) to check the quality of subsamples on a regular basis. In many cases stages (i) and (ii) need to be applied only once. However, we believe that if invertebrates are to be routinely incorporated into inventories, stage (iii) should always be included. We have shown that, with care, the number of species missed when sub-samples are analysed by specialists can be small.

More research is needed to determine if the results presented here can be reproduced by different biodiversity technicians studying different taxa. The data presented suggest that rapid biodiversity assessment is worth investigation. They also suggest that invertebrates, which comprise the greater part of non-microbial biodiversity, may provide more comprehensive and accurate sampling of total biodiversity (see Oliver and Beattie 1993).

Reference

Oliver, I. and Beattie, A. J. (1993). A possible method for the rapid assessment of biodiversity. *Conservation Biology*, **7**, 562–8.

10. Survey and conservation status of Indian plants

S. MANDAL
Department of Botany, Visva-Bharati University, Santiniketan—731235, India

Abstract

Conservation of wildlife and the environment is vital if our society is to survive. However, a large number of species are now endangered. Factors such as sustained pressure on forests result in habitat destruction, and diverse human socio-economic problems have contributed considerably to the deteriorating status of our magnificent wildlife. Fortunately, recent trends in some states of India have created a keen awareness among the general public of the need to conserve wild natural habitats and forest wealth, and also to protect endangered wild animals and plants. Investigations into the population status of certain wild plants and animals in different Indian forests have been carried out. The present paper illustrates the survey and conservation status of the natural plant life of India based on a detailed analysis of vegetation types and other sources of published information.

The Indian subcontinent can support a tremendous diversity of plant and animal species because of its great variety of vegetation habitats. About 15 000 flowering plant species and 80 000 animals have so far been described from India, and another estimated 25–50 000 remain to be described. Of the 15 000 vascular plants found in India, nearly 1275 monocotyledons and about 5000 dicotyledons are endemic to the country. It has also been noted that about 1500–2000 species (i.e. 10–15 per cent of the flora) fall into one or other category of threatened plants.

Introduction

The problem of over-exploitation of natural resources, environmental degradation and declining wildlife populations are most severe in the

Systematics and Conservation Evaluation (ed. P. L. Forey, C. J. Humphries, and R. I. Vane-Wright), Systematics Association Special Volume No. 50, pp. 137–44, Clarendon Press, Oxford, 1994.

developing countries. Preservation of natural landscapes with their indigenous flora and fauna is a responsibility of every nation. The need for conservation of the floristic gene pool has been appreciated in the last 30 years, so that there is a realization that conservation of plants in their native habitats is essential.

According to Raven (1976) two-thirds of the world's plant species are tropical. Regarding the size of tropical floras, Prance (1978) provided some very useful figures. According to him 155 000 out of the 240 000 angiosperm species are tropical. Of these 90 000 are from tropical America (compared to 35 000 in tropical Asia and 30 000 in tropical Africa). It has also been estimated that Colombia, for example, contains 45 000 or 50 000 species (Schultes 1951; Prance 1978). The figures for other plant groups are even more remarkable. According to Prance (1978) 11 000 out of 12 000 ferns are tropical and 9000 out of 12 000 mosses are tropical.

Now, the requirement for information in conservation efforts concerns at least two levels of data gathering. The first is the need to complete the inventory as proposed by Prance (1983), which implies a great deal more basic exploration, the preparation of floristic lists for poorly known areas, surveys of neglected areas, and so on. Secondly, the threatened and endangered species have to be identified with the available published data. That is why there is a need for more modern information so that appropriate conservation measures can be taken.

The present paper will reveal the survey and conservation status of the natural plant life of India on the basis of a detailed analysis of the available vegetation types and other sources of information. The Indian subcontinent has a tremendous diversity of plant and animal species due to its great variety of ecological conditions. About 15 000 species of flowering plants and 80 000 species of animals have so far been described from India (Gadgil and Meher-Homji 1990).

Observations

A large number of species are now endangered. Factors such as sustained pressure on forests resulting in habitat destruction and diverse human socio-economic problems have contributed considerably to the deteriorating status of our magnificent wildlife. Fortunately, recent trends have created a keen awareness among the general public, at least in some states of India, of the need to conserve natural habitats and forest wealth, and also to protect our endangered wild plants and animals.

The Indian region is well-circumscribed geographically, limited by effective barriers like the seas, the Himalayas, and the Thar desert. The variety of its climates and its geographic position is at the junction of important floras and includes Malesian, African, temperate, Mediterranean, and Chinese floristic elements. The natural barriers have helped in the development of a large number of species in the Himalayas. The intervening area of the Indo-Gangetic plain, either because of its dry climate (in the Thar), or due to high population pressure and uniform climatic conditions (in the Gangetic valley) is poor in endemic species. However, there have been several attempts to divide India into biogeographic regions according to the distribution of certain taxa, e.g. Cyperaceae by Clarke (1898), general flora by Hooker (1906) and Chatterjee (1939), and migration of plants by Razi (1975). Blasco (1979) designated the eight divisions of Chatterjee (1939) as ecofloristic zones and enumerated bioclimatic criteria, vegetation types, and floristic peculiarities, but above all he emphasized the characteristic flora (endemic species) of each zone.

The Indian subcontinent is exceedingly fortunate as it is essentially a tropical country with the Himalayas keeping out the cold air masses from northern latitudes. Therefore, it has in common with other tropical regions a greater wealth of living organisms. The country supports a tremendous variety of environmental regimes from the warm, humid tracts of Kerala to cold, dry heights of the Ladakh plateau, and the hot desert of Rajasthan to the swamps of Sunderbans. It is also a country situated at junction of African, European, and south-east Asian realms and derives characteristic elements from each of them. The full diversity of the country's plant life is only imperfectly known. Out of 250 000 species of higher plants known throughout the world, some 15 000 have been described from India (Mani 1974). There are about 1275 species of monocotyledons and around 5000 dicotyledons endemic to this country (Jain 1976). It has also been found that 1500–2000 species (i.e. 10–15 per cent of the flora) fall into one or the other category of threatened plants. Chatterjee (1962) estimated the number of flowering plants of peninsular India as about 2545 species (about 2045 dicots and 500 monocots). Subramanyan and Nayer (1974) reported about 90 endemic species from the Western Ghats. Henry *et al.* (1979) listed 264 species of rare flowering plants in South India of which 54 were reported by Hooker (1872–1897). According to them many of these 54 species are not represented in the Madras Herbarium (MH) today. It has also been found that out of 15 000 vascular plants reported from India, about 3500 are monocotyledons. Out of these, 1260 (about 36 per cent) are endemic to India and 427 of these (about 12 per cent) could be classified as rare or threatened.

Conservational constraints

The varied and rich flora of a vast country like India places severe constraints on conservationists. The existence of only a few herbaria poses limitations in assessing the status of habitats or areas of past distribution of threatened plants (Jain and Sastry 1981). An analysis of literature records, as in Floras, and of herbarium material, together with the study of plants in gardens and in the wild, gives useful clues to the status of species. For example, a study of the orchid genus *Coelogynae* suggested the possible extinction of three species, *C. albo-lutea* Rolfe, *C. assamica* Linden and Reichb., and *C. treutleri* (Das and Jain, 1980). *Arthraxon lanceolatus* (Roxb.) Hochst. is a very rare grass confined to a small region of India. But confusion about its possible identity with *A. prionodes* (Steud.) led to reports of its abundant occurrence all over the country. *Arthraron microphyllus* (Trin.) Hochst. is confined to the eastern Himalayas, but confusion of its identity with *A. lancifolius* (Trin.) Hochst. led to reports of its wide occurrence. This study also led to the discovery of new, endemic, and rare taxa in the genus, as exemplified by *A. deccanensis* Jain and *A. junnarensis* Jain and Hemadri.

Populations of certain selected plants have also been constantly monitored. Methodologies for this are not yet specific, each scientist trying to evolve his own depending on the size and number of populations, vulnerability of habitat, and other factors. The botanists of the Botanical Survey of India at Shillong have been keeping a watch on populations of *Nepenthes khasiana* Hook.f., the endemic and only pitcher species of the country, and found that its populations were shrinking and needed protection. Its type locality has since been declared as a sanctuary. It was heartening to note that these populations have now responded to conservation measures. Monitoring of populations over long periods is impossible as most of the populations remain exposed even to unnatural hazards. A hill-slope in the eastern Himalayas had been under observation for some years for its rich population of *Paphiopedilum fairrieanum* (Lindl.) Pfitz. (Rao and Hajra 1977), until the story ended with the whole slope being cleared for development. However, Mandal (1988) studied the exploration, evaluation, and conservation of rare plants of the Birbhum district, West Bengal, with special reference to *Drosera* sp., another endangered genus of insectivorous plant of the country.

Role of the Botanical Survey of India

The magnitude of the survey needed to make a reasonably complete assessment poses a considerable problem in India due to the vastness of the country, and the very varied and rich natural flora. The Botanical Survey of India is presently the chief agency involved in this activity, but interest is now evident among botanists in other institutions. Very few revisionary studies on Indian genera have been done. Out of about 2800 genera of vascular plants, around one hundred have been subjected to taxonomic revisions. Furthermore, herbarium sheets, particularly the old ones, lack data on the frequency of occurrence. A few old herbaria and gardens exist in India, a country of over 3 million km^2. Only four herbaria, namely the Central National Herbarium, Calcutta (CAL), the herbarium of the Forest Research Institute Dehra Dun (DD), the herbarium of the Botanical Survey of India, Coimbatore (MH), and the Blatter Herbarium at St Xavier's College, Bombay (BLAT), can claim to have any noteworthy collections from before 1900.

Sanctuaries for individual species or groups of threatened species are the best means of conservation. It has been possible to establish sanctuaries for *Nepenthes* in the Khasi Hills, and for *Rhododendron* and *Primula* in Sikkim. An Orchid Sanctuary was also established at Sessa, Arunachal Pradesh for ensuring *in situ* conservation of orchids. Establishment of 'germ-plasm banks' is also important. Considerable literature has been generated recently on the role of botanic gardens in conservation, chiefly due to efforts of botanists of the Royal Botanic Gardens, Kew, and the University Botanic Garden, Cambridge (Synge and Townsend 1979). A start has been made in the experimental gardens of the Botanical Survey of India. For example, several species of *Rhododendron*, *Paphiopedilum*, *Dendrobium*, *Cymbidium*, *Coelogyne*, *Arundina*, and other orchids are being grown in the Experimental Garden at Orchidarium at Shillong; *Nepenthes*, *Psilotum nudum* (L.)Beauv., and *Platycenum wallichii* Hook. are cultivated in the Botanic Garden at Howrah. The National Bureau of Plant Genetic Resources, New Delhi, and the National Botanical Research Institute, Lucknow, are also building germ-plasm collections of selected groups, particularly of the more economically useful genera. Thompson (1975) has reviewed the limitations of botanic gardens and other collections of living plants for conservation of germ-plasm. He has also reviewed the prospects of seed banks as a means of conservation. Very little effort on conservation through seed banks has been possible so far in India.

Forest cover

The forest cover in India has been diminishing rapidly. Particular concern has been expressed over deforestation of the Himalayas because of drastic consequences for adjacent plains (Gupta and Bandhu 1979; Moddie 1981). Deforestation of the Himalayas' reservoirs is believed to be a primary cause of increased soil erosion and consequent rapid siltation in the Himalayan foothills (Kayastha and Juyal 1979; Moddie 1981). Other undesirable consequences attributed to deforestation are an increased incidence of flooding in the adjacent plains and reduced flow of rivers during the dry season (Bhasin *et al.* 1979; Rieger 1981).

At a height of approximately 1100 metres in the Palghat District of Kerala lies the 8950 ha of the rich and almost unpopulated forests of the Silent Valley. These form part of a contiguous forest block of approximately 40 000 ha, the largest such block in the rapidly deforested State of Kerala. These forests contain wild relatives of agricultural and commercial crops as distinct from medicinal plants. Unlike most of the forests of India, the Silent Valley has been free from any village settlements or tribal populations, and in fact the forests appear to have existed with minimal human intervention. Although Silent Valley was selected as an ideal site for a hydro-electric project its completion has not been possible due to the timely intervention of Indian scientists for the conservation of this Valley for its floristic and faunistic wealth.

Aquatic habitats

The Indian subcontinent possesses a rich variety of aquatic habitats ranging from glaciers and streams in the Himalayas, great estuaries on the sea coast, high altitude lakes like Mansarovar, jheels such as Bharatpur, and brackish lakes such as Chilka, to the extensive continental shelf off the west coast of India and coral reef islands of Lakshadweep. This biota has received even less attention and is quite seriously threatened from river valley projects, siltation, and chemical pollution. There are a few aquatic nature reserves, the best known of them being the jheel at Bharatpur. Three sanctuaries, covering several hundred km^2 along the lengths of Chambal, Ken, and Son rivers have been created. Finally, there are proposals to create two marine biosphere reserves, one off Saurashtra and the other in the Gulf of Manaar, to protect coral reef areas. Careful analysis and conservation action relating to the aquatic ecosystems is evidently needed.

Conclusion

Basic information for an overall strategy for the conservation and utilization of resources must include current knowledge of the distribution, status, and coverage of remaining forests, as well as increased knowledge of the biota of these forests. It also requires recognition of priority sites for conservation, and an understanding of what forms and levels of utilization are sustainable, and what further forms of disturbance are justifiable. A complete and accurate appraisal, based on ground, air, and perhaps even satellite surveys of the present distribution of the forest types and the degree to which they are intact or degraded, is an important requirement for future conservation programmes in India. A fundamental objective of all governments should be the preservation of natural and unspoiled areas which contain the complete flora and fauna of the nation. Serving as living museums, these samples of wild landscape should be conserved as national parks or nature reserves.

All conservationists must now dedicate themselves to protect the remaining areas of their country's natural ecosystems through a well thought out programme of biosphere reserves. Scientists as a community need to air their views on conservation more strongly. By working in the conservation field on wholesale biodiversity, threatened species, rare species, crop-plant relatives, and by giving the maximum publicity to their work and results, they can all make major contributions to saving at least part of the plant resources of the planet for the needs of future generations.

References

Bhasin, V., Bhasin, M. K., and Singh, I. P. (1979). People and forests of Bharmaur Tehsil, Chamba, Himachal Pradesh. In *Ecology and Development* (ed. D. Bandhuand V. Bharadwaj), pp.247–254. Indian Environmental Society, New Delhi.

Blasco, F. (1979). Les territoires biogeographiques du Sous-Continent Indien. pp. 25–30. In *Paleogeographie et biogeographie de l'Himalaya et du Sous-Continent Indien. Greco 12. Compte Rendu la Table Ronde I'Univ.* Paul Sabatier, Toulouse.

Chatterjee, D. (1939). Studies on the endemic flora of India and Burma. *Journal of the Royal Asiatic Society, Bengal Science*, **5**, 1–69.

Chatterjee, D. (1962). *Floristic pattern of Indian vegetation.* Proceedings of a Summer School of Botany, Darjeeling (pp. 32–42).

Clarke, C. B. (1898). On the subsubareas of British India, illustrated by the detailed distribution of the Cyperaceae in that Empire. *Botanical Journal of the Linnean Society of London*, **34**, 1–146.

Das, S. and Jain, S. K. (1980) Orchidaceae: *Coelogyne.* Fascicles of flora of India, **5**, 1–33.

Gadgil, M. and Meher-Homji, V. M. (1990). Ecological diversity. In *Conservation in developing countries: problems and prospects* (ed. J. C. Daniel and J. S. Serrao), pp. 175–98. Oxford University Press, New Delhi.

Gupta, K. M. and Bandhu, D. (1979). *Man and forest.* Today & Tomorrow's Printers and Publishers, New Delhi.

Henry, A. N. *et al.* (1979). Rare and threatened flowering plants of south India. *Journal of the Bombay Natural History Society*, **75**, 684–97.

Hooker, J. D. *et al.*(1872–1897). *Flora of British India.* Vol. 1–7. L. Reeve, London.

Hooker, J. D. (1906). *A sketch of the flora of British India*, (2nd edn). Clarendon Press, Oxford.

Jain, S. K. (1976). Rare and endangered plants—the problems and its solutions. *Annals of Indian Science*, **2**, 20–3.

Jain, S. K. and Sastry, A. R. K. (1981). Techniques and constraints in survey and conservation of threatened plants and habitats in India. In *The biological aspects of rare plant conservation* (ed. H. Synge), pp. 59–66. John Wiley, Chichester.

Kayastha, S. L. and Juyal, G. N. (1979). Forests, environment and development. In *Man and forest* (ed. K. M. Gupta and D. Bandhu), pp. 28–43. Today & Tomorrow's Printers and Publishers, New Delhi.

Mandal, S. (1988). Exploration, evaluation and conservation of rare plants of Birbhum district, West Bengal, India. *Proceedings of the international conferences of researches in plants sciences, Delhi*, p. 128.

Mani, M. S. (1974). *Ecology and biogeography in India.* W. Junk. The Hague.

Moddie, A. D. (1981). Himalayan environment. In *The Himalaya: aspects of change* (ed. J. S. Lall), pp. 341–50. Oxford University Press, New Delhi.

Prance, G. T. (1978). Floristic inventory of the tropics: where do we stand? *Annals of the Missouri Botanic Garden*, **64**, 659–84.

Prance, G. T. (1983). Completing the inventory. In *Current topics in plant taxonomy* (ed. V. H. Heywood and D. M. Moore), pp. 365–96. Academic Press, London.

Razi, B. A. (1975). Some observations on plants of the South Indian hilltops and their distribution. *Proceedings of the National Institute of Science of India*, **21B**, 79–89.

Rao, A. S. and Hajra, P. K. (1977). *Paphiopediaum fairieanum* (Lindl.) Pfitz—Habitat, descriptive and cultural notes. *Indian Forestry*, **103**, 29–32.

Raven, P. H. (1976). Ethics and attitudes. In *Conservation of threatened plants* (ed. J. B. Simmons *et al.*), pp. 155–79. Plenum Press, New York.

Rieger, H. C. (1981). Man versus mountains. In *The Himalaya: aspects of change* (ed. J. S. Lall), pp. 351–76. Oxford University Press, New Delhi.

Schultes, R. E. (1951). La riqueza de la flora Colombiana. Revista Acadamia Colomb. Ci. Exact., **8**, 230–42.

Subramanyan, K. and Nayar, M. P. (1974). Vegetation and phytogeography of the Western Ghats. In *Ecology and biogeography in India*, pp. 178–96. W. Junk, The Hague.

Synge, H. and Townsend, H. (1979). *Survival of extinction.* Bentham-Moxon Trust, Kew, England.

Thompson, P. A. (1975). Should botanic gardens save rare plants? *New Scientist*, **636**, 8.

11. The use of phytogeographic data for conservation planning

G. T. PRANCE
Royal Botanic Gardens, Kew, Richmond TW9 3AB, Surrey, UK

Abstract

Patterns of plant distributions in the New World tropics as derived from *Flora Neotropica* are described using a variety of mapping techniques. Examples of some ways in which phytogeographic data can contribute to conservation, particularly in Brazil, are given. Sixteen per cent of all described taxa are widely distributed whilst the remainder show restricted and/or disjunct patterns. Endemism is common in widespread families and prevalent in rain forests. The major phytochoria are described and particularly areas of endemism as these are prime targets for biological reserves. Clusters of endemics occur for various reasons, including edaphic and refuge areas that have remained forest covered throughout periods of cooler and drier climate. Analysis of over one thousand species which occur in the Atlantic coastal forest of Brazil showed that more than 50 per cent were endemic to that region, stressing the importance of conservation of remaining habitat fragments. Disjunct distributions emphasize the need to conserve genetic diversity and preserve populations of different species in the different isolated areas. For example, many species of plants are disjunct between the Guianas and Central America. Distribution of species confined to restricted habitats is most useful to show where particular habitats occur. Examples will be given of widespread savannah and white sand endemics to show where isolated patches of these habitats occur in the Amazon region. Local demographic data from quantitative inventories of the forest are also now becoming available. Examples are given from a 23–ha sample of Chrysobalanaceae near Manaus, Brazil to show that apparently widespread species may be locally abundant in some areas and very rare in others. Good conservation planning takes these factors into consideration and is essential for exploitation of useful species. Priority areas for conservation of plants (and also birds, insects, and fishes) as determined for the Manaus Workshop 90 are illustrated.

Systematics and Conservation Evaluation (ed. P. L. Forey, C. J. Humphries, and R. I. Vane-Wright), Systematics Association Special Volume No. 50, pp. 145–63, Clarendon Press, Oxford, 1994.

Introduction

Monographers of tropical groups of plants and animals have at their disposal, by virtue of their work, a large amount of information on the geographical distribution of organisms. Some monographers have used these data to contribute significantly to the cause of conservation, but there is still much invaluable information that remains hidden within highly technical monographs rather than being interpreted to further the cause of conservation. The aim of this paper is to present some examples of how the *Flora Neotropica* monographs are providing data for conservation planners in order to encourage greater use of the work of systematists in conservation programmes. At this time of ever increasing loss of natural habitats, the usefulness of systematics for conservation is one of the many arguments in favour of increasing rather than decreasing classical, specimen based, systematic biology research. I will deal here only with aspects that use distributional data, and will not consider the other important area in which systematics contributes greatly to conservation through the use of species to differentiate and define the many vegetation types of the world.

Many different approaches to conservation theory and practice are presented in this volume, all of which will further the cause of conservation. While there is a need for this diversity of methods, in the tropics the basis for all of them is the monograph that defines the species. As we develop new techniques there is also a need to continue basic monographic research. This approach will be more acceptable and fundable if the data contained in monographs are applied to conservation.

Patterns of Neotropical plant distributions

While most groups of Neotropical plants that have been monographed contain a few widespread species (Fig. 11.1), one of the most striking features is the number of species that have extremely limited ranges even within a particular type of vegetation (see Fig. 11.2). For example, only 16.6 per cent of the 349 species of Neotropical Chrysobalanaceae could be termed widespread. The remainder are of restricted range or are local endemics. The reasons for the local distribution of species are many, including edaphic conditions, topography, historic changes in vegetation cover, and local climate. Some species are habitat restricted and occur only in such habitats as forest on white sand, floodplain forest, or montane forest, but even within a more widespread and uniform vegetation type, such as rainforest on non-flooded ground, local endemics are common. The development of adequate conservation plans for any area must involve a detailed knowledge of the distribution of the species of restricted range.

Fig. 11.1 Distribution of *Hirtella racemosa* Lam. var *hexandra* (Willd. ex R. & S.) Prance, an example of a widespread species.

Fig. 11.2 Distribution of five locally endemic species of Chrysobalanaceae, each occurring in a different habitat: ◓ *Couepia steyermarkii* Maguire (montane forest), ▲ *Couepia comosa* Benth. (riverine forest), ◘ *Hirtella dorvalli* Prance (white sand), ★ *Couepia glabra* Prance (terra firme forest), ☆ *Licania lanceolata* Prance (savannah).

Another common pattern of distribution is disjunction (Fig. 11.3). For example, there are many species that occur in Central America and the Guianas (e.g. *Licania affinis* Fritsch, Fig. 11.3), and others that occur in Amazonian and Atlantic coastal Brazil (e.g. *Couratari macrosperma* A.C. Smith, Fig. 11.3). The conservation of these species should include the protection of samples from both populations since there are likely to be important genetic differences.

The study of the species distribution of any widespread plant family with a reasonable number of species soon shows that some areas of rainforest have a much higher number of endemic species than others (Fig. 11.2). This has been noticed in the distributions of many different organisms (e.g. Haffer

Fig. 11.3 Two examples of common patterns of species disjunction: ✪ *Licania affinis* Fritsch (Guianas-Amazonia), ▲ *Couratari macrosperma* A. C. Smith (Amazonia-Eastern Brazil).

1969, 1974 for birds; Brown 1976, 1982, 1987 for insects; Prance 1973, 1979, 1982 for plants). These areas have often been equated with forest refugia that remained intact during the drier and cooler periods of climate of the periods of glaciation in more temperate latitudes (Fig. 11.4). Whether or not they were refugia, the fact that these centres of endemism exist has been well established and it certainly makes the conservation of these areas of great importance because without their protection a large number of local endemics will not be preserved.

Another use that has been made of the distribution of endemic species in South America has been for the definition of phytogeographic regions and phytochoria. For example, phytogeographic regions were first proposed for Amazonia by Ducke and Black (1953, 1954) based on the distribution of a

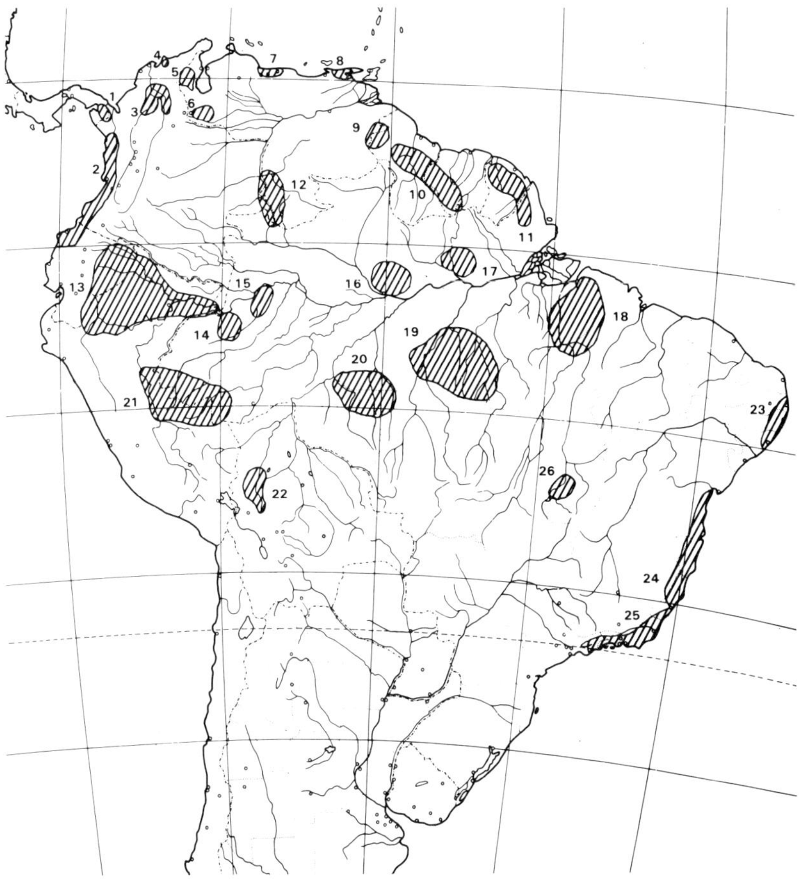

Fig. 11.4 Centres of endemism of lowland rainforest species of trees (after Prance, 1982).

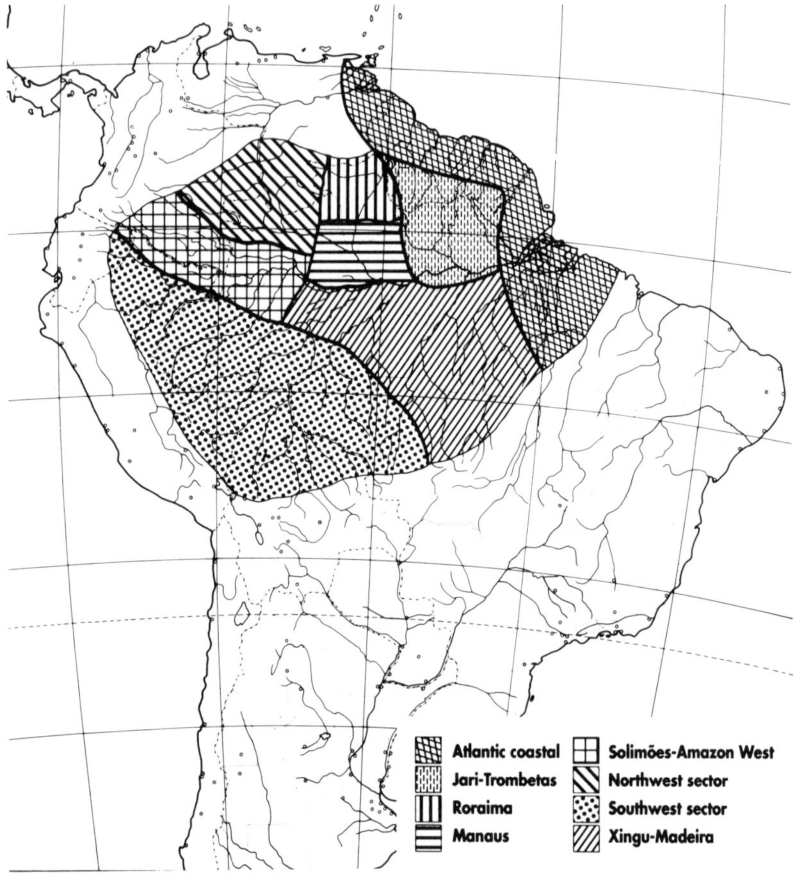

Fig. 11.5 Phytogeographic regions of Amazonia (after Prance, 1977).

large number of locally occurring species in the region. This was refined by Prance (1977), see Fig. 11.5. The conservation recommendations for Amazon proposed by Wetterberg *et al.* (1976) synthesized data from both the centres of endemism of various biogeographers mentioned above and from the phytogeographic regions of Prance (1977). The proposed conservation areas, some of which have been actually declared as reserves (Wetterberg *et al.* 1981), include a part of each phytogeographic region because each one would contain some species of plants that do not occur in other regions.

Figure 11.6 shows the phytochoria of South America as proposed in Prance (1989) slightly modified to include the campos rupestres of eastern Brazil. The phytochoria are defined not primarily by their vegetation type, but mainly as areas within which a large number (over a thousand) of endemic species occur. In addition some areas are either transitional regions between the main phytochoria or mosaics of species from other regions, such as the Pantanal swamp region of Mato Grosso, Brazil. Phytochoria are either continuous areas or they can be broken up into an archipelago-like structure, such as many isolated mountain tops of the Guyana Highland or the campos rupestres. Any good scientifically based conservation policy should also ensure that there are adequate conservation areas within each phytochorion.

An analysis was made by Mori *et al.* (1981) of the distribution of 1127 tree species that occur in the much destroyed Atlantic coastal forests of Brazil. This showed that 53.6 per cent are endemic to those forests, 11.8 per cent to the coastal forest and some part of the adjacent Planalto of Central Brazil, and 7.8 per cent disjunct with the Amazon region, and only 26 per cent were widespread. These figures helped to show conservation planners the high level of endemism of the Atlantic coastal forest. Since they were published in 1981 they have been much quoted because they were the first to quantify the numbers of endemic trees in the regions. The study was based on a simple analysis of the distribution maps of woody species that were available in *Flora Neotropica* monographs up to that date. The tragedy is that so little of the Atlantic coastal forest of Brazil remains today (less than 5 per cent). It is not surprising that many species known only from the types of such nineteenth century collectors as Blanchet and Gardner have not been recollected in recent times.

In addition to plotting a large number of distribution maps based on presence or absence in a grid square, another technique most useful for conservation is isoline maps that surround the entire possible distribution of a species (Fig. 11.7). These are of particular use in large areas of similar vegetation such as within Amazonia. They are useful as long as habitat data are considered. For example, a forest species will not occur in any savannah areas that might be encircled by an isoline for a particular species. In Fig. 11.7, which shows the species of *Licania* subgenus *Licania*, the areas marked a, b, c and d have more species than any other and are centres of diversity for this taxon. To form reserves in these areas would protect a greater number of species than any other area. Figure 11.8 is an isoline diagram for the Caryocaraceae which shows that the region of the Venezuela/Brazil frontier is the centre of diversity for this family. Its origin is therefore, likely to be from the ancient formation of the Guianan shield rather than in the south, and it has migrated into Amazonia and further south from there (Fig. 11.9). The northern area (Fig. 11.9, circle) is the region to protect for the best conservation of this family of considerable economic importance for its timber, edible fruit, oil, and fish poisons.

Figures 11.10–13 show the distribution of species of Chrysobalanaceae which occur in the Ducke Forest Reserve near to Manaus, Brazil, where I have made a detailed local study. Some species are eastern Amazonian

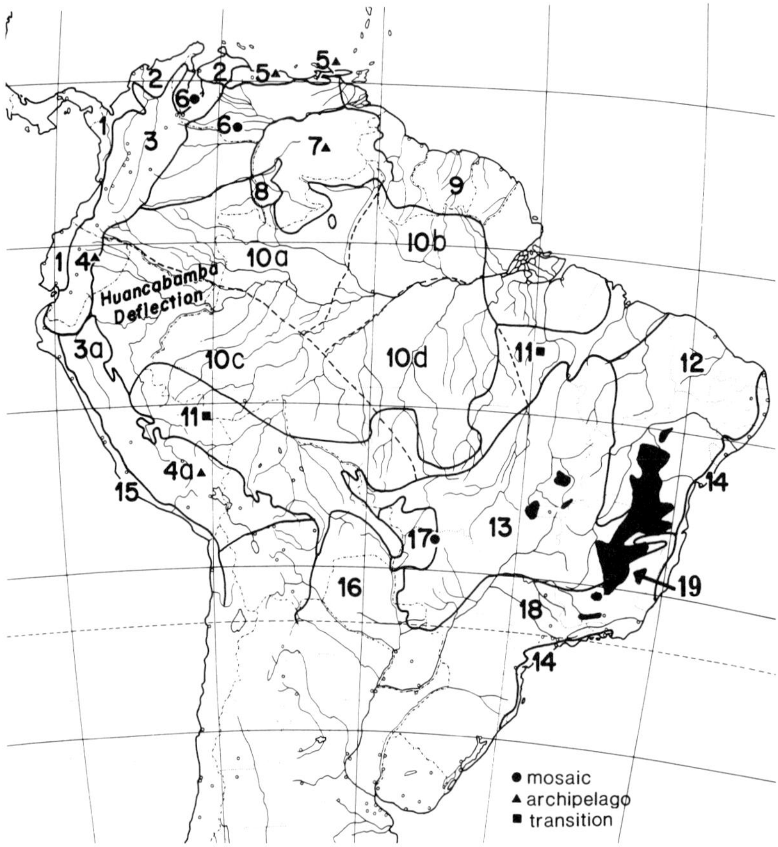

Fig. 11.6 The phytochoria or floristic provinces of South America as defined by the author. These are divided into four types following the categories of White (1976, 1979):

A. Regional centres of endemism with a high percentage of endemism, which are general areas rather than local pockets of endemism such as exist within some regional centres. The regional centres are: 1 = Panamá-Chocó; 2 = Magdalena-Venzuela Gulf; 3 = northern Andean submontane; 3a = southern Andean submontane; 9 = Guiana-eastern Amazonia; 10 = Amazonia; 12 = north-eastern Brazil; 13 = Planalto; 14 = Atlantic coastal region; 15 = southern Pacific; 16 = Chaco; 17 = southern Brazil.

B. Archipelago-like centres (▲) of endemism differ from regional centres in their discontinuous distribution, because of their occurrence on isolated mountain tops. The South American centres are: 4 = north Andean montane; 4a = southern Andean montane; 5 = Central Cordillera Venezuela; 7 = Guayana Highland; 19, shaded area in Atlantic coastal area of Brazil; campo rupestre archipelago.

(*Legend continues on opposite page.*)

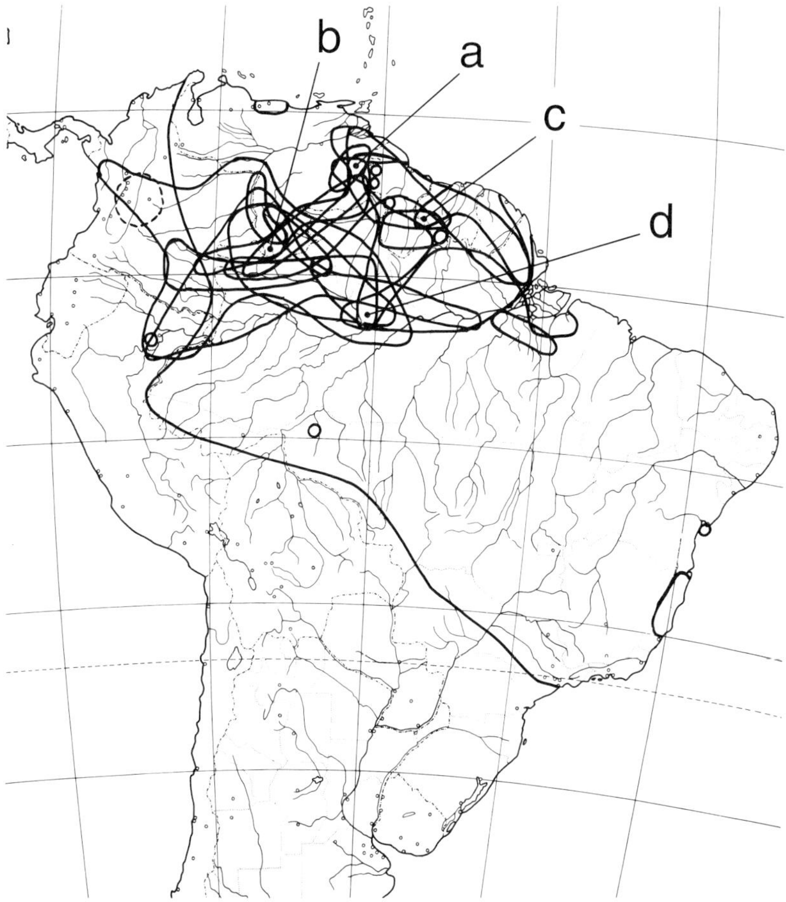

Fig. 11.7 Isoline map of species of *Licania* subgenus *Licania* section *Licania*.

C. Regional transition zones (■) are regions where there is a gradual transition between two centres, with elements from both types and low endemism. In South America the single such zone of significance is: 11 = Amazon Transition Zone, around the south of Amazonia.

D. Regional mosaics (●) are areas that are not well categorized as centres of endemism or as transition zones. They may or may not be rich in endemic species. In South America the Pantanal mosaic (17) is not rich in endemism, but the other areas are: 6 = Catatumbo—Llanos mosaic; 8 = Venezuelan Amazonas—savannas mosaic.

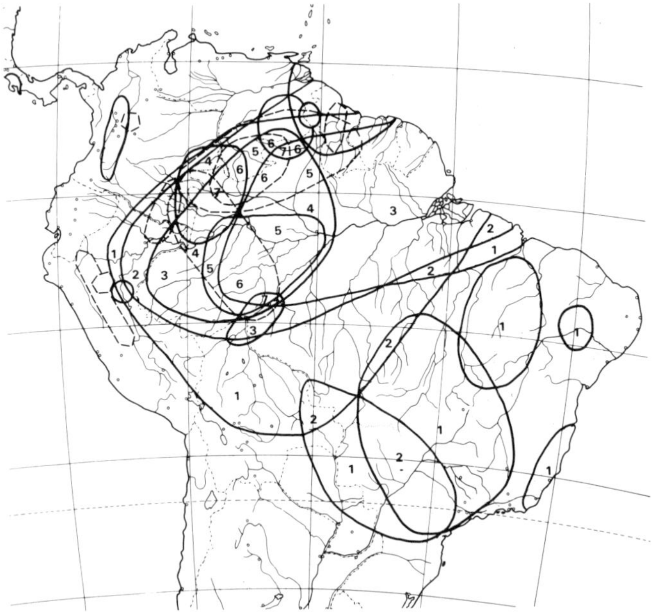

Fig. 11.8 Isoline map showing the distribution of the Caryocaraceae where the greatest diversity is in the Venezuela-Brazil border region.

Fig. 11.9 Map based in Fig. 11.8 showing probable original dispersal of Caryocaraceae.

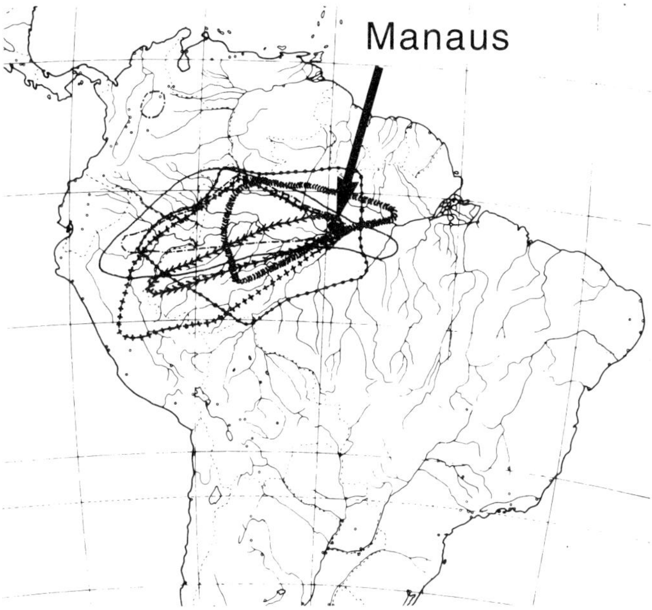

Fig. 11.10 Western Amazonian terra firme forest species of Chrysobalanaceae that also reach as far east as Reserva Ducke.

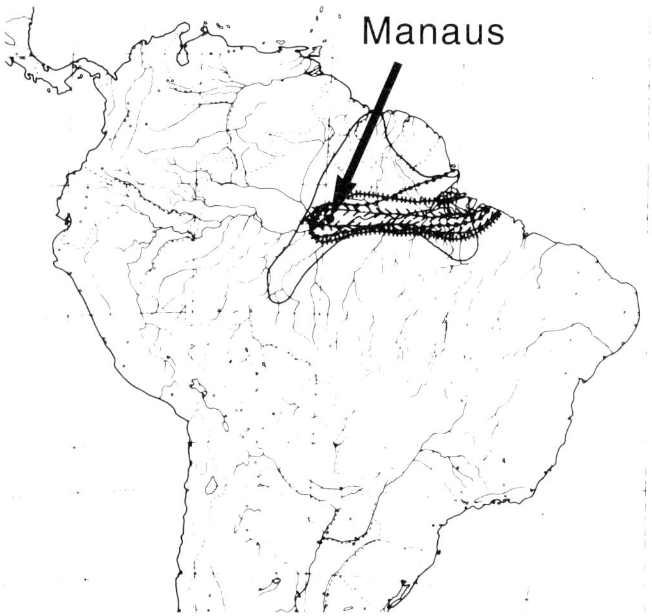

Fig. 11.11 Eastern Amazonian terra firme forest species of Chrysobalanaceae that also reach as far west as Reserva Ducke.

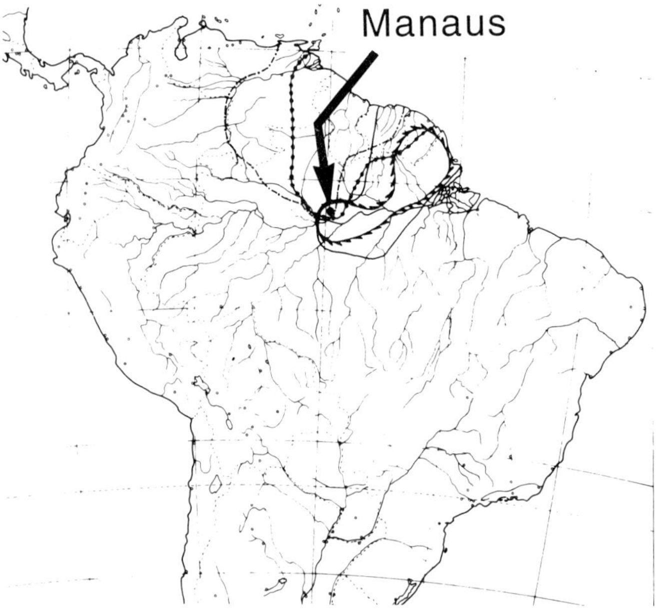

Fig. 11.12 Guianan species of Lecythidaceae that occur in the Reserva Ducke.

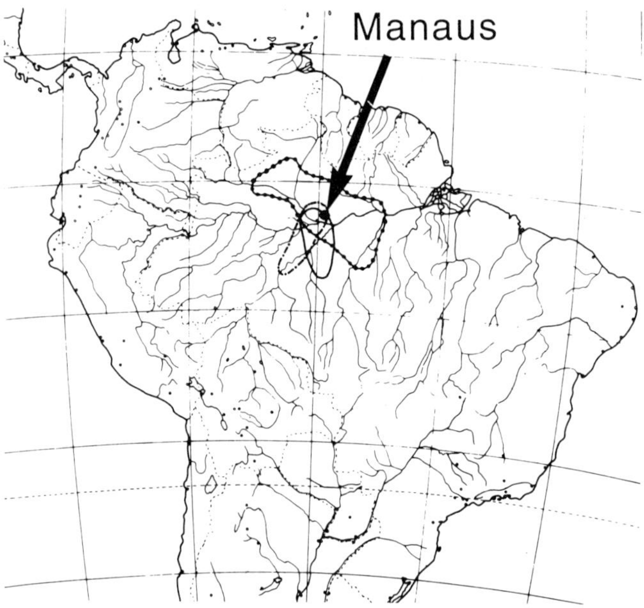

Fig. 11.13 Reserva Ducke species of Chrysobalanaceae confined to Central Amazonia.

whose range extends west just to Manaus, some are western Amazonian with a range extending east to Manaus, and some are Guianan species extending south to Manaus. There is a certain amount of endemism in the vicinity of Manaus, but the diversity is particularly high because the species of several regions come together at the extremes of their ranges in this central Amazonian location.

Figures 11.14–17 are species density maps for four taxa that have been monographed in *Flora Neotropica*: Caryocaraceae, *Nolana* (Nolanaceae), *Hevea* (Euphorbiaceae) and Dichapetalaceae. They show the number of species of each group occurring on each grid square throughout their range. These four groups are chosen to show different patterns of diversity. The maximum numbers of species occur in different places. Dichapetalaceae (Fig. 11.16) is primarily a western family, *Nolana* (Fig. 11.15) is confined to the arid regions of western Peru and Chile, and *Hevea* and Caryocaraceae (Figs. 11.14, 11.17) are predominantly Amazonian. This shows the necessity to base conservation policy on data from a wide range of taxa that cover different geographic areas either for historic reasons or because of the availability of specialized

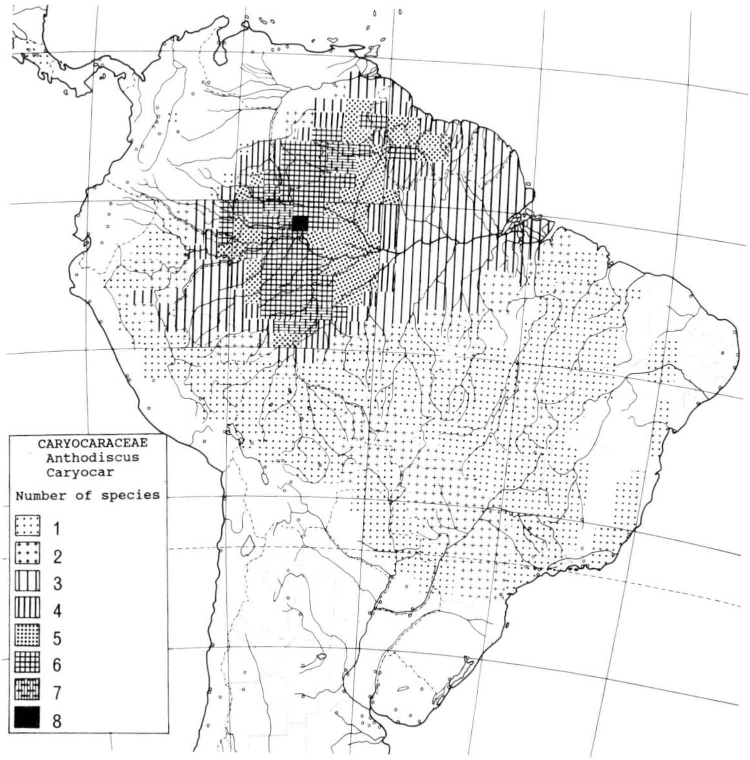

Fig. 11.14 Distribution density map of species of Caryocaraceae.

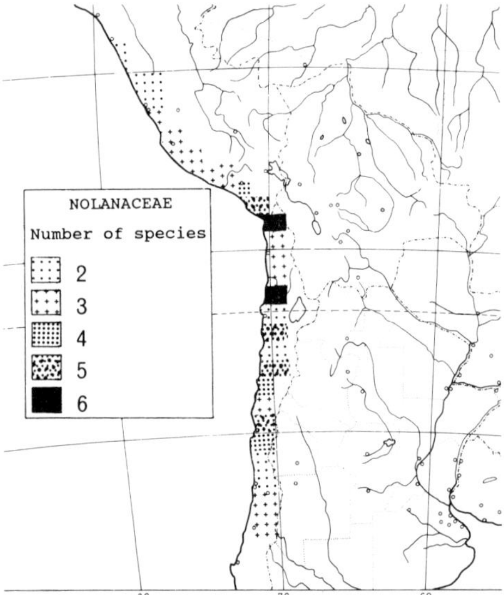

Fig. 11.15 Distribution density map of species of *Nolana* (Nolanaceae).

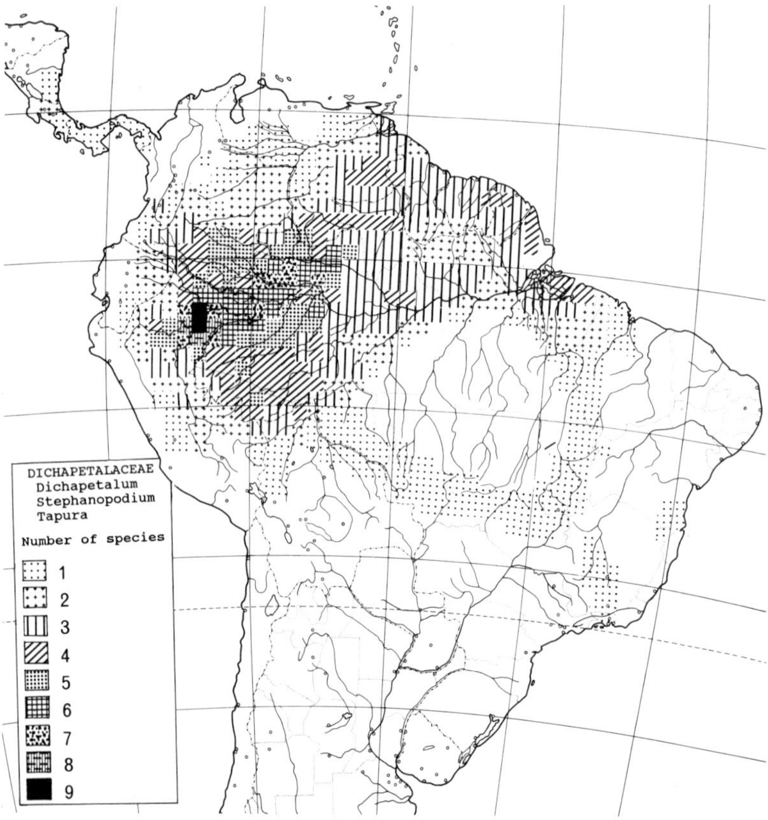

Fig. 11.16 Distribution density map of species of Dichapetalaceae.

habitats. The selection of reserves from the diversity of the four groups in Figs. 11.14–17 would cover some of the most important South American centres of diversity.

The various maps presented in this volume were plotted by hand. Vane-Wright *et al.* (1991) and Williams and Humphries (this volume) explain the theory behind William's WORLDMAP computer program for the selection of areas of maximum species diversity. My data for Lecythidaceae, Chrysobalanaceae, and several other smaller plant families are being incorporated into this system which will enable a greater sophistication of analysis based on both areas of species density and taxonomic relationships.

Figure 11.18 is the result of a workshop held in Manaus in January 1990 where 100 biologists from different fields, including botany, ornithology, entomology, and ichthyology, met together to pool their data to define priority areas for conservation. The conclusions were based on data on centres of diversity, centres of endemism, types of vegetation and soil, and

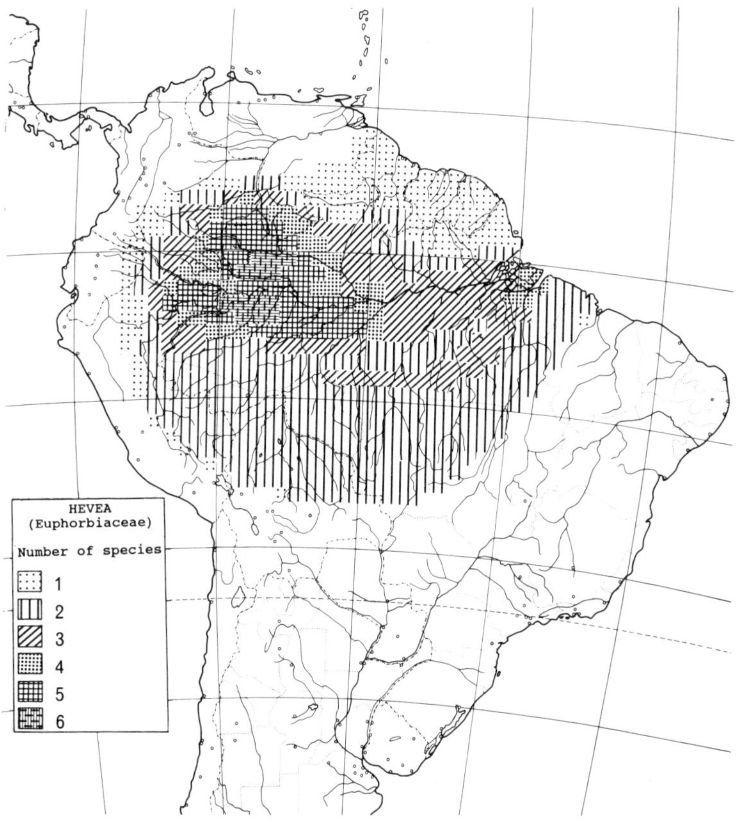

Fig. 11.17 Distribution density map of species of *Hevea* (Euphorbiaceae).

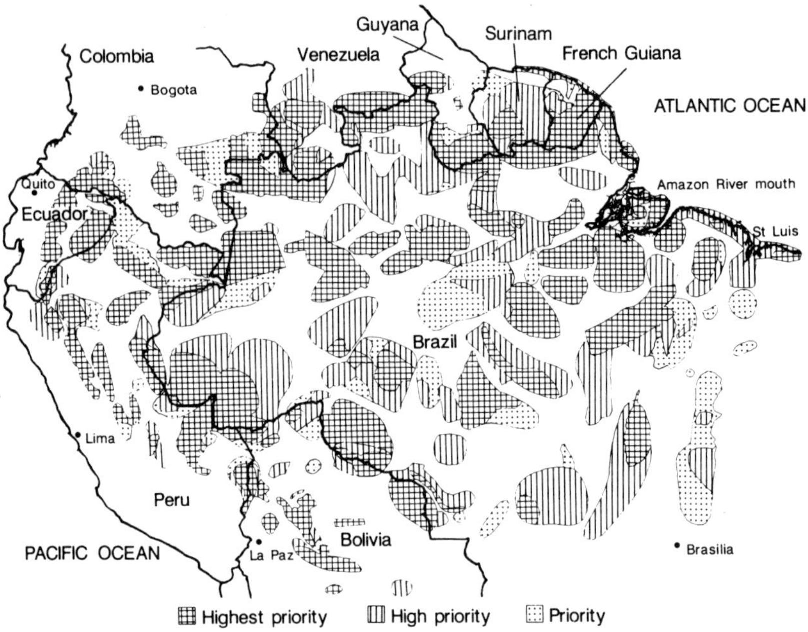

Fig. 11.18 Priority areas for conservation of Amazonia as defined by Workshop 90 in Manaus, Brazil.

greater priority was given to places where data from more than one group of organisms coincided. The end product of a gradual merging of data was a map (Conservation International 1992) which delimited 94 areas evaluated on a scale of 1–5 in importance for conservation. This provides a synthesis of much of the biogeographic knowledge about the fauna and flora of the Amazon region, and is already providing a much more logical basis for future conservation planning throughout Amazonia.

Quantitative inventory

Another line of research which produces much useful information for both taxonomic studies and for conservation is quantitative forest inventory. This provides information about the local distribution and demography of species as well as an understanding about the partitioning of the available habitats within the forest. Some data are presented here: for example, Table 11.1 gives a quantitative inventory of three hectares in the Rio Xingu region of Pará State in Brazil (Campbell *et al.*, 1986). This shows that in this plot some species are abundant, some occasional, and others extremely rare. Every inventory with which I have been involved has shown a similar pattern. It would take a relatively small area in that forest to conserve a

Table 11.1 Summary of data for a 3-ha plot of terra firme forest at O Deserto, Rio Xingu, Pará, Brazil. Importance value is the sum of per cent of relative density, per cent relative frequency and per cent relative dominance

	No. of trees	Importance value
Ctenostigma macrophylum	184	34.69
Orbigyna phalerata	79	19.56
Bertholettia excelsa	4	8.89
Theobroma speciosum	48	7.37
Alexa imperatricia	28	7.23
Neea altissima	45	7.11
Matisia sp.	22	5.92
Sterculia pruriens	20	5.82
Lecythis retusa	31	5.63
Ceiba pentandra	14	4.99
Hirtella piresii	29	4.88
Rinorea juruana	33	4.97
Inga sp 10	29	4.76
Guarea macrophylla	25	4.56
Apuleia lieocarpa	3	4.46
Guatteria schomburgkiana	26	4.16
Protium opacum	23	4.15
Astrocaryum mombaca	20	4.06
Trichilia sp 1	21	3.59
Trichilia sp 3	11	3.33
Sub total	695	150.13
Remaining 245 species	725	149.87
Total trees	1,420	300.00

Number of species represented by 1 individual on 3 ha	99
Number of species represented by 2 individuals on 3 ha	57
	156

viable population of *Orbignya phalerata* Mart. or *Cenostigma macrophylum* Tul. which are abundant. However 57 species are represented by two individuals each and 99 by only one. These scattered species would not be conserved without the protection of an extremely large area of forest. The many inventories that I have carried out show that even a short distance away from a particular area, other species will be abundant, while common species in the first area are often represented by only one individual. An understanding of this clumping is an essential requirement for conservation planning. There is a great need for much further quantitative inventory rather than general collecting in order to understand the local demography as well as the overall distribution of species.

References

Brown, K. S. Jr (1976). Geographical patterns of evolution in Neotropical Lepidoptera. Systematics and derivation of known and new Heliconiini (Nymphalidae: Nymphalinae). *Journal of Entomology*, B, **44**, 201–42.

Brown, K. S. Jr (1982). Paleoecology and regional patterns of evolution in Neotropical forest butterflies. In *Biological diversification in the tropics* (ed. G. T. Prance), pp. 255–308. Columbia University Press, New York.

Brown, K. S. Jr (1987). Biogeography and evolution of Neotropical butterflies. In *Biogeography and Quaternary history in tropical America* (ed. T. C. Whitmore and G. T. Prance), pp. 64–104, Oxford Monographs on Biogeography 3. Clarendon Press, Oxford.

Campbell, D. G., Daly, D. C., Prance, G. T., and Maciel, U. N. (1986). Quantitative ecological inventory of terra firme and várzea tropical forest on the Rio Xingu, Brazilian Amazon. *Brittonia*, **38**, 369–93.

Ducke, A. and Black, G. A. (1953). Phytogeographical notes on the Brazilian Amazon. *Anais Academia Brasileira de Ciências*, **25**, 1–46.

Ducke, A. and Black, G. A. (1954). Notas sôbre a fitogeografia da Amazônia Brasileira. *Boletin Técnico do Instituto Agronomico del Norte, Belém*, **29**, 1–62.

Haffer, J. (1969). Speciation in Amazonian forest birds. *Science*, **165**, 131–7.

Haffer, J. (1974). *Avian speciation in tropical South America. With a systematic survey of the toucans (Ramphastidae and jacamars (Galbalidae)*. Publication of the Nuttall Ornithological Club 14, 390 pp.

Mori, S. A., Boom, B. M., and Prance, G. T. (1981). Distribution patterns and conservation of eastern Brazilian coastal forest tree species. *Brittonia*, **33**, 233–45.

Prance, G. T. (1973). Phytogeographic support for the theory of Pleistocene forest refuges in the Amazon Basin, based on evidence from distribution patterns in Caryocaraceae, Chrysobalanaceae, Dichapetalaceae and Lecythidaceae. *Acta Amazonica*, **3**, 5–28.

Prance, G. T. (1977). The phytogeographic subdivisions of Amazonia and their influence on the selection of biological reserves. In *Extinction is forever* (ed. G. T. Prance and T. S. Elias), pp. 195–213, New York Botanical Garden.

Prance, G. T. (1979). Distribution patterns of lowland Neotropical species with relation to history, dispersal and ecology with special reference to Chrysobalanaceae, Caryocaraceae and Lecythidaceae. In *Tropical Botany* (ed. K. Larsen and L. B. Holm-Nielsen), pp. 58–88, Academic Press, London.

Prance, G. T. (1982). Forest refuges: evidence from woody angiosperms. In *Biological diversification in the tropics* (ed. G. T. Prance), pp. 137–56. Columbia University Press, New York.

Prance, G. T. (1989). American tropical forests. In *Tropical rain forest ecosystems. Ecosystems of the world 14B* (ed. H. Lieth and M. J. A. Werger), pp. 99–132. Elsevier, Amsterdam.

Vane-Wright, R. I., Humphries, C. J., and Williams, P. H. (1991). What to protect?—systematics and the agony of choice. *Biological Conservation*, **55**, 235–54.
Wetterberg, G. B., Jorge Pádua, M. T., de Castro, C. S., and de Vasconcellos, J. M. C. (1976). *Uma análise de prioridades em conservação de natureza na Amazônia.* PNUD/FA0/IBDF/BRA-45 Série Técnica (8), 62 pp.
Wetterberg, G. B., Prance, G. T., and Lovejoy, T. E. (1981). Conservation progress in Amazonia: a structural review. *Parks*, **6**, 5–10.
White, F. (1976). The taxonomy, ecology and chorology of African Chrysobalanaceae (excluding *Acioa*). *Bulletin du Jardin Botanique National de Belgique*, **46**: 265–350.
White, F. (1979). The Guineo-Congolian region and its relationships to other phytochoria. *Bulletin du Jardin Botanique National de Belgique*, **49**, 55.

12. Assessing local distribution of threatened species: a case study considering the threatened plants of Italy

R. I. MILLER*, J. R. PRESS[†], M. WEST[†], and
R. M. BALDINI[‡]
**P.O. Box 2731 Cambridge, Massachusettes 02138, USA*
[†]*European Plant Information Centre, Department of Botany, The Natural History Museum, London SW7 5BD, UK*
[‡]*Museo Botanica, University of Florence, Italy*

Abstract

Conservation actions are based on data collected and stored in scientific institutes. Our project was designed to link two museums with an organisation directly concerned with conservation. Italy was chosen as a European pilot study for assessments of distributions of rare and threatened vascular plants listed in the revised appendix of the Berne Convention. We discuss criteria for defining species rarity and selection of appropriate plant groups. The establishment and development of a computerized system for recording data and for monitoring threatened plants, now operational at the Natural History Museum, London, is linked to a Geographical Information System which enables analyses of a wide range of geo-referenced data and production of high quality maps. Finally, a preliminary analysis of distribution patterns of rare plants throughout Italy is given.

Introduction

Systematic data serve as the basis for conservation action. The majority of systematic data relevant to conservation planning are collected and stored in scientific institutions not necessarily networked into conservation activities. This study was designed to link three collaborating institutions to improve gathering of information relevant to biological conservation: the World

Systematics and Conservation Evaluation (ed. P. L. Forey, C. J. Humphries, and R. I. Vane-Wright), Systematics Association Special Volume No. 50, pp. 165–75, Clarendon Press, Oxford, 1994.

Conservation Monitoring Centre (WCMC), The Natural History Museum (NHM), and the Botanical Museum of the University of Florence (FI). This project is part of a European Plant Information Centre initiative currently underway at the Natural History Museum. The plants of Italy were the subject of this first pilot study.

This research was underpinned by the development and establishment of a computerised monitoring system for threatened plants, which is now operational at the World Conservation Monitoring Centre and at The Natural History Museum. The system consists of a group of networked databases linked to a Geographical Information System (GIS), and it can produce maps depicting threatened plant species distribution patterns. This paper details the development of the database/mapping system and illustrates current monitoring capabilities of the system with regard to some threatened Italian plants.

Threatened species in the computerized monitoring system

The computerized system developed during this study was designed to handle Italian threatened plant species distribution data. The first step was therefore to select a method to identify the threatened Italian plant species to be considered during this study.

Distribution data on a global scale are available for few species (e.g. Gaston 1990). IUCN threat categories are based upon current expert scientific information and serve as the best available gauge of the global threat to a species. Therefore IUCN global threat categories serve effectively as standardized yardsticks to measure the global degree of threat to a plant or animal species.

Various lists of threatened plants are currently available. The Berne Convention, a development of EC legislation which categorizes threatened plants across Europe, includes species with IUCN status categories of Vulnerable, Endangered and Extinct. Several species that are declining very rapidly, but not yet identified within these categories, are also included in the Berne Convention listing. For this study, the Berne Convention list was selected because it is a widely circulated and internationally accepted convention and the Italian government is a signatory.

Development of a database system

A network of databases for the storage, import, and export of threatened species distribution data has been developed. The three available sources of distribution data used in this study were museum specimens, published literature, and field observations. The database network consolidated all these data into a single processing system.

Two of the most important considerations associated with species distribution observations are the definition of the species taxonomic identity and of the geographic location of the record. It became necessary to link these data to standard lists of taxonomic names, geographic locations, and estimates of accuracy. Data categories were considered as follows.

1. *Taxonomic registration of species data*

The vital role of taxonomy in both the scientific and legal realms of conservation has recently been highlighted (Geist 1992). There is a strong taxonomic component in this project. Accurate identification of taxa is vital for anchoring all data in the databases. Therefore, in this study, the names adopted are those used in the Berne Convention Appendix 1 (1990).

2. *Geographic registration of species data*

Many diverse ecological factors influence the disposition of species ranges across the globe (e.g. Whittaker 1975; Rabinowitz *et al.* 1986; Gaston 1990; Pagel *et al.* 1991). However, present-day species ranges are often expressed in terms of political boundaries (Gaston 1990). This is particularly true for threatened species. Therefore, in this study the appropriate political/administrative regions were assigned to each record in addition to specific location data, and when available, the precise latitude–longitude co-ordinates.

Difficulties in abstracting accurate data are due to many different factors (e.g. uninterpretable handwriting or imprecise location data). The availability of experienced curators and of secondary sources (e.g. original collecting books, manuscript itineraries, etc.) can solve individual problems, but detailed information is often irretrievable (e.g. from specimen labels). Similarly, for recorded location data, a collector may cite a local name absent from a gazetteer or replicated numerous times across a country. The omission of coarsely-detailed data records can lead to significant data loss, particularly when the total number of specimens is low, as for many threatened plants and animals. Therefore, each species record needs to be registered at the finest level of resolution derivable from the source data.

Geographic data derived from museum specimens, field observations, and literature citations are very variable and are interpretable at many different levels of resolution. To accommodate this range of data resolution, the database system uses a hierarchical set of geographic fields which cover the spectrum from broad to fine scales of resolution. The point locations are registered on a map as coordinate values (Miller 1986). The inclusion of the appropriate political and administrative regions together with exact co-ordinate data allows these data to be geographically registered based upon their resolution status.

Table 12.1 Current threats to Italian vascular plants. This information was derived from the Berne Convention Appendix 1.

Nature of threat	Number of species threatened by this factor
Abandonment of woodland grazing and coppicing	1
Afforestation	4
Agricultural developments	8
Building developments	14
Canalization	2
Competition from other plant species	4
Deforestation	3
Ditch/stream clearance	2
Drainage	12
Drying out of small water courses	1
Embankment	2
Eutrophication	7
Fertilizer application	3
Fire	8
Firewood collection	1
Forestry practices	4
Gravel extraction	1
Habitat destruction (unspecified)	6
Herbicide run-off	1
Land reclamation for rice-growing	1
Natural succession of lakes to peat bogs	1
Over-collecting	18
Over-grazing	7
Poor regenerative capacity	1
Recreational use/human trampling and disturbance	6
Reproductive collapse due to small population size	1
River bank modification	1
Road construction	3
Rubbish dumping	3
Ski resort development	2
Soil erosion	2
Tourism	13
Undergrazing	4
Uranium mining	1
Water flow reduction caused by irrigation	1
Water pollution	11
Winter sports	1
No/uncertain information	20

The monitoring facility

1. *Broadscale conservation applications*

It has been suggested that preservation of areas rich in rare species will have the greatest effect on the preservation of biological diversity in the twenty-first century (Scott *et al.* 1987). In this case, collective protection of nodes of regional species richness would capture the vast majority of continental biological diversity (Scott *et al.* 1987, p. 787). The system produced in this pilot study provides the data for making conservation decisions based upon known threatened species distribution patterns.

In this study, the data collection and processing techniques for mapping foci of rare plant species richness in Italy requires significantly less time and effort than would a complete survey of species richness. In a comprehensive survey of the fauna and flora in the Hawaiian islands (Scott *et al.* 1988), six years were required for a complete survey of species richness in Hawaii. This time saving will particularly aid conservation planning by providing immediate guidance for the design of future development.

2. *The mapping approach and the GIS link-up*

The distribution patterns of species depicted on range maps are usually produced at very coarse levels of accuracy and resolution. The development of Geographical Information System (GIS) technologies has permitted the production of maps depicting species distribution patterns at much finer levels of resolution and accuracy.

This project linked the data storage and handling capabilities of an advanced database technology to the mapping capacity of a modern GIS. The data stored in the database system can then be automatically transferred to and utilized by the GIS to produce appropriate species distribution maps.

3. *An example illustration: monitoring* Aldrovanda vesiculosa *in Italy*

Aldrovanda vesiculosa is a monospecific genus of the family Droseraceae, the sundews, and like all genera in the family is carnivorous. It is a submerged aquatic with short, simple, and rootless stems lying just below the water surface. It is an Old World species occurring from Europe, central and southern Asia to Australia, but it is best represented in the sub-tropical parts of eastern and southern Africa. In the non-African parts of its range, *A. vesiculosa* is rather scattered and often transitory, capable of producing luxuriant growth, but also prone to rapid decline and disappearance. These population changes seem to be rapid responses to changes in the habitat, similar to those found in other aquatics, for example species of *Chara* and *Tolypela*. The distribution of *A. vesiculosa* is well documented in Europe, its range thoroughly studied and mapped by Walters (1979). A reproduction of Walter's map was produced with the GIS component of this monitoring facility (Fig. 12.1). The data presented clearly show that the species has been observed across the entire European continent. The records were then displayed on a 1: 1 000 000 scale map (Fig. 12.2). This map highlights the two confirmed records of this species in north-western Italy which were known at the time of the original study. Records from The Natural History Museum in London were then gathered, compiled, and input to the database system. The map produced from these records depicts a contrasting distribution pattern (Fig. 12.3). The distinctions between the two maps may simply represent differences in the composition of the Walters and The Natural History Museum collections. The nature of the precise differences between these two collections has not been identified. However, a comparison between the maps in Fig. 12.2 and Fig. 12.3 clearly demonstrates the potential use of this system to monitor changing distribution patterns. This capability of monitoring and updating observed distribution changes is one of the important advantages provided by this database/GIS monitoring system.

4. *A data management system*

A conservation data centre should provide the capability to produce maps depicting changes in both threatened habitat and species distribution patterns over time. This capability has been established in the database/GIS system developed in this study. The complete database management system can store, validate, and network species occurrence data at WCMC and in the Botany Department of the NHM. Taxonomic, geographic, environmental, bibliographic, biological, and source data are stored and updated, and data tables and distribution maps can be produced for publication. This system is currently being used to manage data related to threatened species distributions in Italy.

The database/GIS system can then be used to integrate knowledge obtained from field records, museum specimens, and published materials for the production of media with a conservation focus. Important areas of plant biodiversity can now be identified in Italy based upon the vascular plant records processed during this study. We believe that this database/mapping computer system may provide a useful tool for handling data necessary for conservation action around the world.

Conclusions

The impacts of human civilization define many of the boundaries which limit threatened species ranges. In most areas of the world, conservation must therefore focus upon species ranges set within the context of human land-use patterns. Maps which present the current patterns of threatened species

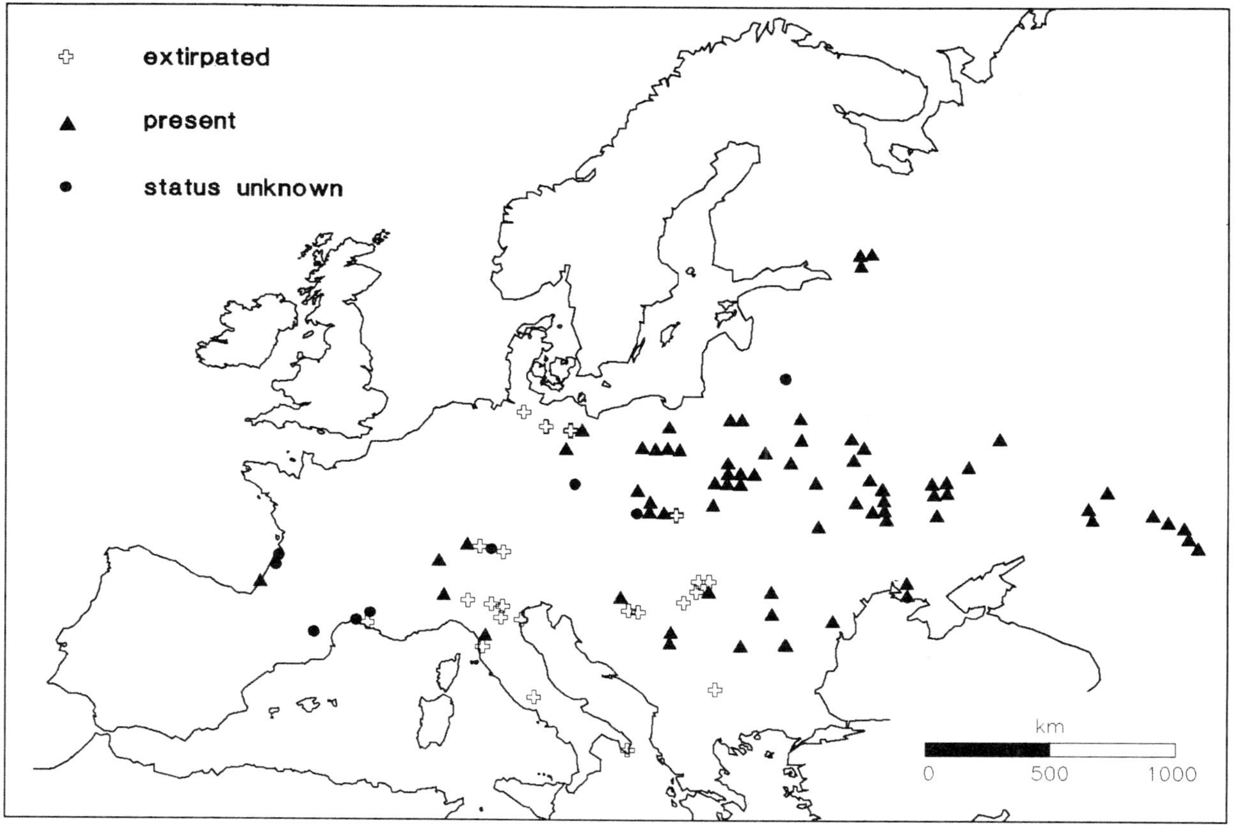

Fig. 12.1 The distribution of recorded occurrences of *Aldrovanda vesiculosa* across Europe. This map was reproduced from the original mapped data published in Walters (1979).

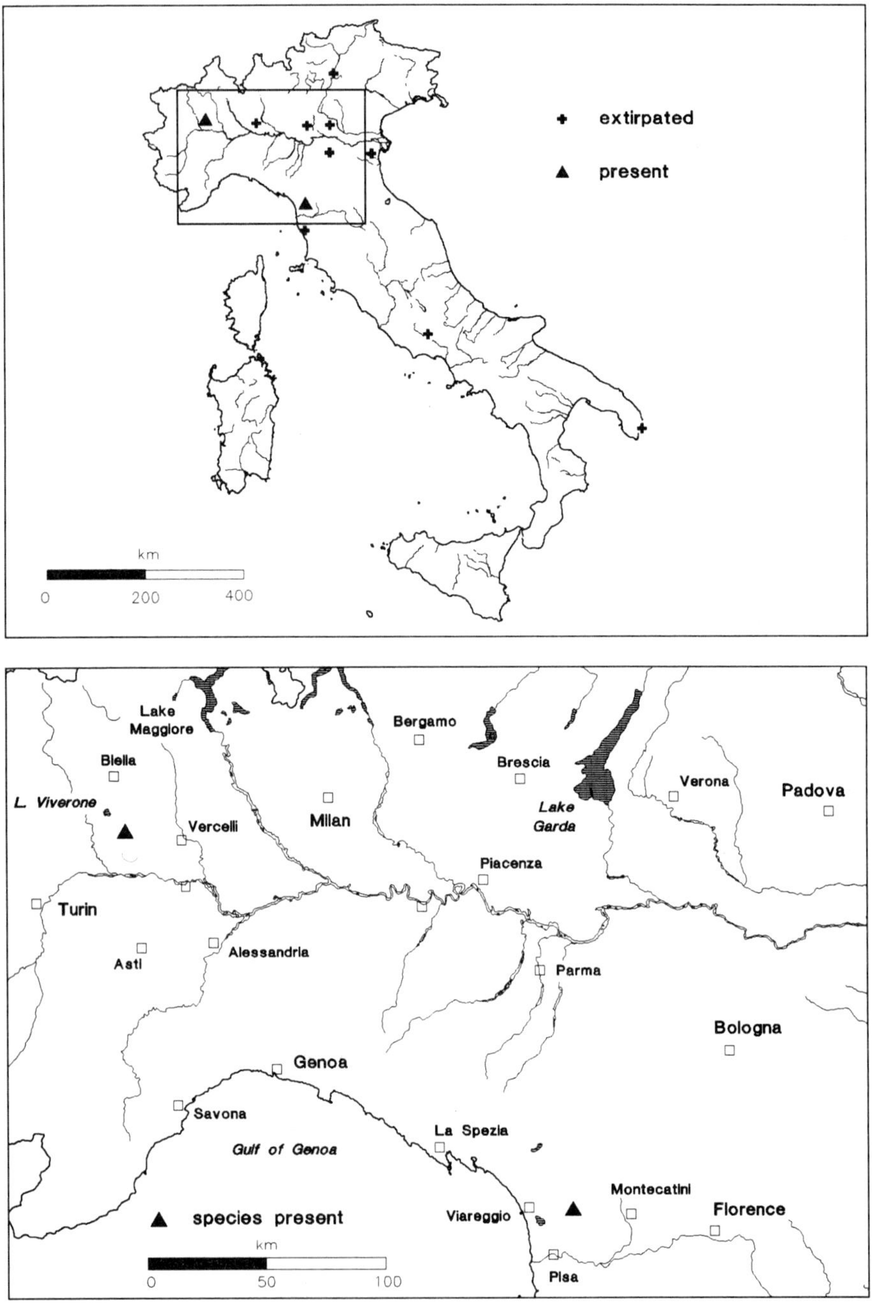

Fig. 12.2 The region of Italy in the Fig. 12.1. map was enlarged and the *Aldrovanda vesiculosa* records for Italy were overlain into the ONC Chart at the 1:1 000 000 scale represented here. The plant records shown are taken from Walters (1979).

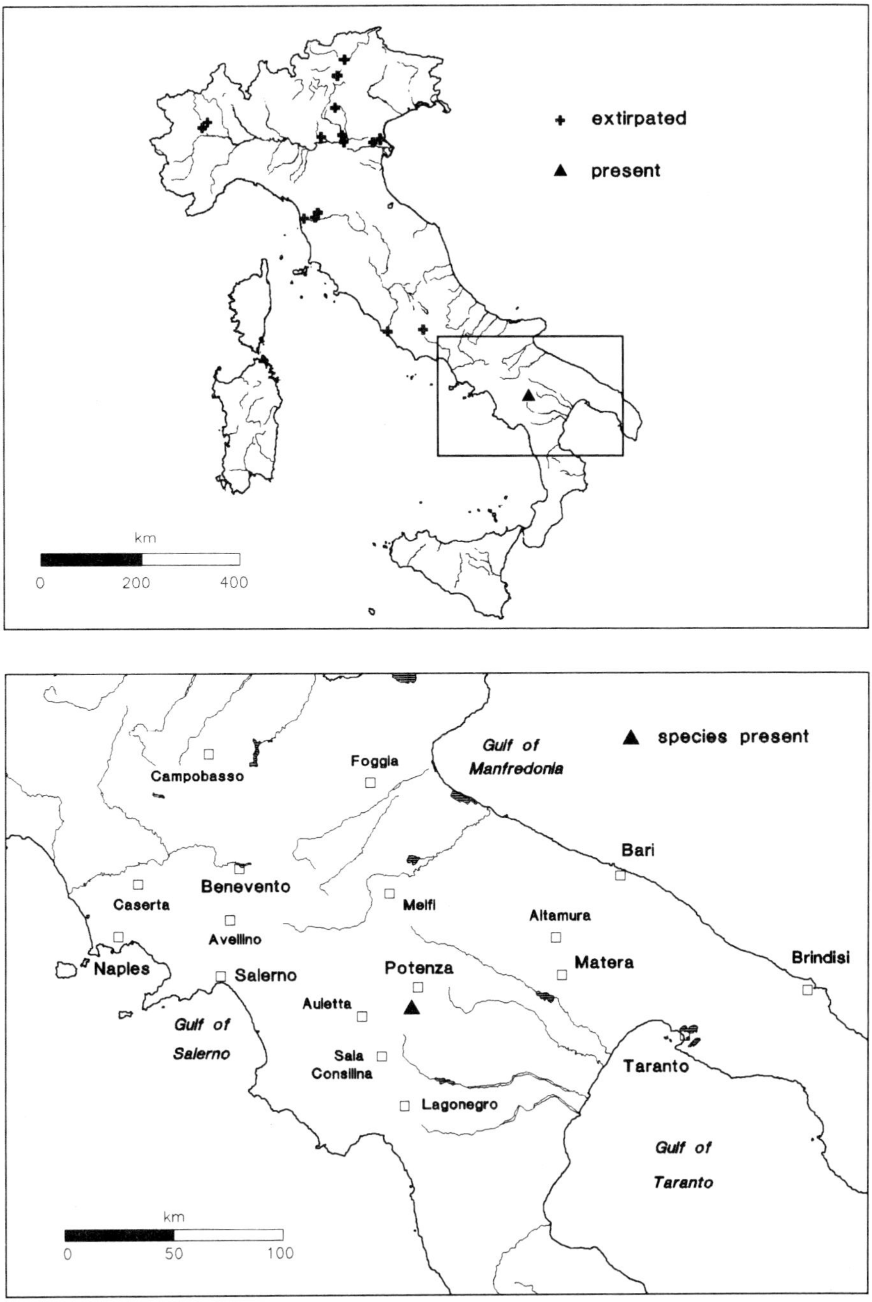

Fig. 12.3 This is an identical 1:1 000 000 scale map of Italy as in Fig. 12.2. The *Aldrovanda vesiculosa* records depicted here represent all the records for this species held at The Natural History Museum as of 1992.

distributions may provide effective media instruments to influence positively the preservation of biodiversity in the future.

Mace and Lande (1991) point out that in the negotiations surrounding development and conservation policy there is a necessary political conflict between expediency and scientific credibility. In the case of national conservation, political expediency requires planning agencies to gain access to species data on a country-wide scale. However, suitable data are not generally available on a comprehensive and accurate basis at this scale. For example, in this study the extant threatened plant data for Italy were laboriously extracted from records in The Natural History Museum in London and the University of Florence herbarium. These records represented the patterns of past botanical collecting, but did not necessarily represent comprehensively the current spatial or temporal patterns for most plant species. However, these records do represent the best vascular plant species data readily available for Italy. The system developed during this study provides the capability of producing high quality, practical maps for the use of interested national and international agencies.

Acknowledgements

This study was able to access the wealth of scientific data and individual expertise at all three home institutions.

References

Berne Convention (1990). Part III. Report of the Council of Europe Project to Revise Appendix 1 (Strictly protected flora species).

Gaston, K. J. (1990). Patterns in the geographical range of species. *Biological Reviews of the Cambridge Philosophical Society*, **65**, 105–29.

Geist, V. (1992). Endangered species and the law. *Nature*, **357**, 274–6.

Mace, G. M. and Lande, R. (1991). Assessing extinction threats: toward a reevaluation of IUCN threatened species categories. *Conservation Biology*, **5**, 148–57.

Miller, R. I. (1986). Predicting distribution patterns of vascular plants in the southern Appalachians of the southeastern United States. *Journal of Biogeography*, **13**, 293–311.

Pagel, M. D., May, R. M., and Collie, A. R. (1991). Ecological aspects of the geographical distributions and diversity of mammalian species. *American Naturalist*, **137**, 791–815.

Rabinowitz, D., Cairns, S. V., and Dillon, T. (1986). Seven forms of rarity and their frequency in the flora of the British Isles. In *Conservation biology, the science of scarcity and diversity* (ed. M. E. Soulé), pp. 182–204. Sinauer, Sunderland, MA.

Scott, J. M., Csuti, B., Jacobi, J. D., and Estes, J. E. (1987). Species richness: a geographic approach to protecting future biological diversity. *Bioscience*, **37**, 782–8.

Scott, J. M., Kepler, C. B., van Riper III, C., and Fefer, S. L. (1988). Conservation of Hawaii's vanishing avifauna. *Bioscience*, **38**, 238–53.

Walters, S. M. (1979). Conservation of the European flora: *Aldrovanda vesiculosa* L., a documented case history of a threatened species. In *Systematic botany: plant utilisation and biosphere conservation* (ed. I. Hedberg), pp.76–82. Almquist and Wiksell International, Stockholm.

Whittaker, R. H. (1975). *Communities and ecosystems.* (2nd ed.). Macmillan, New York.

13. The numbers and spatial distributions of species: analyses of British data

J. H. LAWTON*, J. R. PRENDERGAST*, and
B. C. EVERSHAM†
*NERC Centre for Population Biology, Imperial College at Silwood Park, Ascot, Berkshire SL5 7PY, UK
†Institute of Terrestrial Ecology, Monks Wood, Abbots Ripton, Huntingdon, Cambridgeshire PE17 2LS, UK

Abstract

Britain has one of the best known, but most heavily modified floras and faunas of the world. Data on distributions of many taxa allow the presence and absence of species to be mapped in each 10 × 10 km square of the national grid, which can be used to answer questions of interest to both theoretical ecologists and conservationists. For example: (i) Where are the most species-rich areas of the country for particular taxa? (ii) Do diversity 'hot-spots' coincide for different taxa? (iii) How much turnover (beta-diversity) is shown by different taxa along geographic transects? (iv) What proportion of the national pool of species in a particular taxon might be conserved by setting aside reserves in a limited number of locations? Superimposed on well-known north-south and east-west gradients in alpha diversity, are significant departures in local species richness. Some 'hot-' and 'cool-spots' are relatively easy to explain, but others are not. Furthermore, 'hot-spots' for the taxa examined so far appear not to be coincident. The oft-repeated hope that a reserve in an area that is rich in one or two well known groups will also be rich in other poorly-known taxa is not supported by the British data. Beta diversity in the British flora and fauna is low. This means that for any particular taxon, a few species-rich areas will encompass a large share of the total diversity. It is shown that whilst this message may be true for other temperate regions, it is unlikely to be true in the tropics, where species appear to have smaller geographic ranges compared with temperate taxa. Finally, we show that it is unlikely that conservation strategies in most parts of the world can be devised using data that are anything like as good as those available for Britain.

Systematics and Conservation Evaluation (ed. P. L. Forey, C. J. Humphries, and R. I. Vane-Wright), Systematics Association Special Volume No. 50, pp. 177–95, Clarendon Press, Oxford, 1994.

Introduction

Britain has one of the best known, albeit most heavily modified, floras and faunas in the world. Data on the distributions of many taxa (ranging from higher plants and birds to mosses and molluscs) are held by the Biological Records Centre (Institute of Terrestrial Ecology), described in detail below and hereafter referred to as the BRC. These data allow the presence and absence of species to be mapped in each 10 × 10 km square of the national grid, a finer scale of resolution for a broader range of species than is available for any national biota anywhere else in the world, including North America (e.g. Scott *et al.* 1987).

Patterns in the numbers and spatial distributions of species can be described in a variety of ways. Whittaker (1960) pioneered the concept of partitioning regional (gamma) diversity into its local (alpha) and turnover (beta) components. Usage varies among authors. Here, 'regional diversity' refers to the British Isles excluding Ireland. We use alpha diversity to describe species richness at the level of single 10 km squares within Britain, and beta diversity to denote turnover in species between 50 km squares on transects across Britain (see Harrison *et al.* 1992 for further discussion).

Armed with these definitions and data, we can proceed to answer questions of interest to both theoretical ecologists and conservationists. For example: (i) Where are the most species-rich areas of the country for particular taxa; that is, where are the regions of highest alpha diversity? (ii) Do diversity 'hotspots' coincide for different taxa? (iii) Do 'hotspots' contain a significantly higher proportion of rare species, or just more of the commoner species? (iv) How much turnover (beta-diversity) is shown by different taxa along geographic transects? (v) What proportion of the national pool of species in a particular taxon might be conserved by setting aside reserves in a limited number of locations?

We explore these and related questions in this paper, as an illustration of how conservation strategies might draw on databases of the number and spatial distributions of species. We warn, however, that conservation programmes for the majority of taxa in most parts of the world cannot hope to have available data that are anything like as good as those available for Britain. In most cases, if we wait for comparable data, there will be little or nothing left that is worth conserving.

To avoid possible confusion with different messages emerging from other papers in the symposium, it is useful to clarify exactly what we are, and are not, saying. We are dealing with patterns of species distribution at a much finer scale than a number of other surveys, some of which treat areas roughly as large as the British Isles as one unit. Answers to the questions outlined above are unlikely to be independent of scale. Second, we define 'hotspots' solely in terms of species richness. We are not concerned with areas of high endemism—which is an alternative, but quite different way of defining hot spots. Third and last, we have made no attempt to construct phylogenies for the taxa under investigation, and hence we cannot weight our results by the 'taxonomic distinctness' of species.

The Biological Records Centre

The BRC at ITE Monks Wood was set up in 1964 and is effectively the UK's national biogeographic data centre. It comprises a database of information on the occurrence of plants and animals in the British Isles, and an archive of record cards and manuscript sources from which the database derives.

The database currently holds over five million records, covering more than 9000 species, and with on-going surveys covering some 16 000 species (Harding and Sheail 1992). It is the largest and taxonomically most wide-ranging body of such information in the world, with detailed distribution data on vascular plants, cryptogams, birds, mammals, and a wide range of invertebrate groups. Tabulations of available data are given by Harding and Sheail (1992).

Many of the data are site-relatable, and geographically referenced to 100 m precision. For visual presentation, mapping in the 10 × 10 km squares of the Ordnance Survey national grid is most often used, as it offers fine enough resolution to reveal most landscape-scale patterns in Britain, while permitting comprehensive coverage of all 2910 squares.

1. *Vetting and validating records*

Data are collected by people interested in the natural history of particular groups. Most of the recorders are volunteers. Data collection is administered through some 50 national recording schemes, each managed by a volunteer specialist who is an acknowledged expert in his/her particular group. The scheme organizer is responsible for rigorous taxonomic vetting of incoming data, for seeking extra information in support of records of rare or critical taxa, and for encouraging surveys in more poorly studied parts of the country. The geographic component of the data is also verified, with locality details and statutory conservation site data cross-checked at BRC.

Draft species maps are often circulated throughout the operation of a recording scheme (one of the best incentives to voluntary recorders to keep working), and are ultimately published, usually as an atlas with detailed ecological and descriptive text, and some preliminary analysis of the records (for a list, see Harding 1989).

2. *Evaluation of data sets*

In the early stages of any recording project, species distribution maps reflect the distribution of recorders rather than of species. As records accumulate however, the broad pattern of species distributions gradually becomes apparent, until for well-worked groups the picture is clear enough for absences to be reliable (i.e. a gap in a species' range is due to genuine absence rather than to a lack of recorders).

The view that 'the distribution of a taxon is that of its students and the diversity of a site is a matter of serendipity' (Bullock 1991), is thus partially true of the most obscure groups of micro-organisms and invertebrates in Britain. It is less true of better-known taxa (in Britain this includes bryophytes, macrofungi, and many invertebrates, for example most Coleoptera, many Diptera, and aculeate Hymenoptera), and is completely inapplicable to the most popular such as vascular plants, birds, butterflies, and dragonflies. For the best-known groups although the level of *recording* continues to rise, the curve showing the accumulation of records in new 10 km squares is almost asymptotic. Indeed, coverage is so thorough for butterflies and Odonata, for instance, that real changes in range can now be distinguished from artefacts of recorder behaviour; see Pollard (1991) and Merritt *et al.* (in press) for examples.

Geographic trends in alpha-diversity

In common with most parts of the globe (Schall and Pianka 1978) the flora and fauna of Britain shows a strong (but not universal) decline in species richness at higher latitudes. For example, the south and south-east of the country harbour the highest number of butterflies with over 50 species having been recorded in some 10 km squares over the last 30 years (Fig. 13.1a). Insolation has been cited as a major determinant of species number (Hamilton-Wright 1983; Turner *et al.* 1987) although the precise mechanism is elusive. Dragonflies (Fig. 13.1b) show a similar, but more patchy northward and westward decline in species richness. However, in common with a minority of taxa elsewhere in the world (e.g. Schall and Pianka 1978; Rabenold 1979; Janzen 1981) some groups go against general geographic trends. Liverworts and pteridophytes (Figs. 13.1c and 13.1d), for example, have higher alpha-diversities in the north and west of Britain, where rainfall is highest.

1. *'Hotspots'*

It would be surprising if we did not see local departures superimposed on national trends of species richness. Analysis of BRC data (indeed of any recording scheme based on volunteer input) at local scales introduces special problems caused by variation in recording effort between individual squares; local patterns therefore require cautious interpretation. Nevertheless, it is still possible to identify squares or clusters of squares which differ substantially from their surroundings in terms of species richness. These departures show up particularly clearly on transects across the country. For example (Fig. 13.2a), Wiltshire is depauperate in Odonata whilst there is a large hotspot for this group in east Norfolk, centred on the Norfolk Broads. A second, north-south transect (Fig. 13.2b), at right angles to the first, confirms the national SE-NE trend in Odonata richness, but also shows smaller scale fluctuations, particularly reduced alpha-diversity across London, Lancashire, and the Southern Uplands of Scotland, and higher than average levels in the Lake District, the lowland areas of western Scotland, and the Weald of Kent.

2. *Defining hotspots*

Many of the 'hotspots' of species richness, particularly the larger ones, can be readily explained (like that for Odonata in the Norfolk Broads) in terms of land-use history, geology, etc. However, close examination has shown that several 'hotspots' exist for many different groups for which explanations are less obvious. To some extent, it depends on how a 'hotspot' is defined. Five possible options are currently under investigation:

(a) Absolute species richness. This method simply counts the number of species present in each square for all taxa. The main weakness of this approach arises when depicting total (inter-taxon) species richness. Large groups comprising many species tend to swamp smaller groups; for example, an extra 20 species of butterflies would barely register against a background species number of (theoretically) over 800 species of macro-moths (the number of species for which the BRC has records).

(b) Percentage of total. Problems with different absolute species richness in different taxa can be overcome by using the percentage of the national pool of each species in each square. This method gives equal weight to different sized groups, but causes its own particular problems—not least that some taxa may be of higher conservation priority than others.

Additional measures include:

(c) Percentage of maximum richness. Identify those squares containing at least x per cent (where x will usually be > 50) of the maximum possible number of species in the national pool for a particular taxon.

(d) The most species-rich squares for individual taxa. Identify the top n per cent of squares for particular taxa, where n is an arbitrary number; we have used 20 per cent in our analyses.

Methods a–d, based as they are on maximum values, show 'hotspots' for nearly all taxa to be in the south and east of England since, for most groups, this is where the majority of species reside.

Another approach is to compare squares with their neighbours:

(e) Local 'hotspots'. Identify those squares which exceed the mean species richness value of their surrounding y squares ($y = 8$) by a given factor

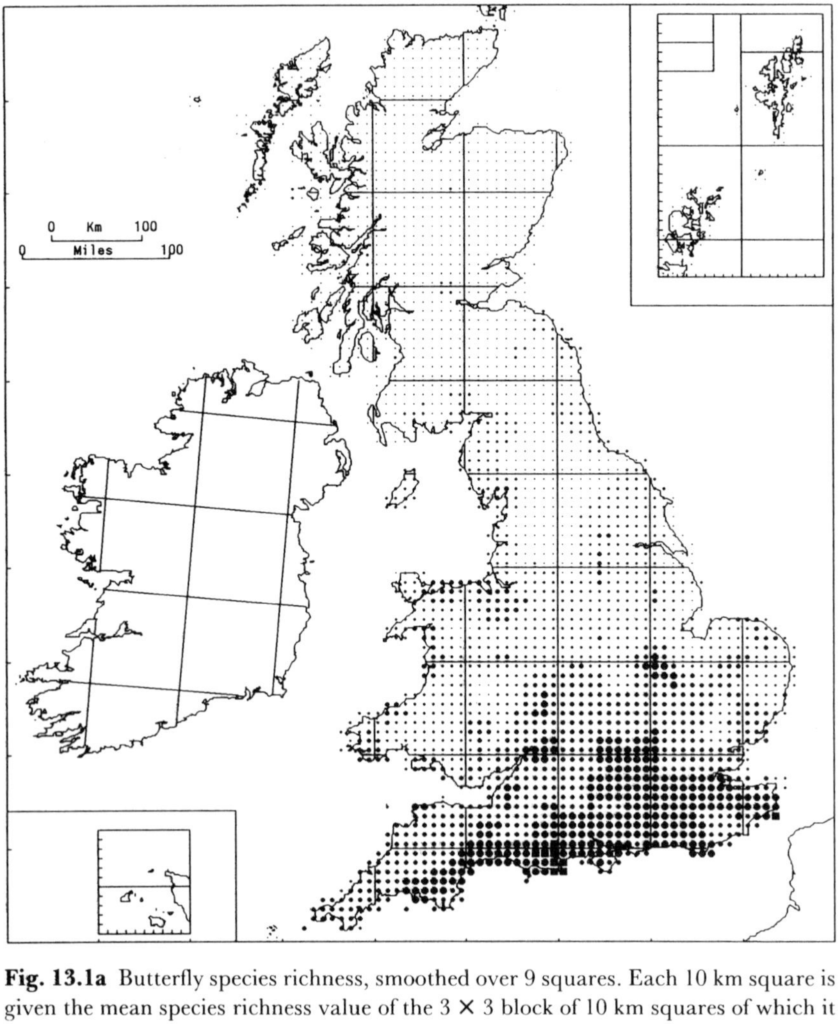

Fig. 13.1a Butterfly species richness, smoothed over 9 squares. Each 10 km square is given the mean species richness value of the 3 × 3 block of 10 km squares of which it forms the centre. Symbols represent, in increasing size: 1–15, 16–25, 26–35, 36–45 and > 45 species. Note the belts of elevated species richness which follow the chalk bands of the North and South Downs.

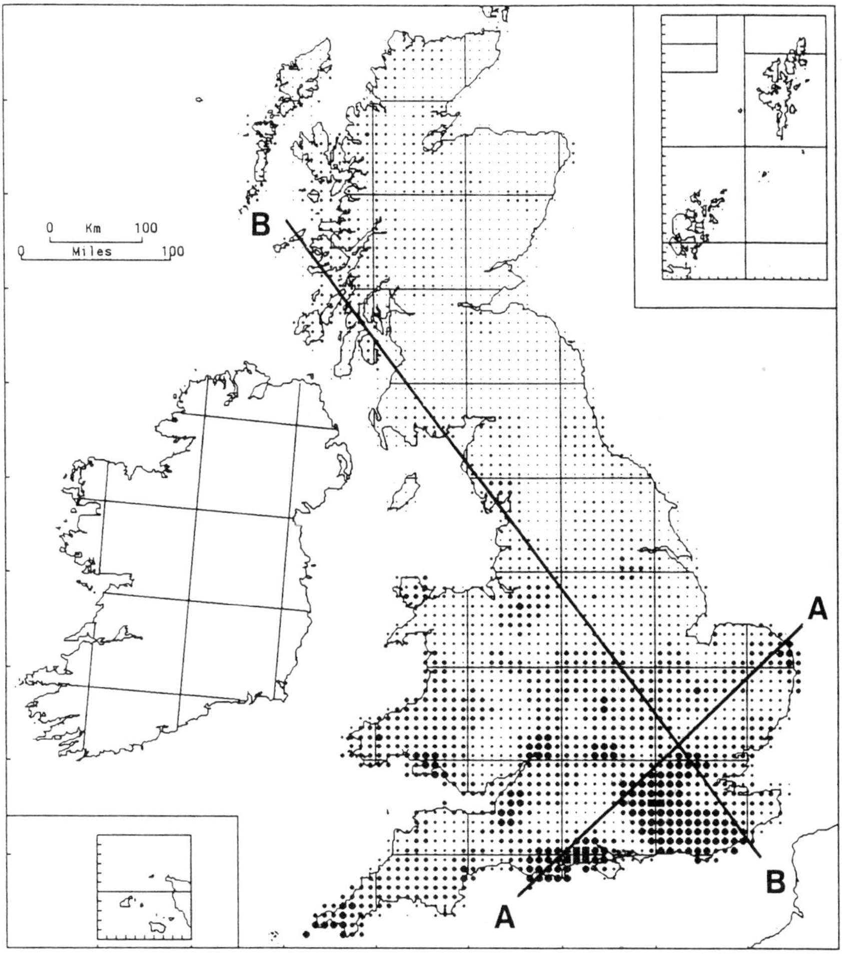

Fig. 13.1b Odonata species richness, smoothed over 9 squares. Symbols represent, in increasing size: 1–5, 6–10, 11–15, 16–24 and > 24 species. Although adopting the same general pattern as butterflies, dragonflies are notably absent from some areas where butterfly species richness is high. Wiltshire and Hampshire are depauperate due to the absence of standing water. Transect lines refer to Figs. 13.2a and 13.2b.

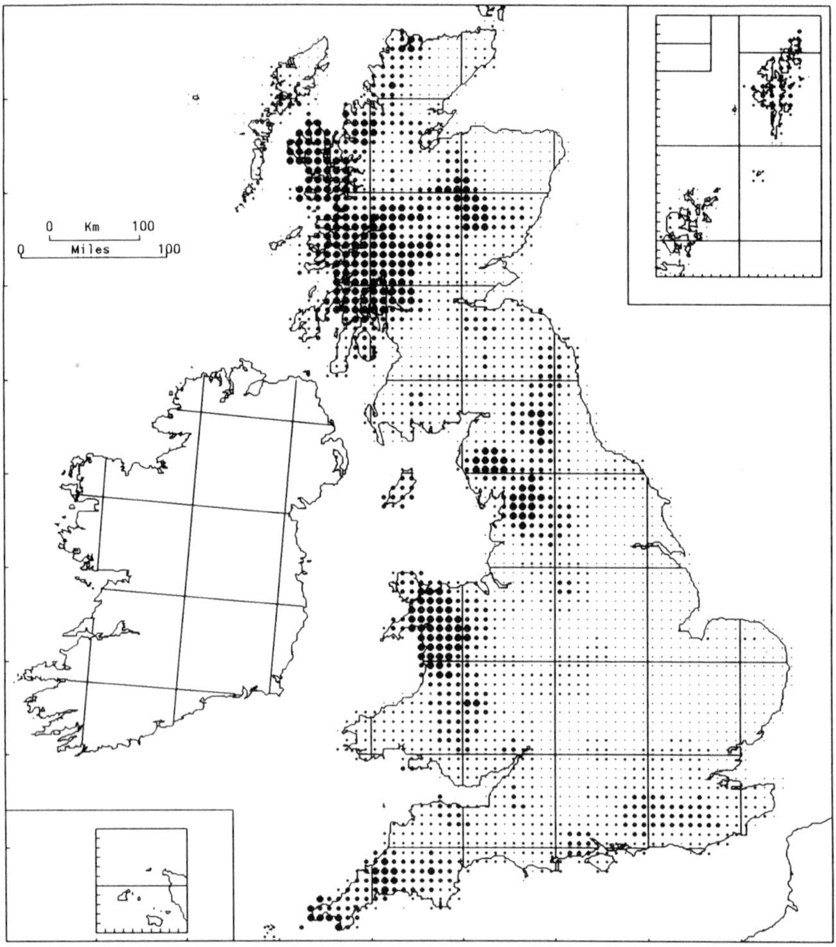

Fig. 13.1c Liverwort species richness, smoothed over 9 squares. Symbols represent, in increasing size: 1–20, 21–40, 41–60, 61–120 and > 120 species. In contrast to butterflies and dragonflies, liverworts are characterised by a marked northern and western distribution, generated mainly by rainfall patterns.

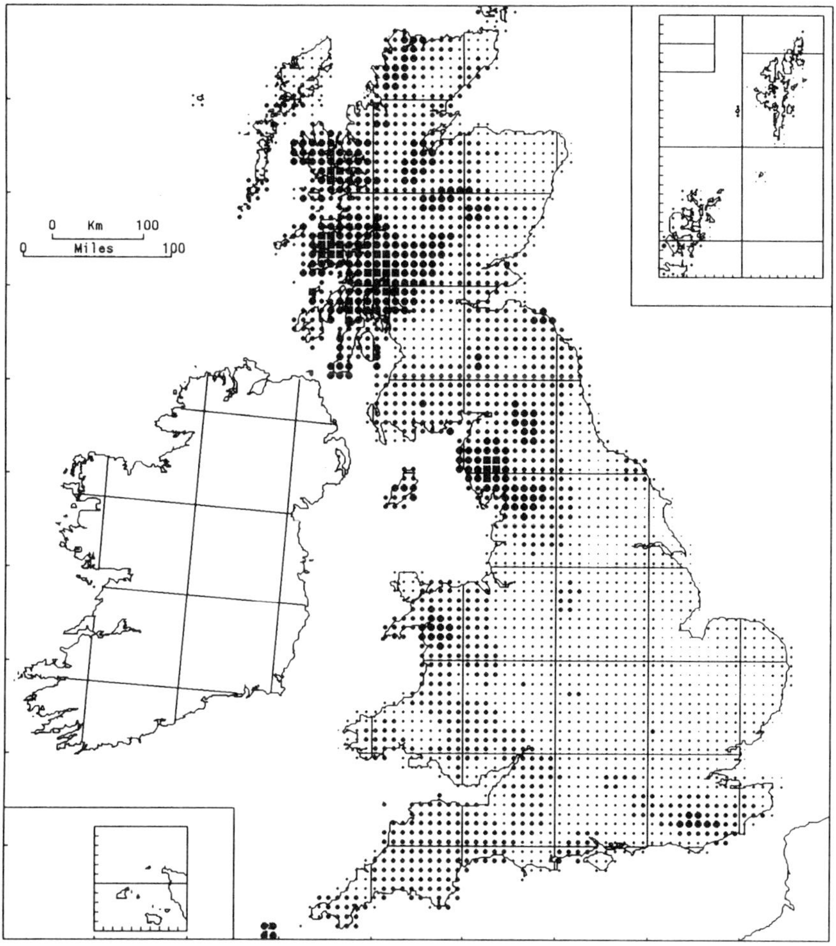

Fig. 13.1d Pteridophyte species richness, smoothed over 9 squares. Symbols represent, in increasing size: 1–7, 8–14, 15–20, 21–25 and > 25 species.

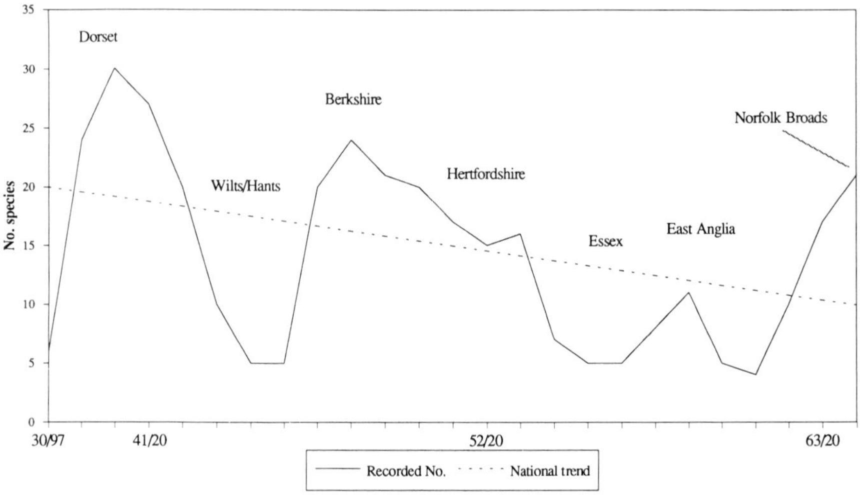

Fig. 13.2a Local variation in Odonata species richness along transect A–A across south-eastern England. Numbers on the x-axis refer to 10 km squares through which the transect passes. Note the paucity of species in the largely agricultural areas of Essex and East Anglia, and the hotspot in the Norfolk Broads.

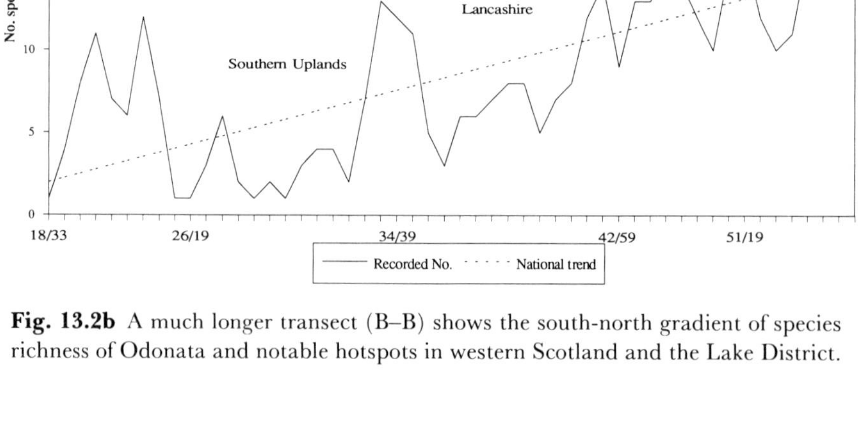

Fig. 13.2b A much longer transect (B–B) shows the south-north gradient of species richness of Odonata and notable hotspots in western Scotland and the Lake District.

thus defining local 'hotspots'. In preliminary analyses, we have found local hotspots that exceed local species richness by a factor > 3.

Although this method readily picks out squares with special characteristics relative to their surrounding area it is scale-limited to one 10 km square. At the present stage of data analysis we do not know whether the properties which make a square a local hotspot are reiterated at larger and/or smaller scales.

It is important to recognize that the squares which are identified as 'hotspots' depend on which criterion is used. There is no definitive 'best' method. Decisions about which methods to use must ultimately depend on the question(s) to be answered, and upon conservation priorities.

3. *Characterizing 'hotspots'*

By way of illustration we present analyses based on method (e) to depict local 'hotspots' for all the animals (Fig. 13.3a) and all the plants (Fig. 13.3b) in the BRC database.

The most notable feature of this type of approach is that the south-east of Britain, which generally supports the largest total number of species in most taxa, has rather few squares which stand out from their neighbours. Instead, local 'hotspots' for animals (18 taxonomic groups) are seen mainly in the west and north. For plants (6 groups) there is a remarkable degree of regularity, with very few individual 10 km squares being in any way special in comparison with their neighbours. We emphasize that these are preliminary analyses. It may be that local 'hotspots' on the animal map are artefacts of the recording process, since plants have been subject to more systematic (i.e. more even) recording than many animal taxa. Alternatively there may be real difference between plants and animals in their response to environmental variables, and hence real differences in the tendency for plants and animals to create local 'hotspots' at the scale of single 10 km squares.

4. *Do 'hotspots' for different taxa coincide?*

An important assumption of many conservation strategies is that 'hotspots' for different taxa coincide (Scott *et al.* 1987; Myers 1990; Gentry 1992) If they do, then creating reserves in areas that are 'good' for a few well known taxa (such as birds or butterflies) should ensure that less well studied taxa will also be effectively conserved. Our work on the BRC database does not encourage confidence in this crucial assumption. At a very coarse scale it is obviously true, because as we have already pointed out, there is a concentration of many taxa in southern and eastern Britain. But, as Fig. 13.1 makes plain, setting aside a reserve in an area that is rich in butterflies by no means guarantees that large numbers of dragonfly species will also benefit,

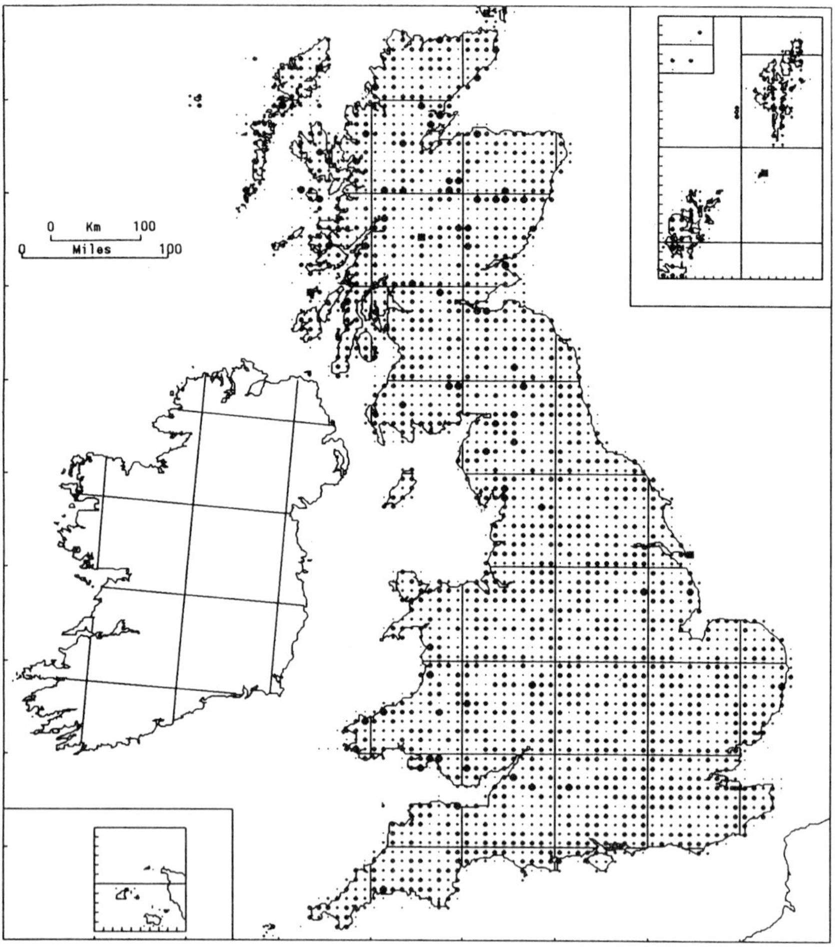

Fig.13.3a Hotspots for 18 groups of animals held in the BRC database. The number of species, summed over all 18 groups is first entered for each 10 km square. Using method (e), each square is compared with the mean of its eight surrounding squares (see text) and the ratio of central square to the surrounding eight squares plotted here. Symbols represent, in increasing size, ratios of: < 0.5 (i.e. a depauperate square), 0.5–1, 1–2, 2–3 and > 3. A total of 80 squares are more than twice as species-rich as their surroundings. Many of the hotspots are in the north and west of the country, possibly due to increased habitat heterogeneity resulting from relatively high topographical relief in these areas.

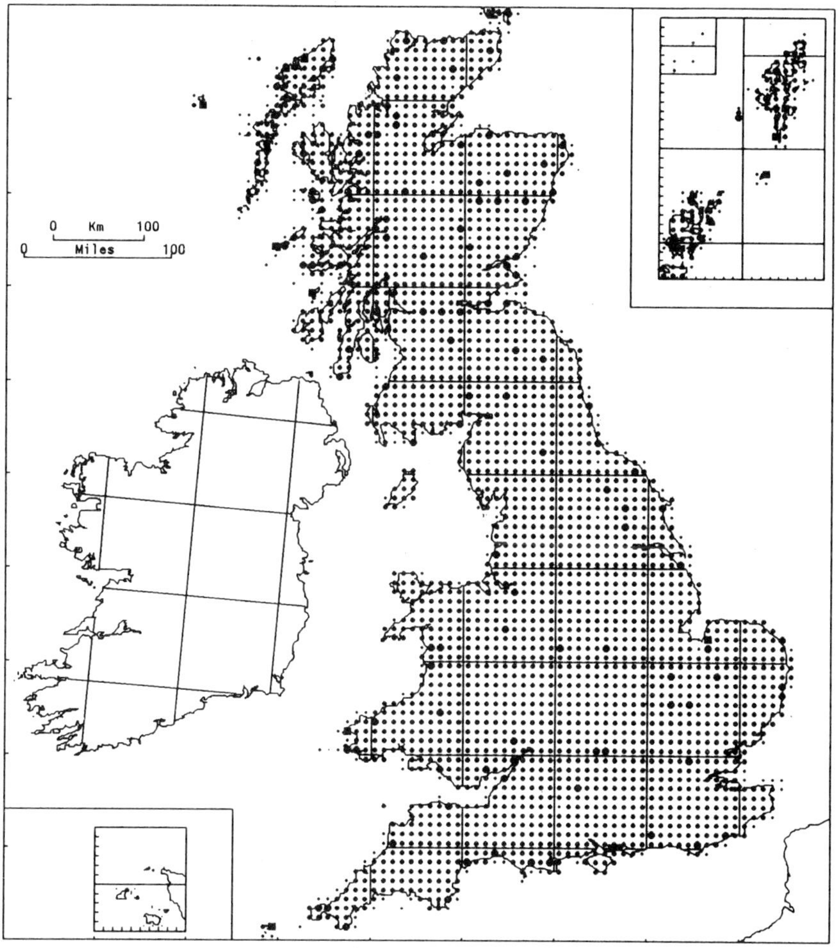

Fig. 13.3b The same process as in Fig. 13.3a applied to 6 groups of plants (higher and lower). Symbols represent, in increasing size, ratios of: < 0.5, 0.5–1.5, 1.5–2 and > 2. Greater regularity is seen in plant patterns with only 12 squares more than twice as rich as their surroundings.

and such a policy would clearly offer little protection to many bryophytes and pteridophytes.

In Britain, the statutory conservation bodies do not explicitly design conservation strategies by setting aside reserves for birds or butterflies, or any other group and hope that bryophytes will also benefit. The BRC data is here being used for illustrative purposes only. We are, however, currently working on much more detailed analyses of national and local 'hotspots', taxon by taxon, in an attempt to answer the question: how often do national or local 'hotspots' for different taxa coincide? Initial results suggest that the answer is much less often than many people hope or assume will be the case for less well studied parts of the world.

5. *Does conserving sites with the most species also conserve the rarest or most specialized species?*

In recent years there has been a shift of emphasis from focusing conservation effort on rare species, particularly those which are rare *and* declining, towards an approach which places priority on the preservation of areas of high biodiversity. This raises an important question about the distribution of rare species. If diversity 'hotspots' are selected for conservation, will this also conserve rare species, or just more, relatively common species?

To test this, we made a simple comparison to reveal congruence of squares containing rare butterflies and dragonflies, and those containing the largest number of species in those groups. Many definitions of rarity exist (Rabinowitz *et al.* 1981); for the purposes of this exercise we have taken the number of occupied 10 km squares ('area of occupancy' *sensu* Gaston 1991) as a measure of rarity irrespective of local abundance within squares or the geographic spread of those squares. This measure, therefore, includes species which are genuinely restricted or which may simply be on the edge of their range in Britain.

Twenty-seven 10 km squares in Britain contained at least 80 per cent of the maximum number of butterfly species found in any one square (40 species) between 1979 and 1982. For the same time period those squares occupied by the 15 rarest UK butterfly species were located. For seven species there was no overlap with the top 27 squares and for none of the other eight did more than one-third of their squares coincide with the species-rich squares. For Odonata the scenario is no better; the maximum number of species to be recorded in one square between 1985 and 1990 was 25, and 21 squares recorded 80 per cent or more of this value. Of the 15 rarest resident species, six have no overlap and only four have more than a 2-square overlap with the list of top squares. We currently do not understand why rare butterflies and dragonflies show these patterns, but the initial conclusion from the analyses is not encouraging. For butterflies, selecting species-rich squares as conservation sites may benefit about half of the rarest UK species. The other half are not represented at all in species-rich squares. For dragonflies however, selecting species-rich squares leaves out nearly three-quarters of the rarest species.

It remains to be seen whether similar patterns emerge for other taxa, and whether the fragmented and highly modified UK flora and fauna is representative of the situation in other parts of the world.

Beta-diversity

Patterns in beta-diversity for selected animal and plant taxa, along north-south and west-east transects of mainland Britain using BRC data, are reported in Harrison *et al.* (1992). Here we summarize their main results.

Harrison *et al.* (1992) examined eight animal and seven plant groups: breeding birds, butterflies, moths (ten families), bees (*Bombus* and *Psithyrus*), dragonflies, non-marine molluscs, woodlice (terrestrial Isopoda), freshwater fishes, mosses, and selected vascular plants [ferns, orchids (Orchidaceae), umbellifers (Apiaceae), composites (Asteraceae), roses (Rosaceae), and native trees]. As we have already pointed out, the BRC data can be examined at many spatial scales, and answers to the same question may be different at different scales of resolution. Harrison *et al.* (1992) examined beta-diversity by measuring species turnover between 50 × 50 km grid squares (i.e. each sample consisted of a list of species recorded in a group of twenty-five 10 km squares, arranged in a 5 × 5 block). It is unfortunate that similar analyses have not been carried out using individual 10 km squares, making the results more directly comparable with those in the first part of the present paper. They examined two transects, one of seventeen 50 × 50 km squares running from northern Scotland to the south coast of England, and another at virtual right-angles to the first, consisting of eight 50 × 50 km squares, extending from southern Wales (Pembroke) to the east coast of England in Norfolk.

Harrison *et al.* (1992) used two measures of species turnover (see their paper for details). The first, termed beta-1, is Whittaker's (1960) original measure, modified slightly to allow comparison between two transects of unequal size. Beta-1 ranges from zero if two squares have identical species, up to 100 if they have no species in common (i.e. bigger values of beta mean bigger floral or faunal differences between regions). One potential drawback of beta-1 is that it was created by Whittaker for comparing sites with relatively constant values of alpha-diversity. A strong north-south gradient in alpha-diversity, such as exists in Britain (see above), can generate misleading values of beta-1 if species simply drop out northwards along the gradient, with few new species being added. Harrison *et al.* (1992) therefore defined a new index, beta-2, which reflects real taxonomic turnover. Beta-2 measures the amount by which regional diversity exceeds the maximum diversity attained locally (in this case locally means within one 50 × 50 km square). Like beta-1, it ranges between 0 and 100; it converges on beta-1 when the sites being compared have similar alpha-diversities. In the context of the present paper, the message from both measures of species turnover (beta-1 and beta-2) is identical, although their numerical values differ.

The results of the study by Harrison *et al.* are relatively simple. Beta diversity in all these taxa, however measured, is very low at the chosen spatial scale. Birds, for example, have representative beta-1 values of 3.3 and 5.7 on the north-south and west-east transects respectively. For comparison, beta-1 was 38.1 for the native land birds of the six largest Hawaiian Islands, which span a considerably smaller range of distances (Harrison and Thomas, unpublished). On the north-south transect, total species richness was 1.5–3 times the average value for each 50 m square, and 1.1–1.6 times the maximum local value. On the west-east transect, total richness was 1.2–1.9 times the average or 1.1–1.4 times the maximum local richness. Combining results from both transects, these correspond to values of beta-1 of 3–13 and beta-2 of 0.6–6.

Not only is beta-diversity very low in Britain, it is not obviously related to species' dispersal abilities. We might expect turnover to be higher in taxa with poor dispersal abilities than in highly mobile taxa, but comparisons of beta-diversity for plants vs. animals, flying vs. non-flying animals, etc. do not suggest any effect of dispersal ability on beta-diversity. Nor does species turnover (specifically beta-2) inevitably increase with distance between localities. There were significant distance effects for only six taxa on the north-south transect, and only one (birds) on the west-east transect.

1. *Conservation implications*

What do these data have to say about conservation strategies? If the UK data are typical of those from other temperate regions of the globe, the message is encouraging, because low beta-diversity implies that setting aside a few, species-rich areas will encompass a large share of a region's total diversity. For example, all 56 surviving species of UK butterflies can be found in just five complementary sites in: Norfolk (33 species), Isle of Wight (another 13 species), West Scotland (6 more), Hampshire (3 more), and Kent (the last 1) (Vane-Wright in Krebs 1992). (Note that these results do not contradict our earlier remarks about hotspots. Quite the reverse. Sites have to be carefully targeted for *particular* species.) Five sites alone, would not, of course, be sufficient to guarantee survival of viable populations of all UK butterfly species for the foreseeable future (approaches to determining a complementary set of sites for a taxon are addressed elswhere in this volume), but the general message is clear enough. Indeed, one does not need fancy statistics to show what natural historians have long known about the British Isles; careful recording in one locality yields a significant fraction of the total national species pool for virtually any taxon. Writing about Wytham Hill near Oxford, Charles Elton noted: 'the estimated fauna of this calcareous south midland hill may be as much as a sixth or a fifth of the British fauna—on an area that is only 1/25,000 of England and 1/60,000 of the British Isles' (Elton 1966).

Unfortunately, we do not know with any degree of confidence that Britain is representative of other temperate regions of the globe; the data are not available. Moreover, Britain is certainly not a good model for expected levels of species turnover in the tropics and sub-tropics. We expect beta-diversity to be much higher in virtually all tropical biomes. Two related observations make this inevitable: first, many species-rich areas show high degrees of endemism (Gentry 1992), and second, compared with related taxa from temperate regions, many tropical species have very small total geographic ranges (Stevens 1989; France 1992). Higher beta-diversity in the tropics means that more reserves are required, in more places, to encompass the same fraction of a tropical compared with a temperate biota. How many more reserves in how many more places remains to be resolved.

Conclusion

The need for effective conservation strategies is growing apace; what conclusions for species preservation can we draw from this limited excursion into the BRC database?

First, we have seen that those parts of the country which contain the most species are different for different taxa. We have also seen that far from being purely a national phenomenon this disparity is repeated at local scales and that 'hotspots' for different taxa are located in different places. Rare species, too, pose a particular problem in that, for the taxa we have studied so far, they do not predictably appear in areas which might intuitively be singled out for special conservation effort, i.e. species-rich areas.

Given the considerable variation in alpha-diversity shown by the groups analysed, the low beta-diversity is somewhat surprising (although the scales of examination are different), but it does imply that for a temperate biota, a small number of (large) areas might embrace a considerable fraction of a region's total species number. Unfortunately this simple, encouraging message is extremely unlikely to be true in tropical areas, where endemism is high, and geographical ranges characteristically small.

Perhaps the clearest message to emerge from the work done so far is that not only are patterns of diversity in Britain complex and scale dependent, but also that our present understanding is very limited.

Many of the assumptions that hard pressed conservation managers are forced to make are not supported by analysis of the British data. In conservation terms this means that there are no easy solutions to species protection.

The most pressing conservation problems are in the tropics, and it is important to put these data for Britain in their proper context. Tropical biomes tend to contain more species, smaller habitat patch sizes, a greater degree of endemism, and higher beta-diversity than temperate regions. We have no evidence that British diversity patterns are in any way representative of those in the tropics (or indeed anywhere else). With the poor availability of reliable data from other parts of the globe for anything other than a handful of species such as birds, the prospect of being able to develop conservation strategies for most taxa based on good ecological information is extremely remote. Saving species will have to rely on luck and intuition; we cannot wait until all the data are in, because if we wait there will be nothing left worth conserving.

References

Bullock, J. A. (1991). The distribution of a taxon is that of its students, and the diversity of a site is a matter of serendipity? *Antenna*, **15**, 6–7.

Elton, C. S. (1966). *The pattern of animal communities*. Methuen, London.

France, R. (1992). The North American latitudinal gradient in species richness and geographic range of freshwater crayfish and amphipods. *American Naturalist*, **139**, 342–54.

Gaston, K. J. (1991). How large is a species' geographic range? *Oikos*, **61**, 434–8.

Gentry, A. H. (1992). Tropical forest biodiversity: distributional patterns and their conservation significance. *Oikos*, **63**, 19–28.

Hamilton-Wright, D. (1983). Species-energy theory: an extension of species-area theory. *Oikos*, **41**, 496–506.

Harding, P. T. (1989). *Current atlases of the flora and fauna of the British Isles*, Biological Records Centre, Huntingdon, UK.

Harding, P. T. and Sheail, J. (1992). The Biological Records Centre: a pioneer in data gathering and retrieval. In *Biological recording of changes in British wildlife* (ed. P. T. Harding), pp. 5–19. HMSO., London.

Harrison, S., Ross, S. J., and Lawton, J. H. (1992). Beta diversity on geographic gradients in Britain. *Journal of Animal Ecology*, **61**, 151–8.

Janzen, D. H. (1981). The peak in North American Ichneumonid species richness lies between 38° and 42°. *Ecology*, **62**, 523–7.

Krebs, J. R. (1992). *Evolution and biodiversity: the new taxonomy*. Report of the Committee set up by the Natural Environment Research Council, NERC, Swindon, UK.

Merritt, R., Moore, N. W., and Eversham, B.C. (in press). *Atlas of the dragonflies of Britain and Ireland*. HMSO, London.

Myers, N. (1990). The biodiversity challenge: expanded hot-spot analysis. *The Environmentalist*, **10**, 243–56.

Pollard, E. (1991). Changes in the flight period of the hedge brown butterfly (*Pyronia tithonus*) during range expansion. *Journal of Animal Ecology*, **60**, 737–48.

Rabenold, K. N. (1979). A reversed latitudinal diversity gradient in avian communities of eastern deciduous forests. *American Naturalist*, **114**, 275–86.

Rabinowitz, D., Cairns, S., and Dillon, T. (1986). Seven forms of rarity and their frequency in the flora of the British Isles. In *Conservation biology: the science of scarcity and diversity* (ed. M. Soulé), pp. 182–204. Sinauer, Sunderland, MA.

Schall, J. J. and Pianka, E. R. (1978). Geographic trends in numbers of species. *Science*, **201**, 679–86.

Scott, M. J., Csuti, B., Jacobi, J. D., and Estes, J. E. (1987). Species richness: a geographical approach to protecting future biological diversity. *BioScience*, **37**, 782–8.

Stevens, G. C. (1989). The latitudinal gradient in geographic range: how so many species coexist in the tropics. *American Naturalist*, **133**, 240–56.

Turner, J. R., Gatehouse, C. M., and Corey, C. A. (1987). Does solar energy control organic diversity? Butterflies, moths and the British climate. *Oikos*, **48**, 195–205.

Whittaker, R. M. (1960). Vegetation of the Siskiyou Mountains, Oregon and California. *Ecological Monographs*, **30**, 279–338.

14. The relative vulnerabilities of moth higher taxa to habitat change in Borneo

J. D. HOLLOWAY
International Institute of Entomology, Department of Entomology, The Natural History Museum, Cromwell Road, London SW7 5BD, UK

Abstract

Biogeographic and ecological data for about 1200 Bornean species in 15 higher taxonomic groups of moths are analysed to assess the relative vulnerabilities of these groups to habitat change. The numbers of taxonomic decisions required to establish these data are calculated. The results illustrate a spectrum of vulnerability amongst the groups to different types of habitat change, with those most at risk having a high proportion of endemic species dependent on primary lowland rainforest. The implications for environmental management and conservation policy are discussed.

Introduction

Biogeographic methods for identifying areas of the world critical for the conservation of biodiversity offer a sound scientific basis for developing strategies for the establishment of protected areas and habitats. This paper comments on the potential consequences of not adopting such a strategy, in terms of the predicted loss of biodiversity in the event of changes in key habitat types in a tropical area.

In any area, different higher taxonomic groups will exhibit different degrees of endemism, and these endemics will show fidelity to different habitat types in a diverse manner. A sample of about 1200 species in 15 higher taxonomic groups of the larger moths (macrolepidoptera) are used to illustrate this differential vulnerability to habitat change. The analysis is presented in more detail by Holloway and Barlow (1992).

Systematics and Conservation Evaluation (ed. P. L. Forey, C. J. Humphries, and R. I. Vane-Wright), Systematics Association Special Volume No. 50, pp. 197–205, Clarendon Press, Oxford, 1994.

The Lepidoptera

The Lepidoptera are a high-diversity, taxonomically tractable group of herbivorous insects exhibiting a moderate to high degree of dietary specificity and therefore potential habitat specificity. The macrolepidopteran moths have been relatively well-collected and hence biogeographic data are mostly reliable. Sampling with light-traps on transects provides a rapid means of assessing habitat preference for a high proportion of the fauna. Extensive sampling work of this nature has been undertaken by surveys in Borneo (Holloway 1989) and elsewhere in the Indo-Australian tropics. Data from these surveys can be placed in a biogeographic context by reference to the outstanding collections of Lepidoptera material from the Indo-Australian tropics in The Natural History Museum and elsewhere. These resources have also been essential for establishing the composition of the higher taxonomic groups through revisional work.

Methods

Two-way tables for the 15 higher taxonomic groups were prepared from data in Holloway (1983–1989) and work in progress on the Geometridae: Ennominae. The groups are as follows: 1. Cossoidea (43 species); 2, Limacodidae (97 species); 3, Noctuidae: 'lower trifines' (Acronictinae, Agaristinae, and Amphipyrinae from *Sasunaga* onwards in Holloway (1983–1989) (104 species); 4, Noctuidae: 'higher trifines' (Noctuinae, Heliothinae, Hadeninae, and Amphipyrinae up to *Sasunaga*) (75 species); 5, Noctuidae: Euteliinae (71 species); 6, Noctuidae: Stictopterinae (90 species); 7, Arctiidae: Arctiinae (39 species); 8, Notodontidae: (122 species); 9, Lasiocampidae: (62 species); 10, Sphingidae: (93 species); 11, Geometridae:

Fig. 14.1 The biogeographic and habitat categories of the two-way tables. The heavy lines on the map of the Indo-Australian tropics (after Holloway 1987) indicate major distributional discontinuities, and the light dashed lines delimit frequently encountered lepidopteran distribution patterns as applied at the left-hand side of the two-way tables: Wallacea (D); Sundaland (B); Wide Oriental (A+B+D(+C)); Indo-Australian (A to E (+F)). The diagram of diversity with altitude (from Holloway and Barlow 1992) illustrates the contribution to total diversity made by overlapping associations of species in the habitat categories of the two-way tables: lowland forest (Low), lower montane forest (LMF), upper montane forest (UMF), and the Radio Sabah and summit associations (RS, S) that are only developed to any great extent on Mt Kinabalu. The tables also include categories for species that extend over a wider range of altitude (LMF+UMF; lowland + montane), and for species characteristic of open habitats.

Ennominae tribe Hypochrosini (60 species); 12, Geometridae: Ennominae tribe Baptini (48 species); 13, Geometridae: Ennominae, *Cassyma* group of genera (39 species); 14, Geometridae: Ennominae tribe Boarmiini (in a broad sense) (154 species); 15, Geometridae: Ennominae, residue (106 species).

The tables indicate the percentages of species falling within each of 49 habitat/biogeographic categories (Figs. 14.1 and 14.2). The habitat categories for each species were established in part by R-mode classifications of

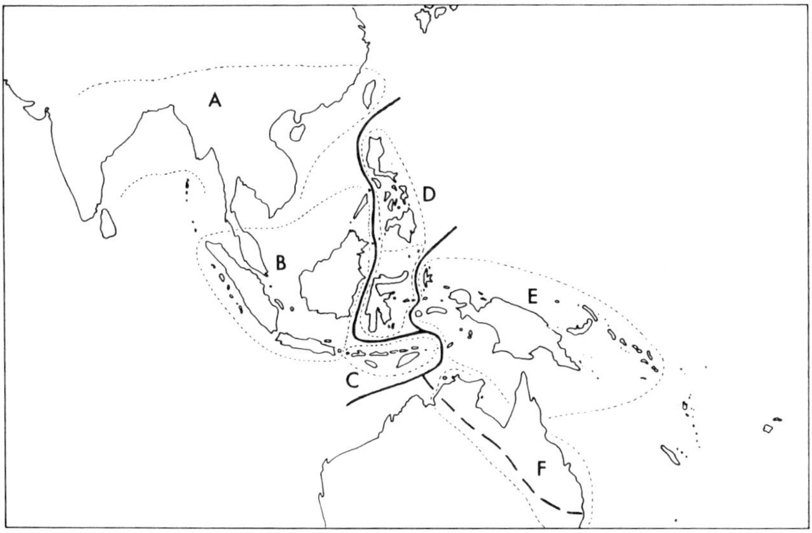

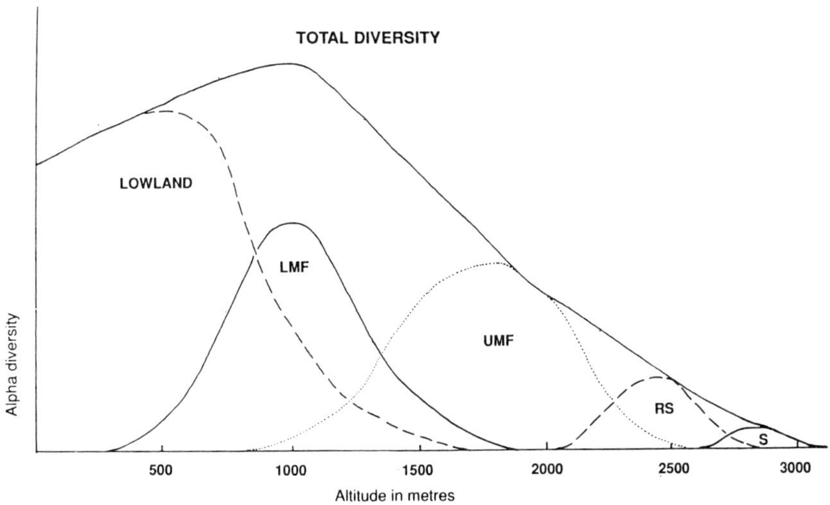

	Low	LMF	UMF	RS + S	LMF + UMF	Low + Mont	Open
Endemic	20.93						
Borneo + Wallacea							
Sundaland	27.90				4.65		
Sundaland+ Wallacea	13.95				2.33	2.33	
Himalaya+ Sundaland	4.65						
Wide Oriental	16.28						2.33
Indo-Australian							

1 Cossoidea

	Low	LMF	UMF	RS + S	LMF + UMF	Low + Mont	Open
Endemic	22.58	4.30	6.45		1.08	1.08	
Borneo + Wallacea	1.08						
Sundaland	40.86	2.69	2.69		1.08	2.15	
Sundaland+ Wallacea							
Himalaya+ Sundaland	4.30				1.08	1.08	
Wide Oriental	6.45						
Indo-Australian	1.08						

2 Limacodidae

	Low	LMF	UMF	RS + S	LMF + UMF	Low + Mont	Open
Endemic	8.25		5.15		7.22		
Borneo + Wallacea					1.03		
Sundaland	11.34	1.03	4.12		1.03	8.25	2.06
Sundaland+ Wallacea	4.12					1.03	
Himalaya+ Sundaland	3.09	1.03	1.03			4.12	
Wide Oriental	7.22	1.03	2.06			5.15	1.03
Indo-Australian	2.06	2.06	2.06			3.09	10.31

3 Lower trifine Noctuidae

	Low	LMF	UMF	RS + S	LMF + UMF	Low + Mont	Open
Endemic			21.13	11.27	0.70		0.70
Borneo + Wallacea			1.41	1.41			
Sundaland	2.82		2.82			1.41	
Sundaland+ Wallacea			4.23				
Himalaya+ Sundaland							
Wide Oriental	2.82		2.82	4.23	0.70	1.41	3.52
Indo-Australian	1.41		4.23			1.41	28.16

4 Higher trifine Noctuidae

	Low	LMF	UMF	RS + S	LMF + UMF	Low + Mont	Open
Endemic	7.46		1.49		1.49	1.49	
Borneo + Wallacea	2.99						
Sundaland	22.39	1.49	1.49		2.99	5.97	
Sundaland+ Wallacea	4.48				1.49		
Himalaya+ Sundaland		1.49					
Wide Oriental	4.48		1.49		1.49	1.49	
Indo-Australian	16.42		1.49		4.48	11.94	

5 Euteliinae

	Low	LMF	UMF	RS + S	LMF + UMF	Low + Mont	Open
Endemic	3.53		8.24			2.35	
Borneo + Wallacea			1.18				
Sundaland	8.24		5.88		3.53	12.94	
Sundaland+ Wallacea	2.35	1.18	1.18		2.35	3.53	
Himalaya+ Sundaland		1.18	1.18		1.18	3.53	
Wide Oriental	2.35		4.71			4.71	
Indo-Australian	3.53	1.18	2.35		3.53	14.12	

6 Stictopterinae

	Low	LMF	UMF	RS + S	LMF + UMF	Low + Mont	Open
Endemic	5.26	5.26	15.79		10.53	2.63	2.63
Borneo + Wallacea	2.63						
Sundaland	10.53	2.63			2.63	2.63	2.63
Sundaland+ Wallacea						2.63	
Himalaya+ Sundaland					2.63		
Wide Oriental					5.26		10.53
Indo-Australian						5.26	7.89

7 Arctiinae

	Low	LMF	UMF	RS + S	LMF + UMF	Low + Mont	Open
Endemic	8.33		5.83			1.67	
Borneo + Wallacea							
Sundaland	30.00		8.33		1.67	5.83	1.67
Sundaland+ Wallacea	6.67					4.17	1.67
Himalaya+ Sundaland	3.33					1.67	
Wide Oriental	8.75	1.25	3.33		0.83	3.33	
Indo-Australian			0.83			0.83	

8 Notodontidae

	Low	LMF	UMF	RS + S	LMF + UMF	Low + Mont	Open
Endemic	9.68		6.45		8.06	1.61	
Borneo + Wallacea							
Sundaland	41.93		6.45		1.61	8.06	
Sundaland+ Wallacea							
Himalaya+ Sundaland					1.61	3.23	
Wide Oriental	8.06				1.61	1.61	
Indo-Australian							

9 Lasiocampidae

	Low	LMF	UMF	RS + S	LMF + UMF	Low + Mont	Open
Endemic						1.10	
Borneo + Wallacea							
Sundaland	15.38	1.10			2.18	6.59	1.10
Sundaland+ Wallacea						1.10	
Himalaya+ Sundaland	6.59		1.10				
Wide Oriental	16.48				3.30	10.99	4.40
Indo-Australian	4.40		2.18			12.09	9.89

10 Sphingidae

	Low	LMF	UMF	RS + S	LMF + UMF	Low + Mont	Open
Endemic	15.83	0.83	9.17	2.50	1.67	1.67	
Borneo + Wallacea					1.67		
Sundaland	25.83	1.67	1.67		1.67	6.67	0.83
Sundaland+ Wallacea							
Himalaya+ Sundaland	6.67		1.67		3.33	1.67	
Wide Oriental	5.00		1.67		3.33	1.67	
Indo-Australian		1.67	1.67				

11 Hypochrosini

	Low	LMF	UMF	RS + S	LMF + UMF	Low + Mont	Open
Endemic	18.75	2.08	12.50	2.08		4.17	
Borneo + Wallacea		2.08					
Sundaland	17.70	11.46	4.17		2.08		
Sundaland+ Wallacea	4.17		2.08				
Himalaya+ Sundaland						2.08	
Wide Oriental	2.08	2.08	2.08			4.17	
Indo-Australian	2.08				2.08		

12 Baptini

	Low	LMF	UMF	RS + S	LMF + UMF	Low + Mont	Open
Endemic	29.73	1.35	4.05			5.41	
Borneo + Wallacea	2.70						
Sundaland	21.62	2.70	2.70		2.70	8.11	
Sundaland+ Wallacea	2.70						
Himalaya+ Sundaland		2.70			2.70		
Wide Oriental	5.41					5.41	
Indo-Australian							

13 Cassyma group

	Low	LMF	UMF	RS + S	LMF + UMF	Low + Mont	Open
Endemic	8.78	4.05	11.49	2.03	2.70	2.03	
Borneo + Wallacea	0.68		0.68				
Sundaland	16.22	4.05	5.41		0.68	6.08	0.68
Sundaland+ Wallacea	2.70				1.35	0.68	
Himalaya+ Sundaland	2.70	2.70	1.35		1.35	0.68	
Wide Oriental	4.05	2.03	4.05		1.35	2.03	0.68
Indo-Australian	4.73					2.03	

14 Boarmiini

	Low	LMF	UMF	RS + S	LMF + UMF	Low + Mont	Open
Endemic	3.96	3.96	7.92		1.98	1.98	
Borneo + Wallacea							
Sundaland	15.35	2.48	1.98		2.97	6.93	
Sundaland+ Wallacea	4.95		1.98		0.99	2.97	0.99
Himalaya+ Sundaland	0.99	1.98	3.96			2.97	
Wide Oriental	10.89	0.99	2.97		2.97	1.98	
Indo-Australian	1.48	0.50				5.94	0.99

15 Other Ennominae

Fig. 14.2 The 15 two-way tables, with habitat and biogeographic categories as defined in Fig. 14.1.

commoner species in terms of distribution over samples in the quantitative surveys (Holloway 1989), with assignation of rarer species to the most likely of the categories so recognized. The biogeographic categories are those common patterns of distribution recognized by R-mode phenetic classification of species ranges (e.g. Holloway 1987).

A classification of the tables can be obtained by applying cluster analysis (single-link supported by a linkage diagram to assess cluster structure) to similarity coefficients that express the extent to which two tables overlap in representation in the 49 categories. The single-link dendrogram and the linkage diagram are shown in Fig. 14.3. In the linkage diagram thickness of line indicates different degrees of similarity—the thicker, the more similar.

Taxonomic input

Generation of the data for the 1200 species required numerous taxonomic decisions.

For the ecological categorizations, field samples had to be checked to ensure species series were not mixed, then the species concerned could be associated with a named taxon to bring in any additional data from collections and literature. The first process involved 20 000–30 000 taxonomic decisions, and the second obviously 1200 positive decisions, and many more negative ones.

Establishment of the geographical ranges of these 1200 named taxa involved anything from a single decision of exclusiveness (endemism) built on any number of negative (different-from) decisions, to the ten or so positive decisions necessary to recognize a widespread Indo-Australian tropical range. A rough estimate of the total for the database of 1200 Bornean species came to well over 5000 positive decisions and probably somewhat fewer negative ones.

Definition of the higher taxonomic categories has involved extensive revisional work associated with production of a series of monographs on the fauna (Holloway 1983–1989). The parts so far produced or in current production have involved several thousand genitalic dissections and examinations.

Results

The clustering in the dendrogram and linkage diagram indicates there is a major grouping of higher taxa with a high proportion of endemic and Sundaland species dependent on lowland rain forest for their survival. At the other end of the continuum is a smaller group of taxa where a higher proportion of localized montane forest specialists occurs, and peripheral to that are several groups that contain a high proportion of geographically

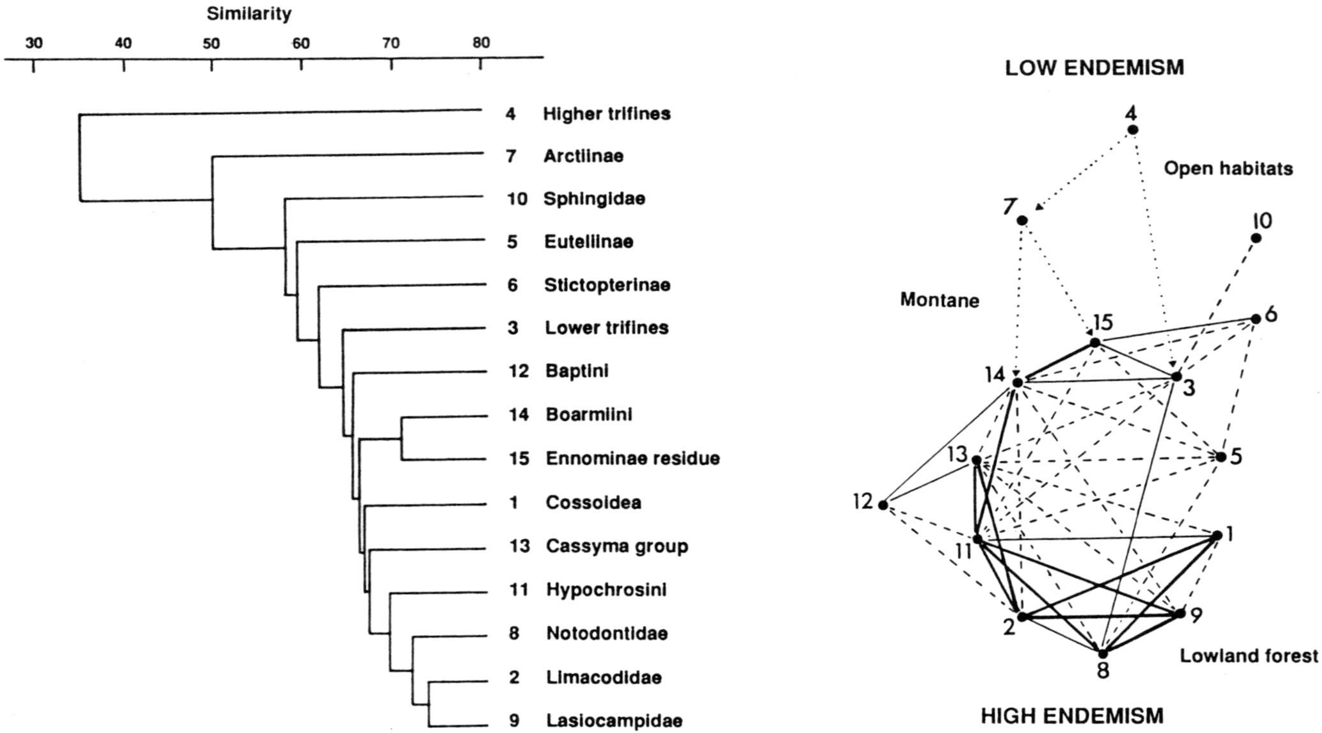

Fig. 14.3 Single-link dendrogram and linkage diagram, clustering the two-way tables illustrated in Fig. 14.2 in terms of similarity of percentage representation of species in the habitat/biogeographic categories. The linkage diagram illustrates high pairwise similarities between tables: 66–100 per cent similarity (heavy solid lines); 60–65 per cent (light solid lines); 55–59 per cent (dashed lines). Dotted lines with arrows indicate the links (lower than 55 per cent similarity) that associate 4 and 7 with the main cluster.

widespread ecological generalists. The three outlying groups also contain a high proportion of widespread taxa, many of which are open-habitat specialists, but group 7 and, to a greater extent, group 4 include a significant number of montane endemics.

It would therefore appear that most higher taxa in tropical South East Asia have a high proportion of species dependent on lowland forest, and most of these are geographically restricted or endemic. Expansion of the sample to include more noctuid subfamilies might add more groups to the peripheral categories, but, in general, diversity in montane habitats or open habitats appears to be a specialized phenomenon.

In the two-way tables, vulnerability to extinction could be considered to increase inversely to geographical range. Similarly, species restricted to a narrow habitat range will be more vulnerable on the whole than those with greater ecological amplitude.

An exception must be made for open-habitat specialist species, dependent on continual environmental perturbation for their survival. Prior to the advent of man, these species and their habitats were probably rather rare in the humid tropics, though geographically very widespread. Man's activities in converting natural habitats to managed systems have benefited them. Indeed, a high proportion are serious agricultural pests.

Many higher taxonomic groups are very much more diverse in tropical latitudes. If this high tropical diversity is associated with high representation in the upper left-hand squares of the two-way table, then the group must be considered particularly vulnerable to habitat modification in the tropics. Examples from the sample include the families Limacodidae and Lasiocampidae, and some of the ennomine geometrid groups such as the Hypochrosini and the *Cassyma* group of genera.

Implications for environmental management and conservation policy

1. *Higher taxa mostly dependent on lowland rain forest*

Groups with a majority of localized species restricted to lowland rain forest are particularly vulnerable to timber extraction, shifting cultivation, conversion to arboreal plantation, agroforestry, and conversion to field crop systems.

Both timber harvest and shifting cultivation can be managed wisely in a sustainable cyclical manner such that diversity is only temporarily depressed, to recover over a period of several decades. But it is necessary now to monitor the effects of different types of perturbation (e.g. the composition of the disturbance mosaic) on vulnerable taxa, particularly rare taxa, to ensure that potential for recovery is optimized. Moth sample data provide one means to assess this.

Conversion to different types of managed forest appears to result in a significant drop in diversity. The contribution of such systems to conservation of biodiversity will depend on the extent to which they can sustain populations of key taxa in vulnerable groups (Holloway *et al.* 1992; Chey *et al.* in press). There are indications that representation of vulnerable groups declines on conversion to plantation, but is maintained in forest regenerating after selective timber extraction. A variety of Limacodidae are frequently recorded in palm crop plantations, sometimes at pest levels, but it is not clear whether, with their 'outbreak' population dynamics, they can be sustained by such plantations alone. In field crop systems only geographically widespread, open-habitat specialist taxa thrive, so all high conservation value taxa are irreversibly lost (Holloway and Stork 1991). Thus, whilst establishment of protected areas of lowland forest will be necessary to guarantee survival of vulnerable groups, buffer zones of forest harvested in natural cycles, or of controlled shifting cultivation, can also make a contribution to conservation.

2. *Conservation in montane systems*

Montane forest systems are vulnerable to conversion to cropping systems such as tea plantation and vegetable growing, where the same considerations apply as in the lowlands. High endemism in such habitats occurs in fewer, more specialized higher taxa.

A particular threat in the Asian tropics is the development of hill resorts and associated transport infrastructure, such as roads. Subsequent erosion of vegetation cover and soils can be serious. Stringent watershed protection policies may serve to ensure conservation of montane ecosystems in such situations, but key areas with forest of exceptional stature and diversity, such as on the Pinosuk Plateau of G. Kinabalu, have already suffered considerable losses through resort development. A number of moth species have been recorded only from the Pinosuk Plateau, and their status must be regarded as imperilled.

References

Chey, V. K., Speight, M. R., and Holloway, J. D. (1992). Comparison of biodiversity between rain forest and plantations in Sabah, using insects as indicators. *Proceedingss of Oxford Conference on Tropical Forests 1992* (ed. F. R. Miller and K. L. Adam), pp. 221–4.

Holloway, J. D. (1983–1989). *The moths of Borneo*. Part 4 (1983), Notodontidae, *Malayan Nature Journal*, **37**, 1–107. Part 14 (1985) Noctuidae (part), *Malayan Nature Journal*, **37**, 157–317. Part 1 (1986), Key to families, Cossidae, Limacodidae etc., *Malayan Nature Journal*, **40**, 1–166. Part 3 (1987), *Bombycoidea*, Southdene, Kuala Lumpur. Part 6 (1988), *Arctiidae* (part), Southdene, Kuala Lumpur. Part 12 (1989), Noctuidae (part), *Malayan Nature Journal*, **42**, 57–226.

Holloway, J. D. (1987). Lepidoptera patterns involving Sulawesi: what do they indicate of past geography? In *Biogeographical evolution of the Malay Archipelago*, (ed. T. C. Whitmore), pp. 103–18. Clarendon Press, Oxford.

Holloway, J. D. (1989). Moths. In *Tropical forest ecosystems of the world* (ed. H. Lieth and M. J. A. Werger) **14B**, pp. 437–53. Elsevier, Amsterdam.

Holloway, J. D. and Barlow, H. S. (1992). Potential for loss of biodiversity in Malaysia, illustrated by the moth fauna. In *Pest management and the environment in 2000* (ed. Abdul S. A. Aziz Kadir and H. S. Barlow), pp. 293–311. CAB International, Wallingford.

Holloway, J. D. and Stork, N. E. (1991). The dimensions of biodiversity: the use of invertebrates as indicators of man's impact. In *The biodiversity of microorganisms and invertebrates: its role in sustainable agriculture* (ed. D. L. Hawksworth), pp 137–62. CAB International, Wallingford, UK.

Holloway, J. D., Kirk-Spriggs. A. H., and Chey, V. K. (1992). The response of some rain forest insect groups to logging and conversion to plantation. *Philosophical Transactions of the Royal Society of London* B, **335**, 425–36.

15. Global patterns of endemism and the conservation of biodiversity

S. J. THIRGOOD and M. F. HEATH
BirdLife International (formerly International Council for Bird Preservation), Wellbrook Court, Girton Road, Girton, Cambridge CB3 0NA, UK

Abstract

The approach to conservation of biodiversity by the International Council for Bird Preservation Biodiversity Project is described. The assumption is made that centres of endemism are priorities for the conservation of biodiversity. Birds were selected as a high priority target group, since most species are known and described, and are present throughout all major ecosystems of the world. Distributions of all species with breeding ranges restricted to 50 000 km^2, or less, were chosen as the study group. The analysis revealed that 2 497 species (26 per cent of all birds) were shown to occur in 221 endemic bird areas (EBAs) which cover about 5 per cent of the world's land area. Two questions were asked of the results: (1) Is there congruence of patterns of endemism between different animal and plant groups? (2) Are endemic birds good indicators for areas of endemism in other plant and animal groups? Through a regional assessment of the 221 EBAs it is shown that there is good congruence between global patterns of endemism in birds and other taxa, although exceptions were noted for the several groups of organisms in the UK and plants in the neotropics. It is argued that the 221 EBAs should be conserved as a first priority, and that the International Council for Bird Preservation Biodiversity Project is justifiable since collection of more sophisticated data will take longer than conservation can wait.

Introduction

The conservation of biodiversity is one of the major global environmental concerns of governmental and non-governmental conservation organizations, particularly as warnings of imminent mass species extinctions are

Systematics and Conservation Evaluation (ed. P. L. Forey, C. J. Humphries, and R. I. Vane-Wright), Systematics Association Special Volume No. 50, pp. 207–27, Clarendon Press, Oxford, 1994.

being recognized as based on good information, and not merely speculation. Biodiversity is not distributed evenly, and therefore an efficient approach to its conservation is to identify areas whose protection would ensure the continued survival and evolution of a high proportion of the world's biota. But how should these areas be selected?

One criterion which could be applied is species richness, exemplified by the 'megadiversity countries' concept of Mittermeier (1990). However, as Lovejoy and Oren (1981) have stated 'pursuit of the objective of maximum species diversity or even maximum species richness could lead to serious negative consequences if taken literally'. Many important components of biodiversity are not represented in countries or areas of high species richness or diversity, whilst a large number of the species they contain may be widespread and highly adaptable, under no immediate threat, and of no great conservation concern. Indeed, as Diamond (1976) stated, 'the question is not which reserve system contains more species, but which contains more species that would be doomed to extinction in the absence of refuges'.

Any strategy aimed at conserving biodiversity will need to consider places which hold concentrations of endemic species, i.e. species with limited distributions, as the loss of these areas would result in global extinctions. Studies which have incorporated both species richness and endemism include the botanical 'hotspots' analysis of Myers (1988, 1990) and 'critical faunas analysis' (Ackery and Vane-Wright 1984; Vane-Wright *et al.* 1991). Given the aim of conserving the full species complement of a certain taxonomic group, critical faunas analysis identifies the minimum set of areas which contain at least one viable population of every species. Like most methods, critical faunas analysis demands full knowledge of species' distributions for a certain taxon. Most of the world's species are not yet described or named, so this type of analysis is currently feasible for only a limited number of taxa.

The International Council for Bird Preservation's Biodiversity Project (ICBP 1992) has applied the idea that centres of endemism are priorities for conservation. The project has identified centres of avian endemism through the compilation of a data set more complete than is currently possible for any other life form. Birds are a group which are both taxonomically and distributionally well known, and additionally have dispersed to, and diversified in all regions of the world and virtually all terrestrial habitat types and altitudinal zones, thus enabling a global perspective. The distributions of all of the world's restricted-range bird species, defined as having breeding ranges below 50 000 km^2, were accurately mapped and analyzed to identify all areas with important concentrations. The analysis showed that 2 497 species, 26 per cent of the world's birds, are restricted to 221 Endemic Bird Areas (EBAs), which cover five per cent of the world's land area.

The ICBP study on birds also addressed two very important and related questions: (1) Is there congruence of patterns of endemism between different animal and plant groups? (2) Are endemic birds good indicators for areas of endemism in other vertebrate, invertebrate, and plant taxa? In other words, is there justification for extrapolating from birds to other life-forms for which the data collection will be a time-consuming challenge? In this paper we examine the general issue of whether patterns of endemism are similar between taxa, and illustrate this with a summary of the review of patterns of endemism in plants, invertebrates, and vertebrates presented in ICBP (1992). We then examine how the areas of endemism identified by ICBP for birds correspond to areas of importance for other groups, and discuss the implications for conservation of using birds as indicators of endemism and hence biodiversity.

Global patterns of endemism: a review

1. *Analytical approach*

During ICBP's research on biodiversity, information was assembled from a review of the available literature on patterns of endemism for terrestrial vertebrates, invertebrates, and plants. Where known, the number of species confined to a discrete geographical range was recorded. In some cases, areas of endemism were identified on the basis of a specialist's knowledge of a particular region and do not include information on the number of endemic species. An attempt was made to incorporate data from several different life-forms in all regions of the world, and for the purpose of data presentation the world was subdivided into six regions: Caribbean islands, Middle, and North America; South America; Africa, Europe and the Middle East; Continental Asia; South-east Asian islands, New Guinea and Australia; and the Pacific islands. More extensive data and complete references are given in ICBP (1992).

The EBAs as identified by ICBP (1992) were used as a global framework to allow a comparison of patterns of endemism between different taxa. These EBAs are defined as areas in which two or more restricted-range bird species are confined. Additional restricted-range bird species may occur within but not be confined to an EBA. EBAs are found on isolated islands or comprise pockets of a particular habitat, especially in tropical forest zones. Regional EBA maps are given in ICBP (1992), for example Africa (Fig. 15.1).

(a) Africa. In continental Africa distributional data comparable to that available for birds are available only for mammals, amphibians, and reptiles. Patterns of endemism amongst these vertebrate groups demonstrate, on the whole, a high level of congruence, with the majority of identified areas of endemism being in the tropical region of Africa (Fig. 15.2).

The Afrotropical mountains are important centres of endemism for

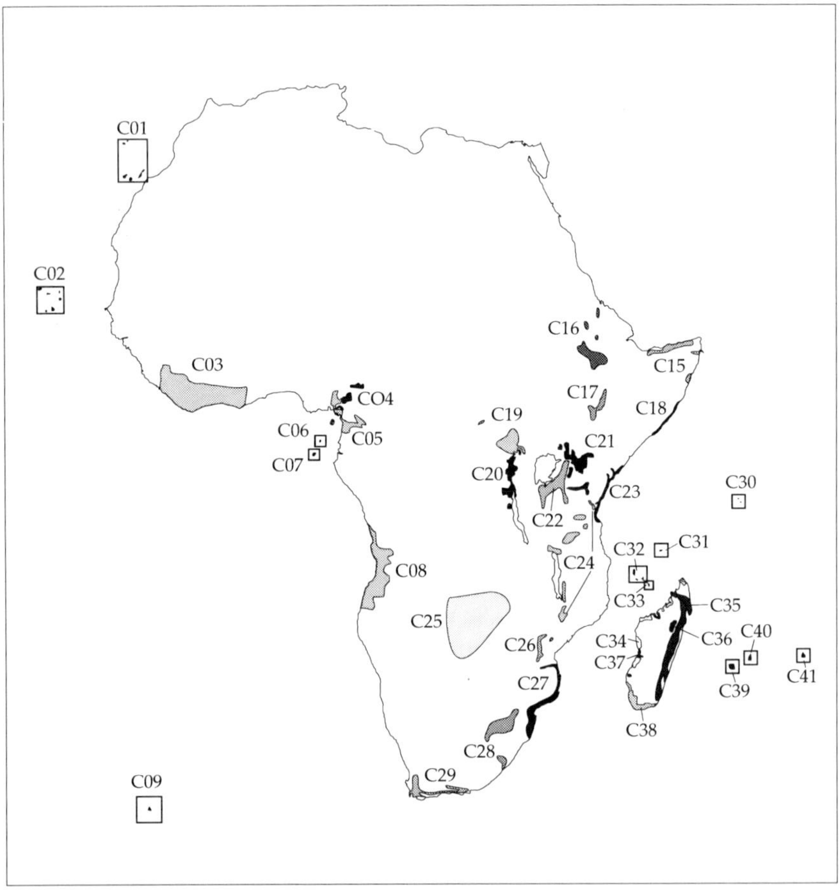

Fig. 15.1 Endemic Bird Areas of Africa. Similar maps are available for other regions of the world in ICBP (1992).

amphibians (Groombridge 1987) as well as birds. Of the four most important areas for birds in terms of restricted-range species richness, three areas, the Cameroon mountains, the Albertine Rift mountains, and the Eastern Arc mountains, also contain the largest concentrations of endemic amphibians. The locations of recognized concentrations of endemic mammals (excluding bats) also correspond closely with the distribution of EBAs. Congruence between birds and mammals is not perfect, however, since the three richest areas for birds are not the same as the three richest areas for mammals, although all six areas are important for both groups (Table 15.1). Additionally, several more widespread bird and mammal species are endemic to the arid habitats found in the north-east and south-west of the continent (Crowe and Crowe 1982; Kingdon 1990; Stuart *et al.* 1990).

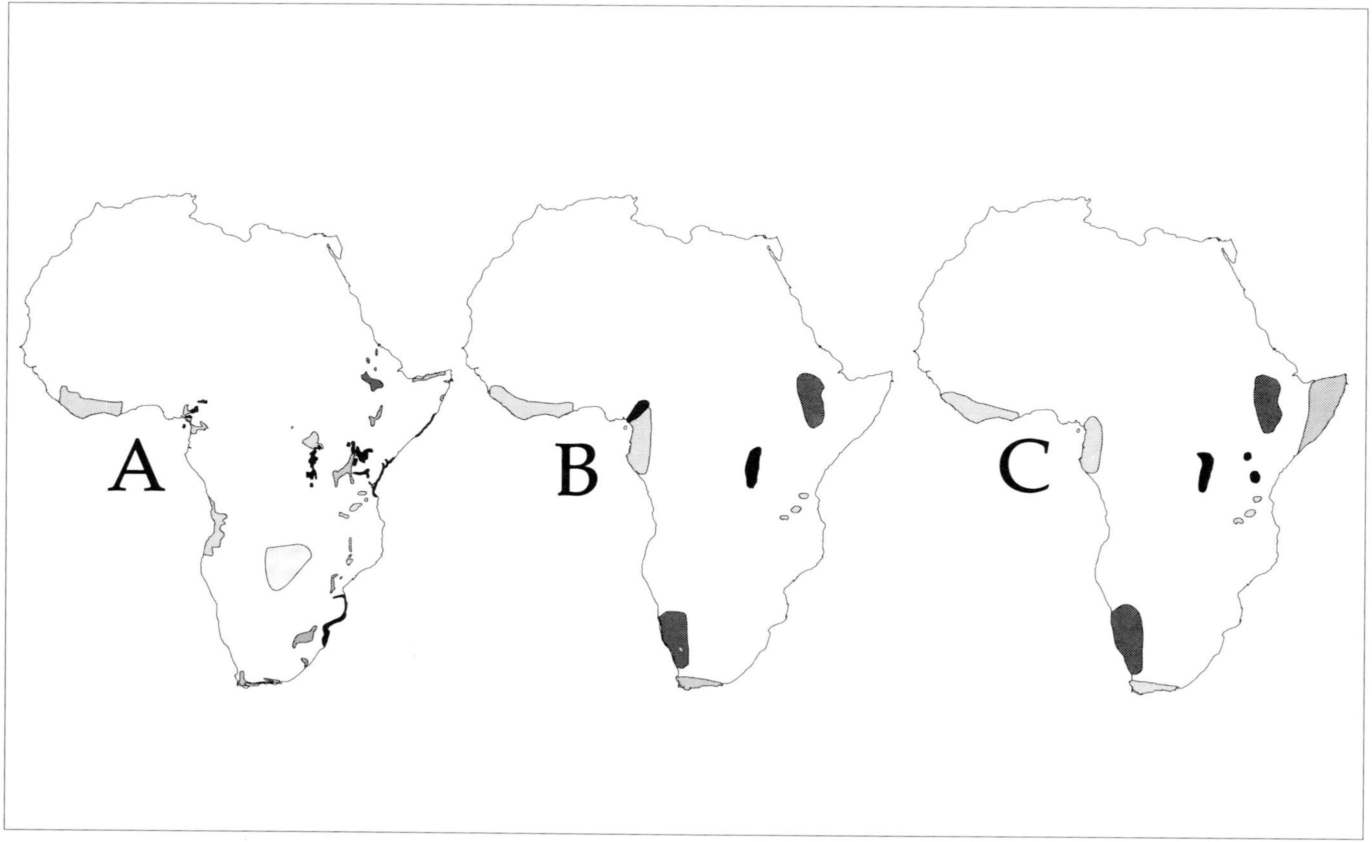

Fig. 15.2 Congruence of areas of endemism for Africa. A = birds; B = amphibians and C = mammals.

The phytogeography of Africa is comparatively well documented (White 1983), but areas of floristic endemism have not been fully identified, and current research initiatives are concentrating on quantifying floristic richness and endemism at local levels (IUCN/WWF in prep.). It is only possible to state in general terms that there appears to be good congruence between botanical and avian endemism. However, in some areas of endemism there are differences in the importance of the area between taxa. For example, the importance of endemism and species richness in the Cape flora, with 6 800 endemic species and 70 per cent endemism, greatly exceeds that of the vertebrates, including birds (Gibbs-Russell 1985).

Table 15.1 Comparison of endemism between birds, other animals, and plants in Africa. Bi = Birds; Ma = Mammals excluding bats; Am = Amphibians; Re = Reptiles; Mo = Terrestrial Molluscs; Bu = Butterflies; Pl = Vascular Plants. Figures refer to numbers of endemic species. Similar data are available for the other regions of the world in ICBP (1992).

	Area code and name	Bi	Ma	Am	Re	Mo	Bu	Pl
C01	Canary Islands and Madeira	10	0			334		631
C02	Cape Verde Islands	4	0	0	12			92
C03	Upper Guinea forest	6	20	17				E
C04	Cameroon mountains	28	3	60				1600
C05	Cameroon and Gabon lowlands	6	11	20				
C06	Príncipe	12	0	5	7			35
C07	São Tomé	21	1					108
C08	Angola	15	0					
C09	Tristan da Cunha Islands	6	0	0	0			40
C14	Socotra	6	0		13			215
C16	Central Ethiopian highlands	4	19	20	8			E
C17	South Ethiopian highlands	4						
C15	North-east Somalia	5	11		85			500
C18	Central Somalian coast	2						
C19	East Zairean lowlands	5	8					E
C20	Albertine Rift mountains	40		34				
C21	Kenyan mountains	7	6	24	14			E
C24	Eastern Arc mountains	30						
C22	Serengeti	3						
C23	Kenyan and Tanzanian coastal forest	8	3	2				100
C25	South Zambia	2	0					
C26	East Zimbabwean mountains	4	0					

Table 15.1 (*cont.*)

	Area code and name	Bi	Ma	Am	Re	Mo	Bu	Pl
C27	South-east African coast	3	0					
C28	South-east African grasslands	2	0					
	Namib-Karoo		11	11	14			2000
C29	Cape region	4	14	18	40			5800
C30	Granitic Seychelles	10	0	11	11			90
C31	Aldabra	4	0	0	0			43
C32	Comoro Islands	13	0	0	11			136
C33	Mayotte	6	0	0				
C34	West Madagascan dry forest	4						
C35	East Madagascan humid forest	18						
C36	Central Madagascan lakes	2	80	148	230	361	182	8000
C37	West Madagascan coastal wetlands	2						
C38	South Madagascan Didiera scrub	8						
C39	Réunion	7	0	0	2			150
C40	Mauritius	10	0	0	10			300
C41	Rodrigues	2	0	0	0			40

The high levels of endemism in the flora and fauna of Madagascar are well known. Phytogeographically, the island has been divided into two regional centres of endemism with the eastern humid forests being particularly rich in endemic plants. Many of the island's endemic birds, mammals, and amphibians are also confined to these forests. Endemism amongst the invertebrate fauna parallels that of the vertebrate groups, with for example some 70 per cent of the butterflies (182 species) and 95 per cent of the terrestrial molluscs (361 species) being unique to the island (Table 15.1: also Jenkins 1987).

In common with most oceanic islands, the vertebrate faunas of the islands of the Atlantic and Indian Oceans are depauperate. Long isolation has, however, resulted in the development of some significant endemic faunas (Stuart *et al.* 1990). Those islands rich in endemic birds, namely the Canaries and Madeira, Príncipe and São Tomé, the Seychelles, the Comoros, and Mauritius, also contain a number of island endemics from other groups, however mammals (excluding bats) are poorly represented on these islands and few endemic species are found (Kingdon 1990).

(b) Caribbean islands, Middle, and North America. Patterns of endemism in the flora and fauna of the Caribbean Islands are relatively well documented (e.g.

Johnson 1988; CCA/IRF 1991). The general trend for birds, reptiles, and amphibians is for the occurrence of a large number of single-island or island-group endemic species on the four Greater Antillean islands of Cuba, Jamaica, Hispaniola, and Puerto Rico, each of these islands being an EBA. Jamaica has the greatest number of single-island endemics with 26 birds and 47 amphibians and reptiles. Whilst the individual island groups of the Lesser Antilles have relatively few single-island endemics, a large element of the Eastern Caribbean avifauna and herpetofauna is endemic to the Lesser Antilles as a whole. Similar patterns were shown in the Caribbean mammal fauna of historic times (Woods 1989), but many of these species are now extinct. Zoogeographic documentation of Caribbean insects is comparatively advanced (Liebherr 1988), and the patterns of insect endemism for single islands and island groups are similar to those identified for the main vertebrate groups.

There is less information available regarding endemism in the flora of the Caribbean, particularly amongst the smaller islands of the Lesser Antilles. The four Greater Antillean islands each contain a number of single-island endemics, with the endemic flora of the islands of Cuba (3000 species) and Hispaniola (1800 species) being of particular note (Davis *et al.* 1986).

There are 21 EBAs in Mexico and Central America. Comprehensive overviews for non-bird taxa encompassing the whole region do not exist, although valuable comparisons between taxa can be made within more limited areas. For instance, there are striking similarities between areas of endemism in the herpetofauna (Savage 1982) and avifauna of Central America and Mexico, with congruence occurring in the Yucatán Peninsula, northern central American highlands, Isthmus de Tehuantepec, northern central American Pacific slope, Costa Rican and Panamanian highlands, Caribbean lowlands of Nicaragua and Costa Rica, Darién highlands, and the North Chocó and Darién lowlands.

Areas of endemism identified in the Central American region for heliconiine and ithominiine butterflies (Brown 1987*a*) correspond reasonably well with concentrations of restricted-range birds. More recent work on the butterfly fauna of Costa Rica suggests similar congruence where areas of butterfly and avian endemism are coincident (Thomas 1991). Importantly, Thomas (1991) further suggests that 'conservation of unmodified habitats based on a thorough knowledge of plant and vertebrate biogeography would also meet the needs of most butterflies'.

The richness and high levels of endemism in the Mexican flora have been highlighted in the 'megadiversity' analysis of Mittermeier (McNeely *et al.* 1988), and in Panama and Costa Rica the level of endemism is expected to exceed 20 per cent of the total flora (Gentry 1986). However, the boundaries of these areas of endemism have not been adequately defined to allow comparison with the EBAs. The Mediterranean habitats of California hold large concentrations of endemic plants (1517 species) (Gentry 1986) and overlap with the only EBA identified in North America.

(c) South America. Most biogeographical studies of South America have concentrated on the lowland biota, particularly of Amazonia, where patterns of endemism have been presented for birds (Haffer 1987), lizards of the genus *Anolis* (Vanzolini and Williams 1970), heliconiine and ithominiine butterflies (Brown 1987*a*), and trees of the families Bignoniaceae and Chrysobalanaceae (Prance 1987). A high degree of concordance between the location of proposed centres of endemism in each of these groups has been suggested (Simpson and Haffer 1978; Brown 1987*b*; but see Bevan *et al.* 1984).

The lowland forests of the Atlantic coast of Brazil are particularly notable for high degrees of endemism in both plant and animal species (Jackson 1978; Mori *et al.* 1981; Duellman 1982; Cracraft 1985; Brown 1987*a*; Mittermeier 1987); indeed five EBAs are located there. The lowland forests in the Chocó region of Colombia and Ecuador are also important being amongst the most species-rich of all known plant communities, with an estimated 10 000 species, of which a quarter are endemic (Gentry 1992). Gentry (1992) documented the flora of isolated ridges in western Ecuador and suggested that the very high local endemism—an estimated 20 per cent of vascular plants below 900 m are endemic to the region—is due to speciation of epiphytes, palms, and shrubs, possibly related to the broken nature of the terrain and the complex interspersion of different climates and vegetation types. There is also a high level of avian endemism in the Pacific lowlands and foothills of the Chocó, with over 50 restricted-range species confined to the area, which is matched in the reptiles and amphibians (Lynch 1979), and butterflies (Brown 1987*a*).

The biota of the submontane and montane habitats along the Andean chain remain one of the most poorly documented in all the Neotropics, but species richness and local endemism are thought to be high. There are 22 EBAs and 309 restricted-range species confined to the Andes, and this complex avian distribution pattern, particularly in the northern Andes, is well matched by the distribution of the locally endemic herpetofauna (Duellman 1979). The flora of the eastern slopes of the Andes has been identified by Myers (1988) as a global floristic 'hotspot'.

(d) Asia. Analysis of endemic congruence in Asia is severely hindered by the lack of reliable distributional data for virtually all taxa other than birds. Comprehensive data-sets are available for most of the island groups in the region, but with few exceptions, equivalent data are not available for continental areas.

Sri Lanka, Taiwan, the archipelagos of the Andamans and Nicobar

Islands, and the Nansei Shoto have all been identified as EBAs, and are important for endemics in other vertebrate taxa (Collins *et al.* 1991). Knowledge of invertebrates on these islands is invariably poor, although the critical faunas analysis of Collins and Morris (1985) identifies the endemic swallowtail butterfly faunas of Taiwan, and the Andamans and Nicobars as of particular importance. Floristically, each of the main island groups contains many endemics (Davis *et al.* 1986), this being particularly well documented in the case of the south-western moist forests of Sri Lanka (Gunatilleke and Gunatilleke 1990).

On the Indian subcontinent there are four principal areas of endemism, all identified as EBAs: the Western Ghats; the state of Assam; and the western and eastern Himalayas (World Conservation Monitoring Centre 1989; Collins *et al.* 1991; ICBP 1992). The evergreen forests of the Western Ghats are rich in endemic fauna, with 16 endemic birds, 16 mammals, 17 reptiles, and 84 amphibians (Groombridge 1983), and also hold several distinct centres of local floristic endemism (Nair and Daniel 1986). Invertebrate distributions in India are generally poorly known, but the Western Ghats contain two endemic swallowtail butterflies (Collins and Morris 1985), and two endemic milkweed butterflies (Ackery and Vane-Wright 1984). Both the western and eastern Himalayas and the state of Assam have been identified as centres of floristic endemism (Myers 1988; World Conservation Monitoring Centre 1989), but the precise geographical boundaries of these areas are unclear, and estimates of the size of the endemic flora vary, thus preventing detailed comparison with data available for endemic birds.

The endemic flora and fauna of China is of global significance, and yet is exceptionally poorly known to Western science (Davis *et al.* 1986). The country has 12 EBAs, of which several are considered to be important for other life-forms. High floristic endemism is suggested in the Sichuan and Yunnan Mountains (Ying Tsun-Shen and Zhang Zhi-Song 1984). The Central Sichuan mountains have several species of endemic primates (Eudey 1987), and the Giant Panda *Ailuropoda melanoleuca*, whose distributions closely correspond to those of the endemic birds of the area.

Patterns of endemism in much of Indochina are also very poorly documented. The level of endemism of flora and fauna of Burma, Thailand, Laos, and Cambodia, whilst incompletely known, is believed to be relatively low (Collins *et al.* 1991). In contrast, endemism in the Vietnamese birds is high, with EBAs in the Annamese lowlands, Da Lat Plateau, and Cochinchina, although reliable distributional data are lacking for virtually all other groups. Floristic endemism is particularly high in the alpine zone of the Hoang Lien Son mountains in northern Vietnam, the rainforests of central Vietnam, and the montane forests of the Da Lat Plateau (Anon 1986).

(e) South-east Asian islands, New Guinea and Australia. Patterns of endemism within the region largely result from insular speciation and much of the available data are in the form of lists of single-island or island-group endemics. Distributional data comparable to the avifauna are available for mammals, reptiles, amphibians, swallowtail butterflies, and vascular plants.

Endemism in the avifauna of the islands of the Sunda Shelf, Wallacea, and the Philippines is high, with 25 EBAs and over 450 restricted-range bird species, which display considerable congruence with the patterns of endemism in other taxa. Single-island or island-group endemic birds, mammals, and reptiles are found throughout the region, particularly in the Philippine archipelago and the larger islands of Sulawesi, Borneo, Sumatra and, to a lesser extent, Java (MacKinnon and MacKinnon 1986; Hauge *et al.* 1986; Cox 1988). Endemism amongst the vertebrates of the smaller islands of the Lesser Sundas and Moluccas is less pronounced both proportionally and in terms of absolute numbers. Comparable invertebrate data-sets are provided for milkweed and swallowtail butterflies by Ackery and Vane-Wright (1984) and Collins and Morris (1985). These analyses highlight the importance of the butterfly faunas of the Philippines, Sulawesi, and the smaller islands in the Lesser Sundas and Moluccas.

Floristic endemism is particularly high on the Sunda Shelf islands of Sumatra and Borneo, the islands of the Philippine archipelago, and in Peninsular Malaysia (Davis *et al.* 1986). The lowland forests of Sumatra and Borneo are the centre of evolution for the Dipterocarpaceae, whilst many mountains of the region are local centres of floristic endemism, with Mount Kinabalu in Sabah the best-known example. The flora of Peninsular Malaysia is particularly well studied and various estimates of endemism have been made: 30 per cent for tree species, 90 per cent for the Bignoniaceae, 80 per cent for the Gesneriaceae, and 50 per cent for the orchids (Earl of Cranbrook 1988). The high floristic endemism is in marked contrast to the peninsula's vertebrate and invertebrate fauna, much of which is shared with Sumatra and Borneo (MacKinnon and MacKinnon 1986).

New Guinea supports a remarkable diversity of animals and plants, and is particularly rich in endemic forms in all taxa for which records exist (Diamond 1986). The island itself, with its complex topographic features and diversity of ecosystems, contains many centres of endemism. These areas can be broadly divided into the western and eastern Papuan islands, the central cordillera, the isolated north coast mountains, and the intervening areas of lowland. Distributional data necessary to identify these centres of endemism are currently available only for birds, although there is qualitative documentation of similar patterns in other taxa (Gressitt 1982).

Patterns of endemism within the Australian continent are already well known, with endemic congruence in all taxa for which information is available (Keast 1981; Heatwole 1987). Restricted-range species of birds, mammals, reptiles, amphibians, and terrestrial molluscs are distributed peripherally in the north-west, north-east, south-west, and south-east of the continent. Endemism in the flora shows broadly similar trends, with the south-west and south-east regions particularly rich in locally endemic forms (Davis *et al.* 1986).

(f) Pacific islands. The Pacific region's importance in terms of global biodiversity conservation has, until recently, received little attention (Dahl 1986). Documentation of the biological resources of the region, in particular of the invertebrates and plants, is very incomplete, thus hampering an assessment of the extent of endemic congruence between taxa. There are 30 EBAs in the region, but the distributional data available for particular island groups for other vertebrates, terrestrial molluscs, Lepidoptera, and vascular plants are limited.

The Bismarck and Solomon archipelagos typify the difficulties of analyzing patterns of endemism in the region. These islands are of exceptional importance in terms of their endemic avifauna, encompassing eight EBAs which support over 150 restricted-range bird species. Unfortunately, information regarding other taxa is scarce. It has been suggested that as many as 23 mammals are confined to the two archipelagos (Musser 1987), although corroboration and further details are not available. Endemism in the reptile, amphibian, terrestrial mollusc, and butterfly faunas is thought to be high in the Solomons (Ackery and Vane-Wright 1984; Groombridge 1985; Dahl 1986), but similar data are unavailable for the Bismarcks. The size and endemicity of the flora are unknown for either archipelago (Davis *et al.* 1986).

More detailed information is available for the island groups of Vanuatu, Fiji, and New Caledonia. Each of these groups are EBAs, containing a rich endemic avifauna and display similar levels of endemism in several other taxa. Whilst the mammal faunas of these oceanic islands are depauperate, high levels of endemism are shown in the herpetofauna (Gibbons 1985; Groombridge 1985; Dahl 1986). The terrestrial mollusc and lepidoptera faunas, though imperfectly known, appear to have undergone extensive radiation and contain many endemics (Holloway 1984; Solem 1984; Wells 1985). In addition, New Caledonia has a particularly distinctive and ancient flora associated with the unusual geology of the island (Holloway 1979), and with 1400 endemic species it has the highest known plant endemism in the world at 89 per cent (Gentry 1992).

Hawaii is especially famous for its endemic flora and fauna with several taxa including birds, terrestrial molluscs, arthropods, and angiosperms having undergone adaptive radiation and displaying uniquely high levels of endemism (Loope *et al.* 1988). Radiation on a smaller scale has been documented on several isolated South Pacific island groups. The terrestrial mollusc faunas of the Society, Austral, Marquesas, and Tuamotu archipelagos are particularly rich and contain many endemics, although recent large-scale extinctions severely depleted several islands (Wells 1985; Dahl 1986).

Birds as indicators of areas of endemism

The 221 EBAs are classified according to the degree of endemic congruence they display for the three major terrestrial life-forms, namely vertebrates (excluding birds), invertebrates, and vascular plants, all based on data extracted from the literature review. Since the global numbers of known species differ by orders of magnitude across taxa, it is necessary to set different thresholds for what is termed 'significant' or 'globally significant' endemism (Table 15.2). Significant endemism refers to 50 endemic species for vertebrates, and 500 species for invertebrates and vascular plants; globally significant endemism refers to 100 endemic species for vertebrates, and 1000 species for invertebrates and vascular plants. These criteria are simply an initial suggestion and remain open to revision. If the insects were more fully known, higher thresholds would almost certainly be appropriate. Knowledge of certain taxa varies considerably in different regions of the globe, but assessment of comparable levels of significance would involve

Table 15.2 Criteria for the scoring of Endemic Bird Areas for plants and animals other than birds.

Score	Criteria
***	Significant endemism in at least two of the three major terrestrial-life forms (vertebrates, invertebrates, plants) or globally significant endemism for one major life-form. ***? is used if this level is probable from described endemism in the region but quantitative data applying to the same boundaries do not exist.
**	Significant endemism in at least one of the three major terrestrial life-forms (vertebrates, invertebrates, plants). **? is used if this level is probable but quantitative data applying to the same boundaries do not exist.
*	Some endemism among other vertebrates, invertebrates, or plants, with levels below the critical thresholds above.
?	Insufficient data to assess levels of endemism amongst other taxa.

unacceptable subjectivity. Quantitative information is not always available for the number of species confined to a particular area of endemism described in the literature, but uncertainty will be reduced by further analysis and publication of data for other groups.

The congruence rating assigned to each EBA was compared to the importance of the EBA for numbers of restricted-range bird species. Three categories of restricted-range bird species richness were calculated by fitting a regression line to the log-log graph of species number against area and seeing how each EBA compared with what would be expected for its area. EBAs with more than twice as many species as expected are classified as the richest (***), and those with fewer than the expected rank as the poorest (*) for bird richness.

The analysis shows that overall there is positive congruence amongst vertebrates, invertebrates, and plants: over 50 per cent of EBAs show globally significant endemism or significant endemism in at least one of these major life-forms (Table 15.3). The lack of data is evident, with 20 per cent of the EBAs having insufficient data to assess levels of endemism for taxa other than birds, though most of these areas have the lowest level of endemic bird species richness (Table 15.3). The locations of the EBAs and the congruence rating assigned to them is shown in Fig. 15.3. The comparison of bird species richness and importance for numbers of species from other taxa shows that although many EBAs important for high numbers of restricted-range bird species are important in terms of numbers of endemics from other taxa, this is not always the case. For example, 19 EBAs are exceptionally important for

Table 15.3 Restricted-range bird species richness scores in relation to other taxonomic group scores for Endemic Bird Areas. See text and Table 15.2 for scoring.

Other group score	Restricted-range bird species richness ***	**	*	Total
/?	24	27	33	84
/?	7	16	14	37
/?	19	19	19	57
?	3	6	34	43
Total	53	68	100	221

Fig. 15.3 Endemic Bird Areas scored for congruence of patterns of endemism for plants and animals other than birds. Square = ***; circle = **; diamond = *; triangle = ? See Table 15.2 for scoring.

a large number of restricted-range birds but less important for other taxa, and conversely 33 EBAs appear to be more important for taxa other than birds (Table 15.3).

Discussion

The literature review suggests that there is good congruence between global patterns of endemism for different life forms, supporting work by others (Simpson and Haffer 1978; Hauge *et al.* 1986; Collar and Stuart 1988; Gentry 1992). Our results show that the majority of areas identified on the basis of their restricted-range birds are also important for plant and other animal groups. Furthermore, on the basis of currently available information, it seems unlikely that many major terrestrial areas of high endemism fall outside of the 221 EBAs. Thus birds are good indicators of areas of endemism in other taxa.

A seemingly conflicting conclusion has been drawn from a recent analysis of species richness in Britain by Lawton *et al.* (this volume), who suggested that congruence between taxa exists only on a very coarse scale, and that national and local hotspots of species richness do not coincide across taxa. Two comments can be made on this apparent contradiction. Firstly, Lawton *et al.* are concerned with species richness and not endemism; and second, their analysis is based on a northern temperate country where restricted areas of endemism are not typically found, and where habitats are extensively modified and the flora and fauna much altered by human impact.

The identification of the EBAs was based on restricted-range criteria, hence they generally occupy relatively small areas (85 per cent are below 50 000 km^2). However, areas of endemism occur at a variety of different scales, therefore there are some larger areas of endemism, to which more widely distributed bird species and other taxa are restricted, that were not considered under the criteria of the EBA analysis. These larger areas include the Amazon and Congo basins, the Namib-Karoo deserts, and the Greater Sundas, where endemics are not as concentrated and vulnerable as in the EBAs, but undoubtedly also need the attention of conservationists.

Conversely there are some taxa, for example plants, that have large numbers of species with extremely localized distribution patterns (Gentry 1986, 1992). In many cases, these small areas of extremely local endemism will be located within larger areas of avian endemism. For example, the cloud-forest at Centinela in Ecuador, documented by Gentry (1986, 1992) as containing levels of local endemism of 10–24 per cent (prior to its destruction), is located within the western side of the Andes of Colombia and Ecuador EBA. Similarly, the results of the Centres of Plant Diversity project (IUCN/WWF in prep.), a global study based on both quantitative and qualitative data, suggests that most areas of local floristic endemism on the

African continent are contained within the framework of the African EBAs, one major exception being the European Mediterranean region (World Conservation Monitoring Centre 1992). Thus, information on locally endemic floras will be of particular importance in identifying the most important areas and sites within EBAs.

The approach followed in the ICBP study, which focused on the identification of aggregations of restricted-range bird species and has been shown to reflect endemism in other taxa, is undoubtedly a sensible and data efficient way of identifying key areas where species should be conserved so as to maintain overall global biodiversity. Gentry (1992) suggests that for Neotropical plants at least, patterns of endemism and diversity are largely non-coincident. In terms of conservation priorities however, concentration on restricted-range or endemic species is extremely important irrespective of the relationship between endemism and overall diversity. The conservation of areas with concentrations of restricted-range species will help to ensure the survival of a unique complement of species, many of which are threatened by virtue of the vulnerability of small areas to extensive habitat modification. Indeed, any strategy that embraces a minimum complementary set of sites to represent the world's biodiversity will have to include coverage of these relatively small centres of endemism. Adequate protection of the 221 EBAs identified by ICBP would contribute to the survival of a disproportionately wide variety of birds and other life. Alternatively, massive extinction could be caused by modification or destruction of habitat in these most valuable places. There is an absence of data from many taxa from several regions of the world, but the collection of this data will take more time than conservation can afford to wait. As Erhlich (1992) has commented, 'it is critical that conservation biologists build on what is already understood about well-known organisms to develop tools that can be used to save biodiversity quickly and en masse'.

Acknowledgements

Our thanks go to our co-workers Colin Bibby, Nigel Collar, Mike Crosby, Adrian Long, Christoph Imboden, Alison Stattersfield, and David Wege at BirdLife International, and Tim Johnson of the World Conservation Monitoring Centre, all of whom have been involved with the Biodiversity Project at various stages, and who made extensive comments on earlier drafts of this paper. Colin Bibby encouraged us to prepare the paper and contributed to data evaluation, and Adrian Long produced the camera-ready artwork.

References

Ackery, P. R. and Vane-Wright, R. I. (1984). *Milkweed butterflies*. British Museum (Natural History), London and Cornell University Press, Ithaca, NY.

Anon. (1986). *Viet Nam: national conservation strategy*. WWF-India, New Dehli.

Bevan, S., Connor, E. F., and Bevan, K. (1984). Avian biogeography in the Amazon basin and the biological model of diversification. *Journal of Biogeography*, **11**, 383–99.

Brown, K. S. Jr. (1987*a*). Biogeography and evolution of neotropical butterflies. In *Biogeography and quaternary history in tropical America* (ed. T. C. Whitmore and G. T. Prance), pp. 66–104. Clarendon Press, Oxford.

Brown, K. S. Jr. (1987*b*). Conclusions, synthesis, and alternative hypotheses. In *Biogeography and quaternary history in tropical America* (ed. T. C. Whitmore and G. T. Prance), pp. 175–96. Clarendon Press, Oxford.

Caribbean Conservation Association/Island Resources Foundation (1991). *Environmental profiles*. Caribbean Conservation Association.

Collar, N. J. and Stuart, S. N. (1988). *Key forests for threatened birds in Africa*. International Council for Bird Preservation, Cambridge, UK.

Collins, N. M. and Morris, M. G. (1985). *Threatened swallowtail butterflies of the world: the IUCN Red Data Book*. International Union for Conservation of Nature and Natural Resources, Gland, Switzerland.

Collins, N. M., Sayer, J. A., and Whitmore, T. C. (ed.) (1991). *The conservation atlas of tropical forests: Asia and the Pacific*. Macmillan, London.

Cox, R. (1988). The conservation status of biological resources in the Philippines. Unpublished report to IUCN-CMC, Cambridge, UK.

Cracraft, J. (1985). Historical biogeography and patterns of differentiation within the South American avifauna: areas of endemism. *Ornithological Monographs*, **36**, 49–84.

Crowe, T. M. and Crowe, A. A. (1982). Patterns of distribution, diversity and endemism in Afrotropical birds. *Journal of Zoology, London*, **198**, 417–42.

Dahl, A. L. (1986). *Review of the protected areas system in Oceania*. International Union for Conservation of Nature and Natural Resources/United Nations Environment Programme, Gland, Switzerland.

Davis, S. D., Droop, S. J. M., Gregerson, P., Henson, L., Leon, C. J., Villa-Lobos, J. L. *et al.* (1986). *Plants in danger: what do we know?* International Union for Conservation of Nature and Natural Resources, Gland, Switzerland.

Diamond, J. (1976). Island biogeography and conservation: strategy and limitations. *Science*, **193**, 1027–9.

Diamond, J. (1986). The design of a nature reserve system for Indonesian New Guinea. In *Conservation biology* (ed. M. E. Soulé), pp. 485–503. Sinauer, Sunderland, MA.

Duellman, W. E. (1979). The herpetofauna of the Andes: patterns of distribution, origin, differentiation and present communities. In *The South American herpetofauna: its origin, evolution, and dispersal* (ed. W. E. Duellman), pp. 371–459. University of Kansas Museum of Natural History (Monograph 7).

Duellman, W. E. (1982). Quaternary climatic-ecological fluctuations in the lowland tropics: frogs and forests. In *Biological diversification in the tropics* (ed. G. T. Prance), pp. 389–402. Columbia University Press, New York.

Ehrlich, P. R. (1992). Population biology of checkerspot butterflies and the conservation of global biodiversity. *Oikos*, **63**, 6–12.

Earl of Cranbrook (ed.) (1988). *Malaysia (key environments)*. Pergamon Press, Oxford.

Eudey, A. A. (1987). *Action plan for Asian primate conservation: 1987–1991*. International Union for Conservation of Nature and Natural Resources (IUCN/SSC Primate Specialist Group), Gland, Switzerland.

Gentry, A. H. (1986). Endemism in tropical versus temperate plant communities. In *Conservation biology* (ed. M. E. Soulé), pp. 153–81. Sinauer Press, Sunderland, MA.

Gentry, A. H. (1992). Tropical forest biodiversity: distributional patterns and their conservational significance. *Oikos*, **63**, 19–68.

Gibbons, J. R. H. (1985). The biogeography and evolution of Pacific island reptiles and amphibians. In *The biology of Australasian frogs and reptiles* (ed. G. Grigg, R. Shine, and H. Ehmann), pp. 125–42. Surrey Beatty, Chipping Norton, NSW, Australia.

Gibbs-Russell, G. E. (1985). Analysis of the size and composition of the South African flora. *Bothalia*, **15**, 613–29.

Gressitt, J. L. (1982). Zoogeographical summary. In *Biogeography and ecology of New Guinea* (ed. J. L. Gressitt), pp. 897–918. Junk, The Hague.

Groombridge, B. (1983). Comments on the rain forests of southwest India and their herpetofauna. Paper prepared for the Centenary Seminar of the Bombay Natural History Society. 6–10 December, 1983 (revised January 1984).

Groombridge, B. (1985). Amphibians and reptiles of Oceania. Unpublished report, World Conservation Monitoring Centre, Cambridge.

Groombridge, B. (1987). Amphibians of Africa: an annotated list of national and regional endemics. Unpublished report, World Conservation Monitoring Centre, Cambridge.

Gunatilleke, I. A. U. N. and Gunatilleke, C. V. S. (1990). Distribution of floristic richness and its conservation in Sri Lanka. *Conservation Biology*, **4**, 21–31.

Haffer, J. (1987). Biogeography of neotropical birds. In *Biogeography and quaternary history in tropical America* (ed. T. C. Whitmore and G. T. Prance), pp. 105–50. Clarendon Press, Oxford.

Hauge, P., Terborgh, J., Winter, B., and Parkinson, J. (1986). Conservation priorities in the Philippine Archipelago. *Forktail*, **2**, 83–91.

Heatwole, H. (1987). Major components and distributions of the terrestrial fauna. In *Fauna of Australia. Vol 1A. General articles* (ed. G. R. Dyne and D. W. Walton), pp. 101–35. Australian Government Publishing Service, Canberra.

Holloway, J. D. (1979). *A survey of the Lepidoptera, biogeography and ecology of New Caledonia*. Junk, The Hague.

Holloway, J. D. (1984). Lepidoptera and the Melanesian arcs. In *Biogeography of the tropical Pacific* (ed. F. J. Radovsky, P. H. Raven, and S. H. Sohmer), pp. 129–70. Bishop Museum, Honolulu.

ICBP (1992). *Putting Biodiversity on the map: priority areas for global conservation.* International Council for Bird Preservation, Cambridge.

IUCN/WWF (in prep.). Centres of plant diversity: a guide and strategy for their conservation.

Jackson, J. F. (1978). Differentiation in the genera *Enyalius* and *Strobilurus* (Iguanidae): implications for Pleistocene climatic changes in eastern Brazil. *Arquivos de Zoologia, São Paulo.*

Jenkins, M. D. (ed.) (1987). *Madagascar: an environmental profile.* International Union for Conservation of Nature and Natural Resources, Gland, Switzerland.

Johnson, T. H. (1988). *Biodiversity and conservation in the Caribbean: profiles of selected islands.* International Council for Bird Preservation, Cambridge.

Keast, A. (1981). Distributional patterns, regional biotas, and adaptations in the

Australian biota: a synthesis. In *Ecological biogeography of Australia* (ed. A. Keast), pp. 1893–997. Junk, The Hague.

Kingdon, J. (1990). *Island Africa: the evolution of Africa's rare animals and plants*. Collins, London.

Liebherr, J. K. (1988). Biogeographic patterns of West Indian *Platynus* carabid beetles (Coleoptera). In *Zoogeography of Caribbean insects* (ed. J. K. Liebherr), pp. 121–52. Cornell University Press, Ithaca, NY.

Loope, L. L., Harmann, O., and Stone, C. P. (1988). Comparative conservation biology of oceanic archipelagos. *BioScience*, **38**, 272–82.

Lovejoy, T. E. and Oren, D. C. (1981). The minimum critical size of ecosystems. In *Forest island dynamics in man-dominated landscapes* (ed. R. L. Burgess and D. M. Sharp), pp. 5–12. Springer-Verlag, New York.

Lynch, J. D. (1979). The amphibians of the lowland tropical forests. In *The South American herpetofauna: its origin, evolution, and dispersal* (ed. W. E. Duellman), pp. 189–215. University of Kansas Museum of Natural History.

MacKinnon, J. and MacKinnon, K. (1986). *Review of the protected areas system in the Indo-Malayan Realm*. International Union for Conservation of Nature and Natural Resources/United Nations Environment Programme, Gland, Switzerland.

McNeely, J. A., Miller, K. R., Reid, W. V., Mittermeier, R. A., and Werner, T. B. (1988). *Conserving the world's biological diversity*. IUCN, WRI, CI, WWF-US, WB, Gland and Washington.

Mittermeier, R. A. (1987). Framework for primate conservation in the Neotropical region. In *Primate conservation in the tropical rainforest* (ed. C. Marsh and R. A. Mittermeier), pp. 305–20. Alan R. Liss, New York.

Mittermeier, R. A. (1990). Wealth of plants and animals unites 'megadiversity countries'. *Tropicus*, **4**, 4–5.

Mori, S. A., Boom, B. M., and Prance, G. T. (1981). Distribution patterns and conservation of eastern Brazilian coastal forest tree species. *Brittonia*, **33**, 233–45.

Musser, G. G. (1987). The mammals of Sulawesi. In *Biogeographical evolution of the Malay Archipelago* (ed. T. C. Whitmore), pp. 73–93. Clarendon Press, Oxford.

Myers, N. (1988). Threatened biotas: 'hotspots' in tropical forests. *Environmentalist*, **8**, 187–208.

Myers, N. (1990). The biodiversity challenge: expanded hot-spots analysis. *Environmentalist*, **10**, 243–56.

Nair, N. C. and Daniel, P. (1986). The floristic diversity of the Western Ghats and its conservation: a review. *Proceedings of the Indian Acadamy of Science, (Suppl.)*, 127–63.

Prance, G. T. (1987). Biogeography of neotropical plants. In *Biogeography and quaternary history in tropical America* (ed. T. C. Whitmore and G. T. Prance), pp. 46–65. Clarendon Press, Oxford.

Savage, J. M. (1982). The enigma of the Central American herpetofauna: dispersal or vicariance? *Annals of the Missouri Botanical Garden*, **69**, 464–547.

Simpson, B. B. and Haffer, J. (1978). Speciation patterns in the Amazon forest biota. *Review of Ecology and Systematics*, **9**, 497–518.

Solem, A. (1984). A world model of land snail diversity and abundance. In *World-wide snails: biogeographical studies on non-marine Mollusca* (ed. A. Solem and A. C. van Bruggen), pp. 6–22. Brill, Leiden.

Stuart, S. N., Adams, R. J., and Jenkins, M. D. (1990). *Biodiversity in sub-Saharan*

Africa and its islands. International Union for Conservation of Nature and Natural Resources, Gland, Switzerland.

Thomas, C. D. (1991). Habitat use and geographic ranges of butterflies from the wet lowlands of Costa Rica. *Biological Conservation*, **55**, 269–81.

Vane-Wright, R. I., Humphries, C. J., and Williams, P. H. (1991). What to protect? Systematics and the agony of choice. *Biological Conservation*, **55**, 235–54.

Vanzolini, P. E. and Williams, E. E. (1970). South American anoles: geographic differentiation and evolution of the *Anolis chrysolepis* species group. *Arquivos de Zoologia, São Paulo*, **19**, 1–298.

Wells, S. (1985). Notes on endemic and threatened Pacific land snails. Unpublished report to World Consevation Monitoring Centre, Cambridge.

White, F. (1983). *The vegetation of Africa: a descriptive memoir to accompany the UNESCO/AERFAT/UNSO vegetation map of Africa*. United Nations Educational Scientific and Cultural Organisation, Paris.

Woods, C. A. (1989). The biogeography of West Indian rodents. In *Biogeography of the West Indies* (ed. C. A. Woods), pp. 741–98. Sandhill Crane Press, Gainsville, FL.

World Conservation Monitoring Centre (1989). India: conservation of biological diversity. Unpublished.

World Conservation Monitoring Centre (1992). *Global Biodiversity: status of the Earth's living resources*. Chapman and Hall, London.

Ying Tsun-Shen and Zhang Zhi-Song (1984). Endemism in the flora of China: studies on the endemic genera. *Acta Phytotaxonomica Sinica*, **22**, 259–68.

16. Species as the 'currency' of conservation: the Karoo/Dune/Red Lark complex of south-western Africa

T. M. CROWE, P. G. RYAN, M. F. ESSOP,
R. K. BROOKE, P. A. R. HOCKEY, and
W. R. SIEGFRIED
FitzPatrick Institute, University of Cape Town, Rondebosch 7700, South Africa

Abstract

Species are considered to be the fundamental units of conservation. This brief account describes a multifaceted approach using morphological, behavioural, ecological, and genetic information to illustrate an appropriate strategy for recognizing species. The Karoo/Dune/Red Lark species complex in south-western Africa is examined as an example of this approach since at least 21 authors have expressed opinions about their natural status. By analysis of qualitative variation in plumage, mensural, behavioural, and genetic characters, together with habitat details, two genera and four species of these larks are recognized as the basic units of conservation.

Introduction

Although biodiversity can be measured from the level of the nucleotide to the biome, species are, in practice, the fundamental units of comparative biology and of conservation evaluation. However, species recognized on the basis of 'biological' species concepts (*sensu* Mayr 1963; Paterson 1985), which set species limits on the basis of the capability (or inferred capability) of interbreeding are not the optimal 'currency' of either discipline. In addition to a range of objections raised against such concepts (e.g. de Queiroz and Donoghue 1988; Templeton 1989; Wheeler and Nixon 1990), the restrictive use of 'biological' species concepts may not accurately recover the historical pattern of evolution of populations which comprise species, and most

Systematics and Conservation Evaluation (ed. P. L. Forey, C. J. Humphries, and R. I. Vane-Wright), Systematics Association Special Volume No. 50, pp. 229–34, Clarendon Press, Oxford, 1994.

certainly does not cater for the obvious, and sometimes essentially qualitative, geographical variation that occurs within many species (Avise *et al.* 1987; Cracraft 1989). Traditionally, qualitative intraspecific variation has been signified through the recognition of races or subspecies within polytypic species (Mayr 1963; Crowe 1978). However, the subspecies category also has been used as a means of describing clinal, quantitative variation within otherwise obviously evolutionarily cohesive entities (see forum in *Auk* 1982, **99**, pp. 593–615). The effect of a 'veritable orgy' (Mayr 1982) of descriptions of clinal subspecies during the first half of the twentieth century has been a proliferation of intraspecific taxa of little or no use to comparative or conservation biologists.

Although there has been much improvement in morphological approaches to the objective identification of species (Smith and Patton 1988) and in the description of intraspecific geographical variation (Zink and Remsen 1986), there are still many borderline cases between apparent species or subspecies and morphologically cryptic, but genetically distinct forms (e.g. sibling species) (Avise *et al.* 1987). The aim of this study is to use a multifaceted approach incorporating morphological, behavioural, ecological, and genetic information to illustrate what we feel is an appropriate strategy for recognizing species that are of use to comparative biologists, as well as to those who are charged with the conservation of biodiversity.

A case history

Starting with Stark and Sclater (1900), at least 21 authors have expressed opinions on the taxonomy of the Karoo/Dune/Red Lark complex for which the earliest name is *Alauda albescens* Lafresnaye (1839). The most extreme classifications have been those of Roberts (1940) who recognize two genera (*Calendulauda* and *Pseudammomanes*), six species and seven subspecies, and Meinertzhagen (1951) who recognized only one species with no subspecies in the genus *Certhilauda*. The most recent revision of this group of south-western African endemic birds (Clancey 1989) also treats them as one species *Certhilauda albescens*, but with eight subspecies. Since these largely plumage-morphometric-based studies have yielded no consensus classification, we decided on a multidisciplinary approach to the problem, involving study of the plumage, morphometrics, vocalizations, habitats, and mitochondrial DNA (mtDNA) of these larks.

Results

1. *Plumage and morphometrics*

Based on discriminant analyses (BMDP-7M, Dixon 1990) of 15 plumage and mensural variables, there appear to be at least seven taxa distinguishable

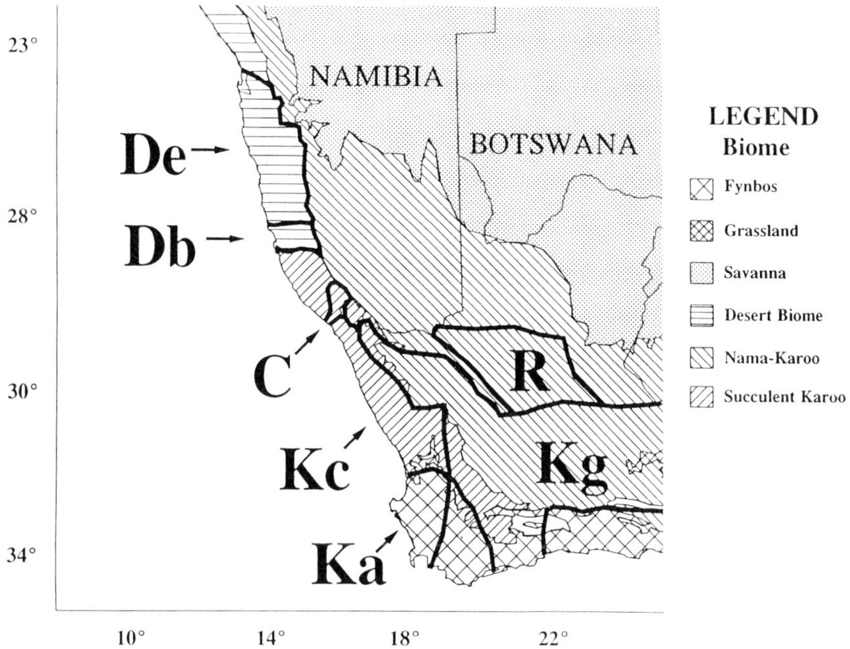

Fig. 16.1 Geographical ranges of the Dune (D: De = *Certhilauda e. erythrochlamys*, Db = *C. e. barlowi*), Caves (C), Karoo (K: Ka = *Certhilauda a. albescens*, Kc = *C. a. codea*, Kg = *C. a. guttata*) and Red (R = *Pseudammomanes burra*) Larks in relation to the distribution of major biomes (after Rutherford and Westfal 1986).

with a 95 per cent probability in the Karoo/Dune/Red Lark complex (Fig. 16.1). Four taxa (the Dune Lark—*erythrochlamys* + *barlowi*, Cave's Lark—*cavei*, the Karoo Lark—*codea* + *guttata* + *albescens*, and the Red Lark—*burra*) are essentially qualitatively distinct (i.e. 100 per cent correct classifications). All of the specimens 'incorrectly' classified in the discriminant analysis were collected within 150 km of the mouth of the Orange River (28° 35′S, 16° 30′E), which may have been an important zoogeographic factor affecting the evolution of these larks.

2. *Vocalizations*

The fundamental structure of the advertisement calls of all seven taxa is similar and consists of three components: a lead-in of 1–12 (median = 4) similar, generally simple, notes followed by a more complex single note, and a trill. The Red Lark has fewer, but relatively complex, elements of longer duration in both the lead-in and trill components, and a shorter call overall, stereotyped within bouts. The Karoo Lark is distinguished by having the highest pitched lead-in and trill components. The Dune Lark differs from the others by having a large number of relatively simple elements in the lead-in call. A single recording of the Cave's Lark exhibits a mosaic of features of Dune and Karoo Larks.

3. *Habitats*

The Dune lark is confined to Desert Biome and the Red Lark to Nama Karoo. The three subspecies of the Karoo Lark are confined to Succulent Karoo (*codea*), Fynbos (*albescens*), and Nama Karoo Biomes (*guttata*) respectively, although *guttata* penetrates well into Succulent Karoo and Fynbos. Cave's Lark is found in Succulent Karoo to the north of the Orange River (Fig. 16.1).

4. *MtDNA*

MtDNA was isolated and purified from at least two individuals of all seven taxa (except *albescens*), and purified using a combination of the methods of Cummings *et al.* (1987) and Smith *et al.* (1971) with modifications. Purified mtDNA of each individual was restricted with 10 six-base-pair-recognizing restriction endonucleases, namely ClaI, XbaI, NheI, SacI, PvuII, PstI, BclI, NcoI, EcoRV, and BamHI. Restriction endonuclease digestions were performed at 37 °C for 2 hours, using a buffer designated KGB (McClelland *et al.* 1988). Restriction fragments were end-labelled using 32P-dCTP and separated by agarose gel electrophoresis. The sequence divergence between mtDNAs was estimated using methods suggested by Nei and Li (1979).

The sequence divergence between the six taxa studied is summarized in Table 16.1. Although no particular degree of genetic distance can be taken as absolute proof of specific distinction, intraspecific comparisons of birds tend to yield divergence values less than 1, whereas the mean divergence values for interspecific comparisons between congeners occur in the range 1.4–6.5 (Shields and Helm-Bychowski 1988; Crowe *et al.* 1992). Using these divergence values as guidelines, like those for our morphometric studies, the mtDNA results suggest that there are four qualitatively distinct taxa in this assemblage of larks (the Dune Lark—*erythrochlamys* + *barlowi*, Cave's Lark—*cavei*, the Karoo Lark—*codea* + *guttata* + *albescens*, and the Red Lark—*burra*).

Conclusions

Based on largely congruent patterns of essentially qualitative variation in plumage, mensural, behavioural, and genetical characters, and in habitat, we believe that the comparative and conservation biology of these larks is best served by recognizing at least four species and two genera. The Karoo Lark *Certhilauda albescens*—subspecies *albescens*, *codea* and *guttata*; Cave's Lark *Certhilauda cavei*—no subspecies; Dune Lark *Certhilauda erythrochlamys*—subspecies *erythrochlamys* and *barlowi*; and the Red Lark *Pseudammomanes burra*—no subspecies.

Table 16.1 Matrix of per cent nucleotide divergence estimates for Karoo, Red and Dune Larks.

	COD	GUT1	GUT2	GUT3	CAV	BAR	ERY
Karoo							
codea (29.15S 16.55E)							
guttata (29.15S 17.01E)	0.0						
guttata (29.15S 17.40E)	0.0	0.0					
guttata (33.55S 20.45E)	0.6	0.6	0.6				
Cave's (*cavei*)	7.2	7.2	7.2	7.2			
Dune							
barlowi	4.2	4.2	4.2	3.4	3.7		
erythrochlamys	7.2	7.2	7.2	6.4	2.0	0.8	
Red (*burra*)	9.9	9.9	9.9	9.9	>15.0	>15.0	>15.0

Acknowledgements

We thank the Cape Department of Nature Conservation and the Namibian Nature Conservation Division for permission to collect the material analyzed genetically. Rob Little and Richard Dean helped with specimen collection. We thank The Natural History Museum (Tring), the South African Museum (Cape Town), the Transvaal Museum (Pretoria), the Durban Natural Science Museum, and East London Museum for access to material in their collections. We acknowledge financial support from the Core and Special Programmes of the Foundation for Research Development, Pretoria, and the Research and Equipment Committees of the University of Cape Town. TMC thanks the organizers of this conference for support which enabled him to contribute towards its proceedings.

References

Avise, J. C., Arnold, J., Ball, R. M., Bermingham E., Lamb, T., Neigel, J. E. *et al.* (1987). Intraspecific phylogeography: the mitochondrial DNA bridge between population genetics and systematics. *Annual Review of Systematics and Ecology*, **18**, 489–522.

Clancey, P. A. (1989). Taxonomic and distributional findings on some birds from Namibia. *Cimbebasia*, **11**, 111–33.

Cracraft, J. (1989). Speciation and its ontology: the empirical consequences of alternative species concepts for understanding patterns and processes of differentiation. In *Speciation and its consequences* (ed. D. Otte and J. A. Endler), pp 28–59. Sinauer, Sunderland, MA.

Crowe, T. M. (1978). The evolution of guinea-fowl (Galliformes, Phasianidae, Numidinae): taxonomy, phylogeny, speciation and biogeography. *Annals of the South African Museum*, **76**, 43–136.

Crowe, T. M., Harley, E. H., Jakutowicz, M. B., Komen, J., and Crowe, A. A. (1992). Phylogenetic, taxonomic and biogeographical implications of genetic, morphological and behavioral variation in francolins (Phasianidae: Francolinus). *Auk*, **109**, 24–42.

Cummings, O. W., King, T. C., Holden, J. A., and Low, R. L. (1987). Purification and characterization of the potent endonuclease in extracts of bovine heart mitochondria. *Journal of Biochemical Evolution*, **262**, 2005–15.

De Querioz, K. and Donoghue, M. J. (1988). Phylogenetic systematics and the species problem. *Cladistics*, **4**, 317–38.

Dixon, W. J. (ed.) (1990). *BMDP statistical software manual*. University of California Press, Los Angeles.

McClelland, M., Hanish, J., Nelson, M., and Patel, Y. (1988). KGB: a single buffer for all restriction endonucleases. *Nucleic Acids Research*, **16**, 364.

Mayr, E. (1963). *Animal species and evolution*. Belknap Press, Cambridge, MA.

Mayr, E. (1982). Of what use are subspecies? *Auk*, **99**, 593–5.

Meinertzhagen, R. (1951). Review of the Alaudidae. *Proceedings of the Zoological Society of London*, **121**, 81–132.

Nei, M. and Li, W. (1979). Mathematical model for studying genetic variation in terms of restriction endonucleases. *Proceedings of the National Academy of Sciences, USA*, **76**, 5269–73.

Paterson, H. E. H. (1985). The recognition concept of species. In *Species and speciation* (ed. E. S. Vrba), pp. 21–9. Transvaal Museum, Pretoria.

Roberts, A. (1940). *The birds of South Africa*. H. F. and G. Witherby, London.

Rutherford, M. C. and Westfal, R. (1986). The biomes of southern Africa—an objective categorization. *Memoirs of the Botanical Survey of South Africa*, **54**, 1–98.

Shields, G. F. and Helm-Bychowski, K. M. (1988). Mitochondrial DNA of birds. *Current Ornithology*, **5**, 273–95.

Smith, C. A., Jordan, M. J., and Vinograd, J. (1971). In vivo effects of intercalating drugs on the super-helix density of mitochondrial DNA isolated from human and mouse cells in culture. *Journal of Molecular Biology*, **59**, 255–72.

Smith, M. F. and Patton, J. L. (1988). Subspecies of pocket gophers: causal bases for geographic differentiation in *Thomomys bottae*. *Systematic Zoology*, **37**, 163–78.

Stark, A. C. and Sclater, W. L. (1900). *The birds of South Africa 1*. R.H. Porter, London.

Templeton, A. R. (1989). The meaning of species and speciation. In *Speciation and its consequences* (ed. D. Otte and J. A. Endler), pp. 3–27. Sinauer, Sunderland, MA.

Wheeler, Q. D. and Nixon, K. C. (1990). Another way of looking at the species problem: a reply to de Queiroz and Donoghue. *Cladistics*, **6**, 77–81.

Zink, R. M. and Remsen, J. V. (1986). Evolutionary processes and patterns of geographic variation in birds. *Current Ornithology*, **4**, 1–69.

17. Basal taxa and the role of cladistic patterns in the evaluation of conservation priorities: a view from freshwater

MELANIE L. J. STIASSNY* and MÁRIO C. C. de PINNA[†]
*Department of Herpetology and Ichthyology, American Museum of Natural History, New York, NY 10024, USA
[†]Departamento de Zoologia, Universidade de São Paulo, Caixa Postal 11461, São Paulo-SP 05422–970, Brazil

Abstract

Taxa basal within their lineages are often notably depauperate in numbers of species when compared with their sister-groups, and frequently have extremely restricted geographical distributions. These two features render them particularly vulnerable to environmental pressures. This phenomenon is strikingly prevalent among freshwater fishes, an assemblage representing upward of 25 per cent of all living vertebrates yet largely ignored in most regional biodiversity estimates and conservation evaluation programmes.

Besides the known fact that basal taxa in some sense have comparative information complementary to that of their more species-rich sister-groups, they also have a crucial importance in resolving the higher- and lower-level intrarelationships of their respective lineages. The phylogenetic relevance of basal clades can be appreciated at three levels. (1) They are invariably the most important element in determining character polarities within their sister groups. (2) In certain instances knowledge of character states shown by these basal groups introduces character conflicts which are the only factor leading to the proposal of a particular phylogenetic hypothesis. (3) They sometimes provide the *only* possible evidence for understanding the evolution of certain character transitions.

Our conclusions indicate that until stable, well-resolved phylogenies are available for most taxonomic groups, basal taxa assume a key importance in phylogenetic analysis. Furthermore for many groups, particularly for those in the tropics where new species are constantly being found, basal taxa will continue to play an important role in determining the ultimate placement of these new species. For these reasons basal taxa should be assigned even higher conservation priority than has been so far realized.

Systematics and Conservation Evaluation (ed. P. L. Forey, C. J. Humphries, and R. I. Vane-Wright), Systematics Association Special Volume No. 50, pp. 235–49, Clarendon Press, Oxford, 1994.

Introduction

There is a growing realization that definitions of biodiversity and the evaluation of conservation priorities should in some way incorporate phylogenetic information. However, the manner in which this information should best be applied has yet to reach a consensus, and recently a number of different measures encoding notions such as 'phylogenetic uniqueness', 'taxonomic distinctiveness', and 'phylogenetic diversity' have been developed (e.g. May 1990; Vane-Wright *et al.* 1991; Nixon and Wheeler 1992; Crozier 1992; Faith 1992, in press; Williams *et al.* 1993). Whether these measures render quantitative indices solely in terms of cladistic branching pattern, or whether they incorporate genetic or morphological distance into their data, they are clearly sensitive to changes in the topology of the trees from which they are derived. The development and exploration of such indices is heuristic and is a useful and progressive exercise. However, we want to emphasize the rather simple point that for most taxonomic groups we are a long way from having well-corroborated and resolved phylogenies, and thus lack the stable tree topologies necessary for fully meaningful analysis (see also Nixon and Wheeler 1991).

For certain groups a lack of phylogenetic resolution poses little problem and some level of conservation priority is virtually self evident. For example, few would argue against a high priority being assigned to measures aimed at saving *Sphenodon*, the last surviving member of the sister-group to the extant Lepidosauria (Fig. 17.1A). Similarly, measures aimed at conserving *Latimeria*, a basal sarcopterygian and sister-group to the remaining extant Sarcopterygii (Fig. 17.1B) are rather noncontroversial. The loss of these taxa is particularly poignant as it would represent the loss of the last living vestige of their lineage. Clearly if there were hundreds of living members of the actinistian clade, conservation priorities would become more complex; it is the marked asymmetry in species numbers between sister clades that emerges as the decisive factor here.

There is another aspect to these examples; both *Sphenodon* and *Latimeria* diverge close to the root in their respective cladograms, i.e. they represent the basal sister-groups in their respective clades. If *Sphenodon* was instead a monotypic squamate genus nested somewhere within the Agaminae for example, no matter how morphologically aberrant or ecologically unusual it may still be, its conservation profile would undoubtedly diminish. In a certain sense, these basal sister taxa have comparative information complementary to that of the entire membership of their sister-group (Vane-Wright *et al.* 1991; Nixon and Wheeler 1992). It is this topological aspect, and its potential significance for the establishment of conservation priorities that we would like to explore further.

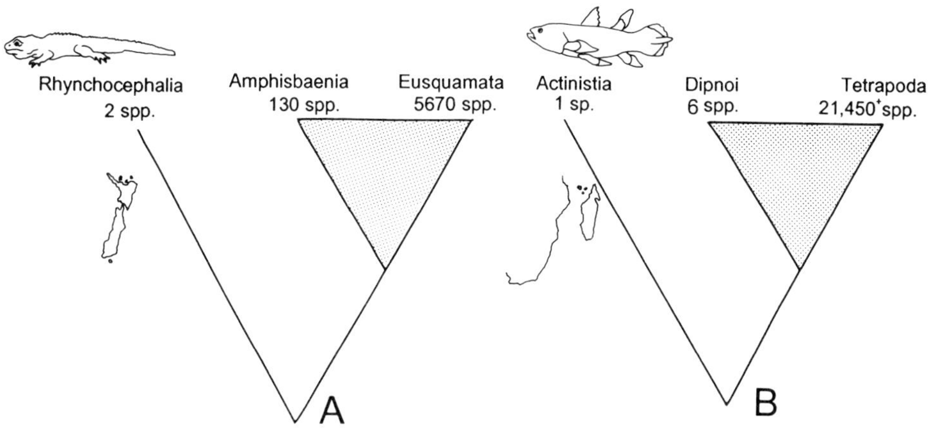

Fig. 17.1 A. Systematic position of *Sphenodon* (see Daugherty *et al.* 1990). Inset indicates geographical distribution of *Sphenodon*, restricted to the islands of New Zealand. B. Systematic position of *Latimeria* (see Rosen *et al.* 1981). Inset indicates the geographical distribution of *Latimeria*, restricted to deep water habitats off the islands of Grande Comore and Anjouan.

Basal teleostean clades

Among teleost fishes, the group with which we are most familiar, we recognize numerous cases of high-level phylogenetic asymmetry where the basal sister-group to a taxonomically diverse clade is a notably depauperate taxon. A selection of such examples is illustrated in Figs. 17.2 and 17.3, and in many of these selected lineages the grey triangles represent what is essentially a profound lack of taxonomic resolution within each of the species-rich sister-groups. If we take, for example, the case of the Siluriformes (Fig. 17.2B), one of the few well-corroborated components of relationships currently available is the placement of the Diplomystidae as the basal sister-group to the remaining 2200+ extant siluriforms (see e.g. Arratia 1987; Grande 1987). Given this limited state of knowledge it is clear that any attempts to resolve the high-level intrarelationships of catfishes will necessarily rely heavily on data derived from the basal taxon Diplomystidae. Similarly, any phylogeneticist working on acanthomorph intrarelationships knows the importance of *Polymixia* in gaining insight into character polarity within the Acanthomorpha (see e.g. Rosen 1985; Stiassny 1986; Stiassny and Moore 1992; Fig. 17.2A). Taxa such as *Diplomystes* and *Polymixia* assume a critical importance when the phylogenetic resolution of their respective sister-groups is in question. Often they are the only piece of information we have regarding the intrarelationships of major lineages and, practically

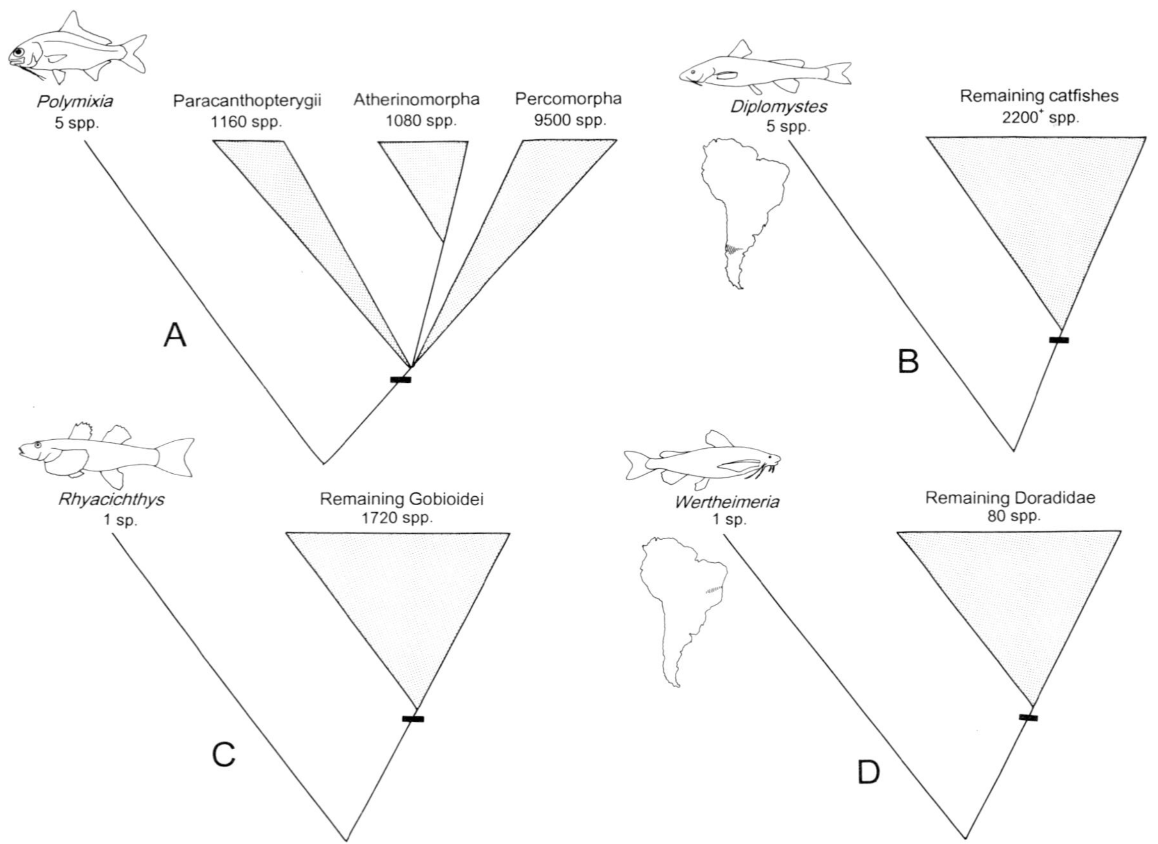

Fig. 17.2 A. Systematic position of *Polymixia* (see Stiassny 1986; Stiassny and Moore, 1992). B. Systematic position of *Diplomystes* (see Arratia 1987; Grande, 1987). Inset indicates geographical distribution restricted to central Chile and Argentina. C. Systematic position of *Rhyacichthys* (see Miller 1972). D. Systematic position of *Wertheimeria* (see Higuchi 1992). Inset indicates geographical distribution in the Rio Jequitinhonha of Brasil.

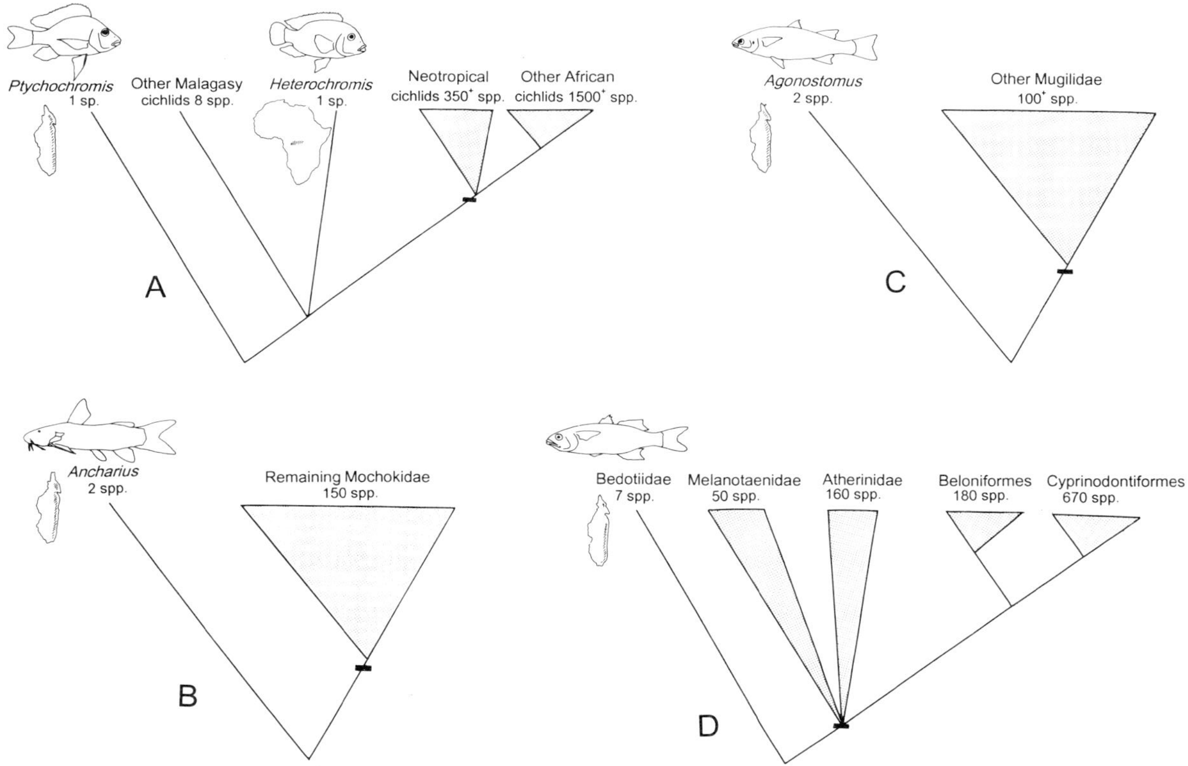

Fig. 17.3 A. Systematic position of *Ptychochromis* and the other Malagasy cichlids (see Stiassny 1991). Inset indicates distribution of *Ptychochromis* in the eastern coastal forests of Madagascar. B. Systematic position of *Ancharius* (see Mo, 1991). Inset indicates distribution of *Ancharius* in the eastern coastal forests of Madagascar. C. Systematic position of *Agonostomus* (see Harrison and Howes 1991; Stiassny, in press). Inset indicates geographical distribution of *Agonostomus telfairii* in the eastern coastal forest of Madagascar, a second species, *Agonostomus monticola* occurs in freshwater habitats in central America. D. Systematic position of the Bedotiidae (see Stiassny 1990). Inset indicates their geographical distribution in the eastern coastal forests of Madagascar.

speaking, without them the chances of ever resolving relationships within their sister-groups are greatly diminished. By virtue of their basal position (as successive outgroups), character state changes in these taxa can profoundly influence hypotheses of relationships among the remaining members of their lineage, regardless of the states observed in more distant outgroups and in more phylogenetically derived taxa (Maddison *et al.* 1984; Clark and Curran 1986). For phylogeneticists, the fact that these critical basal taxa are composed of only one or a few species simplifies analysis, as ingroup resolution and character variation within the basal taxon is necessarily rather muted. In the context of conservation evaluation there is another aspect to these examples that is of considerable importance. It is very often the case that basal taxa, besides being notably depauperate in species composition, also have extremely restricted geographical distributions, rendering them particularly vulnerable to environmental pressures. This phenomenon is strikingly evident among freshwater fishes, an enormous grouping that represents upwards of 25 per cent of all living vertebrate species, but that has been largely ignored in most regional biodiversity estimates and conservation evaluation programmes.

When the geographical distributions of the basal taxa are superimposed onto the cladograms for the primarily freshwater groups (Figs. 17.2B, D and 17.3A, B, C) their fragility becomes readily apparent. While the detrimental effects of deforestation and other human disturbances on endemic freshwater ichthyofaunas are poorly documented, it is clear that regional deforestation results in significant changes in the turbidity, temperature, resource availability, and productivity of associated river drainages (Reinthal and Stiassny 1991). Freshwater endemics, because of often specialized physiological and trophic requirements, are frequently unable to take refuge in downstream brackish waters when their freshwater habitats are degraded. These fishes are therefore particularly prone to regional and, in the case of endemics localized in their distribution to single watersheds, total extirpation. In Madagascan freshwaters for example, the major focus for speciation has been disjunct freshwater habitats. Over 73 per cent of endemics are highly localised freshwater species, versus 21 per cent for euryhaline species (Reinthal and Stiassny 1991), and it is the freshwater endemics that have undergone the most extreme attrition.

At least for a number of the freshwater ichthyofaunal components considered here these phylogenetic data can be used to strengthen arguments for a broad regional conservation plan, in this case recognizing the eastern coastal forests of Madagascar as an area harboring a significant concentration of basal taxa of considerable phylogenetic importance (Fig. 17.3), and undergoing severe habitat perturbation and deforestation (Green and Sussman 1990). As noted by Stiassny (1992), in the case of Madagascar our proposal is more-or-less equivalent to highlighting an area of extreme species endemism as a high-priority region for conservation, but this need not always be the case.

Loricarioid relationships: an example

To further explore the systematic importance of basal taxa and their potential role in conservation evaluation programmes we will outline a specific example taken from a recent study of basal loricarioid catfish relationships (de Pinna 1992). Although highly specific in its details, this example illustrates some important properties of basal taxa of quite general significance.

The loricarioid catfishes are a taxonomically diverse and ecologically ramified grouping of Neotropical catfishes. This large clade (ranked as the superfamily Loricarioidea or suborder Loricarioidei) includes six monophyletic families. (1) The Nematogenyidae, composed of a single species, *Nematogenys inermis* (Guichenot 1848) endemic to a few basins in central Chile. (2) The Trichomycteridae, composed of 200+ species distributed throughout South America. (3) The Callichthyidae, another widely distributed group of about 150 species. (4) The Scoloplacidae, composed of 4 species limited to the Amazon and Parana river drainages. (5) The Astroblepidae, an Andean assemblage of about 40 species. (6) The Loricariidae, a large grouping of over 600 species distributed throughout South America.

The monophyly of the loricarioid catfishes is well established (see e.g. Baskin 1973; Howes 1983; Schaefer and Lauder 1986), and their intrarelationships have traditionally been represented as indicated in Fig. 17.4, with the monotypic *Nematogenys* forming the sister-group to the remaining loricarioid assemblage. The phylogenetic integrity of the non-nematogenyid loricarioids is based on a series of characters (summarized in Schaefer and Lauder 1986), two of which are illustrated in Fig. 17.4. The first of these (Fig. 17.4A) is the derived presence of a superficial ossification on the ventral surface of the Weberian apparatus that extends over the basioccipital bone. The plesiomorphic condition as exemplified by *Nematogenys* and other catfishes, is to have the basioccipital free from the ventral surface of the Weberian apparatus. A second derived character of non-nematogenyid loricarioids is the loss of the intercalarium bone (Fig. 17.4B), one of the Weberian ossicles primitively present in *Nematogenys* and most other catfishes.

In 1983 a new taxon was discovered in a small creek between Rio de Janeiro and São Paulo in Brasil. The fish was determined to belong to a new genus and species of trichomyterid and named *Trichogenes longipinnis* (Britski and Ortega 1983). Phylogenetic analysis places *Trichogenes* quite unambiguously within a monophyletic Trichomycteridae, and more importantly

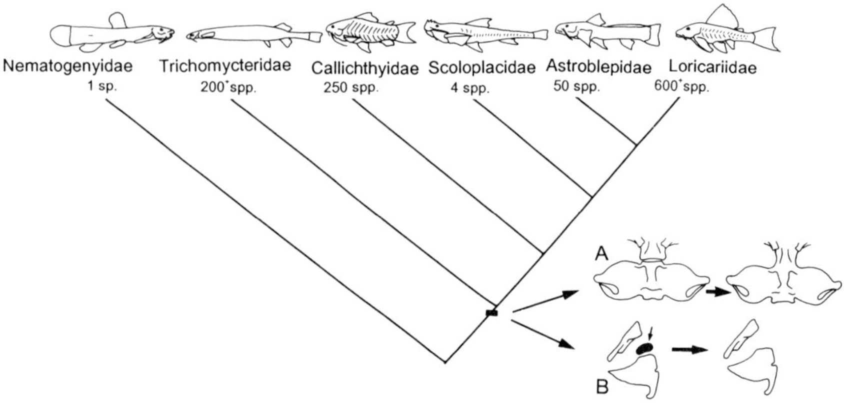

Fig. 17.4 Scheme of loricarioid intrarelationships (after Baskin 1973; Howes 1983; Schaefer and Lauder 1986). A. schematic representation of plesiomorphic (unfused) and derived (fused) states of the basioccipital and ventral surface of the Weberian apparatus. B. Weberian ossicles, small arrow indicates presence of the intercalarium.

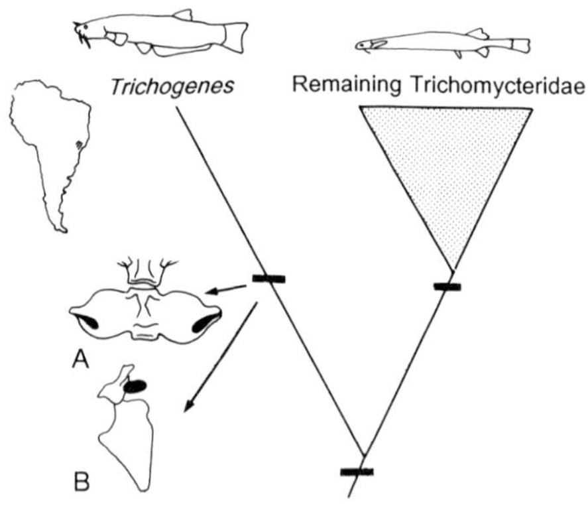

Fig. 17.5 Systematic position of *Trichogenes* as the basal sister-group to the remaining Trichomycteridae (for additional characters supporting this scheme see de Pinna 1992). A. basioccipital not fused with the ventral surface of Weberian apparatus. B. Intercalarium present.

as the basal sister-group to all other trichomycerids (de Pinna 1992); Fig. 17.5).

As a trichomycterid and thus a non-nematogenyid loricarioid, *Trichogenes* would be expected to display the derived states of characters uniting that assemblage (Schaefer and Lauder 1986). This is not the case, and in fact *Trichogenes*, while sharing all of the derived characters of the Trichomycteridae, displays the plesiomorphic states of most of those characters

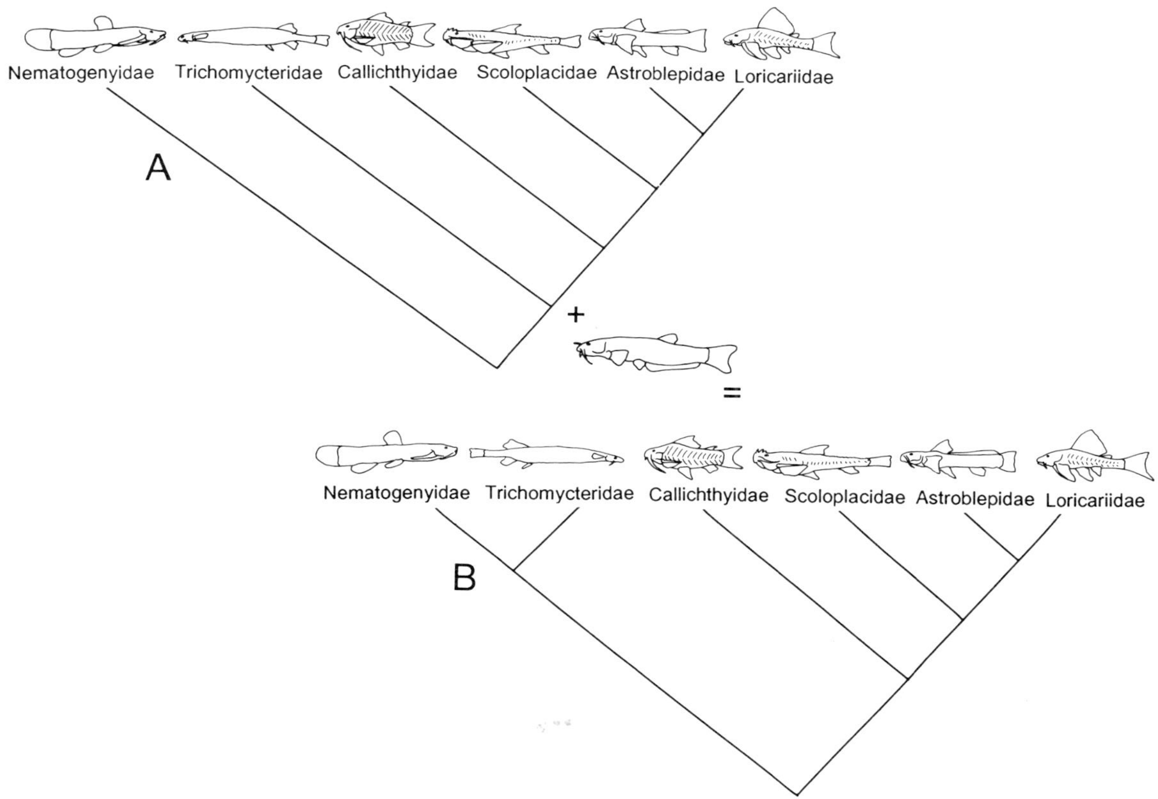

Fig. 17.6 A. Scheme of loricarioid relationships (after Baskin 1973; Howes 1983; Schaefer and Lauder 1986). B. Single most parsimonious scheme of loricarioid relationships after character distributions in *Trichogenes* are introduced into the matrix (see text for discussion).

formerly hypothesized to unite the non-nematogenyid loricarioids. For example, in the two characters illustrated in Fig. 17.4, *Trichogenes* exhibits plesiomorphic states (Fig. 17.5 A,B). The novel mosaic of derived (trichomycterid) and plesiomorphic states exhibited by *Trichogenes* introduces character incongruence that results in a significant alteration in the most parsimonious scheme of relationships among basal loricarioid fishes. The inclusion of *Trichogenes* into the analysis results in a single most parsimonious scheme in which the Trichomycteridae and Nematogenyidae form a monophyletic sister-group to the remaining loricarioids (Fig. 17.6B). The hypothesized monophyly of a non-nematogenyine loricarioid assemblage (Fig. 17.4, 17.6A) is now falsified. It should be noted that this topological change is simply the result of character incongruence caused by the inclusion of *Trichogenes*, no new characters have been introduced into the analysis (de Pinna 1992).

If *Trichogenes* was a derived trichomycterid embedded somewhere high up in the family tree, its unique character combinations would have no effect whatsoever on hypotheses of basal loricarioid relationships. It is by virtue of its basal placement within the Trichomycteridae that character state changes in *Trichogenes* can influence hypotheses of relationships among the remaining loricarioid lineages. de Pinna (1992) observes that *Trichogenes* is playing an analogous role to that played by fossil taxa in the recent studies of Gautier *et al.* (1988) and Donoghue *et al.* (1989). Clearly, the inclusion of basal taxa with previously unobserved character combinations, that introduce character conflict into the analysis, may result in alterations of the most parsimonious scheme of relationships regardless of their fossil or extant status.

The discovery of *Trichogenes* has greatly facilitated our understanding of basal loricarioid interrelationships, yet it is possible that evidence supporting a nematogenyine + trichomycterid clade could have been found in its absence. In fact de Pinna (1992) has found additional derived characters supporting the monophyly of a nematogenyid + trichomycterid clade. However, when we consider character transformations within the Loricarioidei, without *Trichogenes* we would simply be unable to understand unambiguously their evolution. Taking as an example the derived presence of a superficial ossification on the ventral surface of the Weberian apparatus that extends over the basioccipital, in the absence of *Trichogenes* the interpretation of the evolution of this character state within the Loricarioidei is ambiguous and two equally parsimonious interpretations are possible. In the first (Fig. 17.7A) a transition from the plesiomorphic (unfused, 0) state to the derived (fused, 1) has occurred independently in the Trichomycteridae and in the remaining loricarioids. While in the second interpretation (Fig. 17.7B) a transition from the plesiomorphic (unfused, 0) state to the derived (fused, 1) occurs at the base of the Loricariodei, with a reversal from the derived (fused, 1) to the plesiomorphic (unfused, 0) state occurring in the

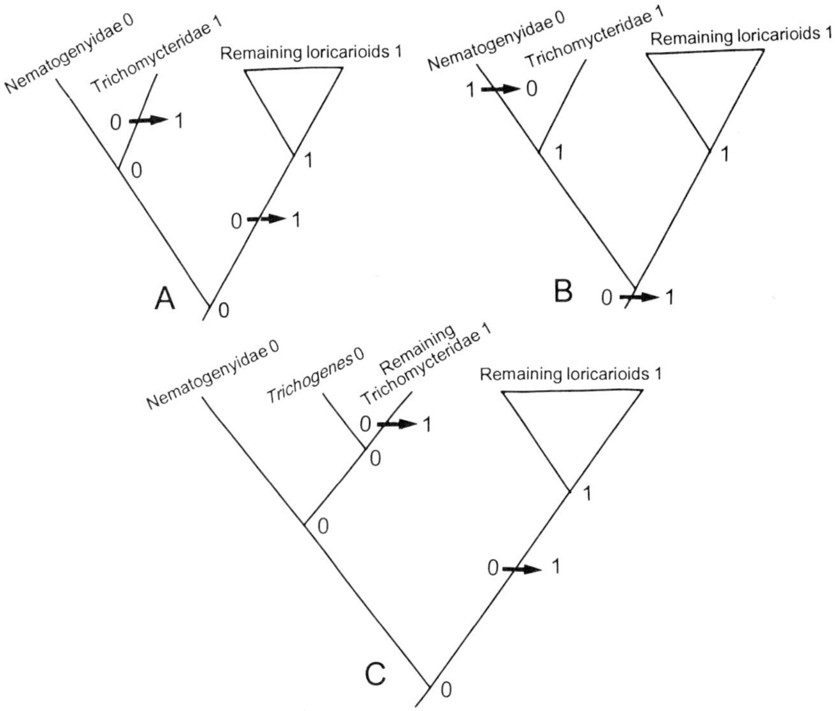

Fig. 17.7 Alternative character optimizations for basioccipital association with ventral surface of the Weberian apparatus. The plesiomorphic state (unfused) indicated by 0, and derived stated (fused) indicated by 1 (see text for explanation).

Nematogenyidae. Either scenario could be true and we have no a priori means of deciding between these alternative interpretations. However, when *Trichogenes* is added to the tree (Fig. 17.7C) the ambiguity is resolved. The distribution of the derived character state (1) within loricarioids is most parsimoniously interpreted as having arisen twice, once within the Trichomycteridae and a second time in all remaining loricarioids. The morphological similarity of the basioccipital/Weberian fusion in the non-trichogenine trichomycterids and the remaining loricarioids is striking, and based solely on morphological criteria there is no evidence against putative homology. It is *Trichogenes*, with its basal position in the Trichomycteridae, that supports the hypothesis that such ossification has originated twice independently in loricarioid fishes.

Of course, few conservationists are going to be overly concerned about obscure bones in the ears of fishes, but what is true for such bones is also true of suites of other biological features, be they anatomical, behavioural, or ecological. The presence of one basal taxon can have a major effect on our understanding of the evolution of certain characters, and these characters can in turn be of great importance to the ecology or functional biology of the taxon.

Conclusions

Current considerations of the importance of phylogenetic information for evaluating conservation priorities have made clear that basal taxa should be assigned high priority. The details and precise quantification of the weight conferred to the 'basalness' of a taxon are still a matter of debate in calculating conservation indices (Williams *et al.* 1993). However, a point that has not been emphasized so far is that basal taxa not only have particular importance in view of their unique comparative value, but they may also determine the outcome of the phylogenetic hypotheses for more distal clades of their lineage. These phylogenetic hypotheses, in turn, will determine conservation priorities for the whole lineage. This situation is first an outcome of the role of basal taxa as first-order source of information for assessing character-state polarities, and consequently a major portion of the derivative phylogenetic hypothesis. Also, basal taxa frequently display character states that have subsequently been modified to unrecognizable degrees in the more distal members of their lineage.

One of the examples discussed in this paper serves to demonstrate the above point. It was shown that the inclusion of a single species of trichomycterid, *Trichogenes longipinnis* in the analysis of relationships among loricarioid catfishes resulted in a major shift of phylogenetic hypothesis for that large clade of South American catfishes. Before the inclusion of *T. longipinnis*, the Nematogenyidae was considered as the sister-group to all remaining loricarioid families. After the inclusion of that species and the consequent increase in homoplasy, the favoured hypothesis changed to unite the Nematogenyidae as the sister-group of the Trichomycteridae. Perhaps more convincingly, the new hypothesis turned out to be the one subsequently supported by an extended character survey (de Pinna 1992). Comparing the two hypotheses (Fig. 17.8), one sees that a measure of conservation weights for each terminal differs markedly between them. In this case, we used the so-called 'Standardized Taxonomic Root Weights' (see Vane Wright *et al.* 1991; Williams *et al.* 1993), but we expect the general conclusions to hold true for any index that relies on cladogram topologies. The striking difference in relative weights for a group composed of over 1000 species is caused simply by the addition or exclusion of a single species, the basal trichomycterid *Trichogenes longipinnis*, in the analysis of loricarioid relationships. This example underlines the fact that basal taxa not only have special relevance because of their unique comparative value, but also because they may directly determine the conservation priorities for taxa that surround them. Basal taxa form the baseline for conservation decisions

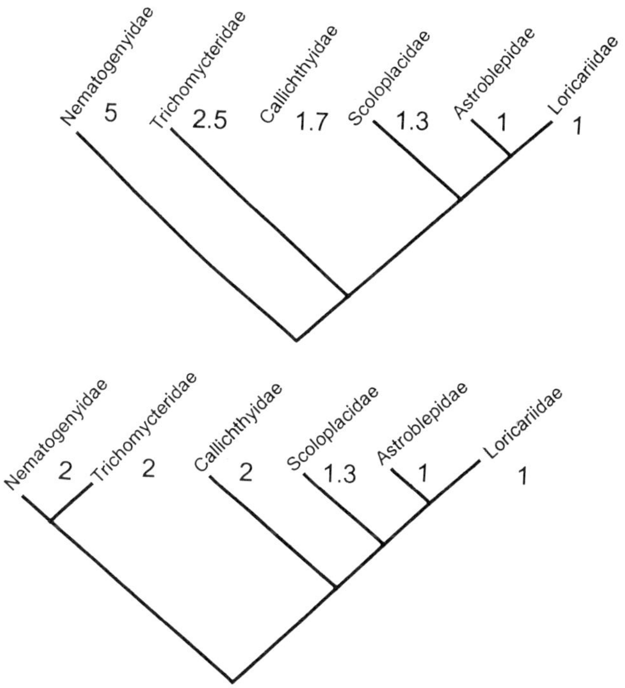

Fig. 17.8 Standardized Taxonomic Root Weights (see Vane-Wright *et al.* 1991; Williams *et al.* 1993) calculated for alternate loricarioid trees.

potentially affecting clades more species-rich than themselves by several orders of magnitude.

Acknowledgements

We would like to thank Peter Forey, Chris Humphries, and Dick Vane-Wright for organizing the 1992 symposium on Systematics and Conservation Evaluation, and for all of their efforts in bringing this volume together. Thanks also to Dave Yeates and members of the American Museum Journal Club for a helpful discussion of some of the ideas presented herein.

References

Arratia, G. (1987). Description of the primitive family Diplomystidae (Silurifomres, Teleostei, Pisces): morphology, taxonomy and phylogenetic implications. *Bonner Zoologische Monographien*, **24**, 1–120.

Baskin, J. N. (1973). Structure and relationships of the Trichomycteridae. Unpublished Ph.D. Dissertation, City University of New York, New York.

Britski, H. A. and Ortega, H. (1983). *Trichogenes longipinnis*, novo gênero e espécie de Trichomycterinae do sudeste do Brasil. *Revista Brasileira de Zoologia*, **1**, 211–16.

Clark, C. and Curran, D. (1986). Outgroup analysis, homoplasy, and global parsimony: a response to Maddison, Donoghue and Maddison. *Systematic Zoology*, **35**, 422–6.

Crozier, R. H. (1992). Genetic diversity and the agony of choice. *Biological Conservation*, **61**, 11–15.

Daugherty, C. H., Cree, A, Hay, J. M., and Thompson, M. B. (1990). Neglected taxonomy and continuing extinctions of tuatara (*Sphenodon*). *Nature*, London, **347**, 177–9.

Donoghue, M., Doyle, J., Gauthier, J., Kluge, A., and Rowe, T. (1989). The importance of fossils in phylogeny reconstruction. *Annual Review of Ecology and Systematics*, **20**, 431–60.

Faith, D. P. (1992). Conservation evaluation and phylogenetic diversity. *Biological Conservation*, **61**, 1–10.

Faith, D. P. (in press). Systematics and conservation: on predicting the feature diversity of subsets of taxa. *Cladistics*.

Gauthier, L. Kluge, A. G., and Rowe, T. (1988). Amniote phylogeny and the importance of fossils. *Cladistics*, **4**, 105–209.

Green, G. M. and Sussman, R. W. (1990). Deforestation history of the eastern rain forests of Madagascar from satellite images. *Science*, **248**, 212–5.

Harrison, I. J. and Howes, G. J. (1991). The pharyngobranchial organ of mugilid fishes: its structure, variability, ontogeny, possible function and taxonomic utility. *Bulletin of the British Museum (Natural History)* Zoology, **57**, 111–132.

Higuchi, H. (1992). A phylogeny of the South American thorny catfishes (Osteichthys; Siluriformes, Doradidae). PhD Thesis, Harvard University.

Howes, G. J. (1983). The cranial muscles of loricarioid catfishes, their homologies and value as taxonomic characters (Teleostei: Siluroidei). *Bulletin of the British Museum (Natural History)* Zoology, **45**, 309–45.

Maddison, W. P., Donoghue, M. J., and Maddison, D. R. (1984). Outgroup analysis and parsimony. *Systematic Zoology*, **33**, 83–103.

May, R. M. (1990). Taxonomy as destiny. *Nature*, 347, 129–30.

Miller, P. J. (1973). The osteology and adaptive features of *Rhyacichthys aspro* (Teleostei: Gobioidei) and the higher classification of gobioid fishes. *Journal of the Zoological Society of London*, **171**, 397–434.

Mo, T. (1991). Anatomy, relationships and systematics of the Bagridae (Siluroidei, Ostariophysii)—with a hypothesis of siluroid phylogeny. PhD thesis, London University.

Nixon, K. C. and Wheeler, Q. D. (1992). Measures of phylogenetic diversity. In *Extinction and Phylogeny* (ed. M. Novacek and Q. D. Wheeler), pp 216–34, Columbia University Press, New York.

Pinna, M. C. C. de. (1989). A new sarcoglanidine catfish, phylogeny of its subfamily, and an appraisal of the phyletic status of the Trichomycterinae. *American Museum Novitates*, **2950**, 1–39.

Pinna, M. C. C. de. (1992). A new subfamily of Trichomycteridae (Teleostei, Siluriformes), lower loricarioid relationships, and a discussion on the impact of additional taxa for phylogenetic analysis. *Zoological Journal of the Linnean Society of London*, **106**, 175–229.

Reinthal, P. N. and Stiassny, M. L. J. (1991). The freshwater fishes of Madagascar: a study of an endangered fauna with recommendations for a conservation strategy. *Conservation Biology*, **5**, 231–43.

Rosen, D. E. (1985). An essay on euteleostean classification. *American Museum Novitates*, **2827**, 1–57.

Rosen, D. E, Forey, P. L., Gardiner, B. G., and Patterson, C. (1981). Lungfishes, tetrapods, paleontology, and plesiomorphy. *Bulletin of the American Museum of Natural History*, **167**, 163–275.

Schaefer, S. A. and Lauder, G. V. (1986). Historical transformations of functional design: evolutionary morphology of the feeding mechanisms in loricarioid catfishes. *Systematic Zoology*, **35**, 489–508.

Stiassny, M. L. J. (1986). The limits and relationships of the acanthomorph teleosts. *Journal of Zoology*, **1**, 411–60.

Stiassny, M. L. J. (1990). Notes on the anatomy and relationships of the bedotiid fishes of Madagascar, with a taxonomic revision of the genus *Rheocles* (Atherinomorpha: Bedotiidae). *American Museum Novitates*, **2979**, 1–33.

Stiassny, M. L. J. (1991). Phylogenetic intrarelationships of the family Cichlidae: an overview. In *Cichlid Fishes: Behaviour, ecology and evolution* (ed. M. H. A. Keenleyside), pp. 1–35. Chapman and Hall, London.

Stiassny, M. L. J. (1992). Phylogenetic analysis and the role of systematics in the biodiversity crisis. In *Systematics, ecology, and the biodiversity crisis* (ed. N. Eldredge), pp. 109–20. Columbia University Press, New York.

Stiassny, M. L. J. (in press). What are grey mullets? *Bulletin of Marine Science.*

Stiassny, M. L. J. and Moore, J. A. (1992). A review of the pelvic girdle of acanthomorph fishes, with comments on hypotheses of acanthomorph intrarelationships. *Zoological Journal of the Linnean Society of London*, **104**, 209–242.

Vane-Wright, R. I., Humphries, C. J., and Williams, P. H. (1991).What to protect? *Systematics and the agony of choice. Biological Conservation*, **55**, 235–54.

Williams, P. H., Vane-Wright, R. I., and Humphries, C. J. (1993). Measuring biodiversity for choosing conservation areas. In *Hymenoptera and biodiversity*, (ed. J. LaSalle and I. D. Gauld), pp. 309–28. CAB International, Wallingford, UK.

18. Phylogenetic diversity: a general framework for the prediction of feature diversity

DANIEL P. FAITH
CSIRO Division of Wildlife and Ecology, P.O. Box 84, Lyneham, ACT, 2602, Australia

Abstract

The phylogenetic diversity of a subset of species is defined as the sum of the lengths of those branches from the estimated phylogeny that are spanned by the species subset. This measure of diversity has useful properties. Its cladistic/phylogenetic basis means that it can be expected to predict underlying feature diversity patterns among taxa. In conservation evaluation, giving priority to that species subset that maximises represented feature diversity is justified as a form of maximizing 'option value'. Further, because the goal or criterion of this measure is explicit, its performance is vulnerable to testing and evaluation. Phylogenetic diversity provides a framework for conservation evaluation that is applicable over a realistic range of information availability, extending from basic information from existing classifications to well-corroborated phylogenetic patterns among taxa.

These properties of phylogenetic diversity contrast with those of other proposed measures, including another classification-based approach based on Pielou's 'hierarchical diversity'.

An example of phylogenetic diversity assessment, for the Orchidaceae, uses existing classification information and produces a ranking of regions that differs from that for species-counting. Furthermore, the set of species with high phylogenetic diversity contrasts with the set of species previously highlighted for conservation effort based on endangered status.

As part of a larger study of the diversity patterns over several groups of invertebrates in Tasmanian rainforest, preliminary results for amphipods provide an example of a case where the phylogeny reveals diversity information that is not available from the existing classification of the species.

Systematics and Conservation Evaluation (ed. P. L. Forey, C. J. Humphries, and R. I. Vane-Wright), Systematics Association Special Volume No. 50, pp. 251–68, Clarendon Press, Oxford, 1994.

Introduction

My involvement in research on the use of taxonomic or phylogenetic information in conservation evaluation was prompted by the work of Hopper and Coates (1990) on *Eucalyptus* conservation in Western Australia. They argued that conservation priorities should be based on maximum representation of the hierarchical pattern of genetic diversity found among populations of a rare *Eucalyptus* species. Greater representation was taken to imply that more genetic resources were protected. A formal definition of phylogenetic diversity (Faith and Cawsey 1990; Faith 1992*a*) has since been developed to provide a general strategy for quantifying the representativeness of such a hierarchy, particularly when the hierarchical pattern could be interpreted as an estimate of phylogenetic relationships.

Vane-Wright *et al.* (1991) presented important theoretical and practical arguments relating to a number of measures of taxonomic diversity, and their use at the global level. This work has lead directly to subsequent refinements of their own suggested measures (Williams *et al.* 1993; Williams and Humphries, this volume), and a profusion of other suggestions in the literature for alternative measures relating to taxonomic uniqueness and/or taxonomic diversity (see below).

The basic idea of placing greater value on taxonomically unique species is an albeit intuitively old idea (IUCN 1980); a species that is the only representative of its family justifiably may receive greater conservation attention. Similarly, an area that has all members of a genus endemic to it is regarded as a focal point for diversity of that group.

These informal, but well-established, rationales (see also Ahern *et al.* 1985 and references within) for taking taxonomic uniqueness into account contrast with some of the mathematical methods recently proposed for quantifying 'weights' or 'ranks' for species based on systematic information. It is the purpose of this paper to argue that:

(1) If the purpose of introducing systematics-related information into conservation evaluation is to better guarantee the representation of the general feature diversity of organisms, then phylogenetic diversity (*sensu* Faith 1992*a*) provides a practical, logical framework that encompasses a range of available taxonomic/systematic information. The established use of current classifications (referred to above) fits comfortably into this framework; at the same time, however, the concept of phylogenetic diversity also reveals how basic classification information can be misleading, with better prediction obtained from phylogenetic (topology and branch length) information, when available.

(2) In contrast to phylogenetic diversity, other proposed methods are shown to be incompatible with conventional classification-based assessments,

generally because they do not necessarily incorporate implied divergence (character difference) information from the classification.

Phylogenetic diversity

The 'phylogenetic diversity' ('PD') of a subset of taxa, *s*, is defined (Faith 1992*a*) as the sum of the lengths of all those branches on a cladogram (or estimate of phylogeny) that are members of the 'minimum spanning path' defined by *s*. The minimum spanning path is made up of the smallest assemblage of branches such that, for any two members of *s*, a path connecting the two can be found that uses only branches in the assemblage.

The phylogenetic diversity of any subset of taxa can be evaluated by simply adding up the branch lengths on the corresponding minimum spanning path, or by using a summary matrix consisting of pairwise distances between taxa taken from the cladogram (or distances from another source assumed to be compatible with an underlying phylogeny). In this matrix, the distance, $D_{a,b}$, between two taxa, *a* and *b*, is the sum of the lengths of the branches on the path between them. If a subset of the taxa, *s*, is already protected (e.g. in a reserve network), then the increase in phylogenetic diversity, I(x), resulting from the addition of a new taxon, x, to the reserve is given by (Faith 1992*a*):

$$\mathrm{I(x)} = \text{Minimum, over all } i, j \text{ in } s, \text{ of } 0.5(D_{x,i} + D_{x,j} - D_{i,j}) \quad (18.1)$$

This formula measures the 'length of the branch or branches' gained through the addition of taxon x to the subset. In practice, the new taxon that implies the largest increase in PD may have highest priority for addition to the reserve network. Similarly, *I* may be calculated for a region, by applying formula (18.1) to those new taxa found in the region.

Species (or other terminal taxa), that are phylogenetically distinctive naturally will be expected to make a large contribution to the overall diversity of any subset of the total set of species. This diversity can be expressed another way: subsets of taxa with higher PD values should have greater underlying *feature diversity*. Thus, phylogenetically distinctive species are expected to contribute more to the diversity of a given subset because they contribute different 'features' (for discussion see Faith 1992*a*). Viewed this way, species are replaced as the basic units of biological diversity by *features* of species; it is this definition that justifies differentially prioritising species based on phylogeny.

McNeely *et al.* (1990) describe diversity as the raw material for adapting to change, so providing 'option value': a safety net of biological variation for responding to unpredictable events or needs. The feature diversity, represented by a subset of species provides option value in ensuring not only that one or more members of the subset can adapt to changing conditions, but also that society may be able to benefit from features of these species in response to future needs (Faith 1992*a*).

There is a practical difficulty in using feature diversity as a basis for evaluating subsets of taxa. *General* feature diversity patterns (rather than those specific to a given set of measured characters) cannot be measured directly; instead, estimates of phylogenetic relationships (e.g. from cladistic studies) provide values for an index of phylogenetic diversity. This implies a form of prediction, in that the subset with maximum phylogenetic diversity is expected, on average, to have greater feature diversity for any new set of features (Faith 1992*a*).

Comparisons of proposed definitions of taxonomic diversity (Faith 1992*b*) suggest that, if underlying feature diversity patterns are to be the basis for setting conservation priorities, then phylogenetic diversity should be used because it predicts these patterns best. The prediction of feature diversity using PD depends on its use of not only topology, but also divergence (branch length) information (Faith 1992*b*).

While other methods have been proposed that use divergence information, the inclusion of such information does not in itself guarantee a method that effectively predicts feature diversity patterns among taxa. Faith (submitted) shows that Altschul and Lipman's (1991) proposed measure results in weights for taxa that are in fact opposite to those required if feature diversity is to be predicted. Weitzman (1992) also uses divergence information, but adopts an unrealistic approach in using an underlying model that is tied to an assumption of equal rates of character change in different lines of descent. [His work does, however, show how systematic diversity measures may be linked, in a quantitative way, with economic considerations.] However, one possible justification for his approach is that the goal is not representation of feature diversity as defined here, but rather representation of combinations of different features (Weitzman, pers. comm.).

These alternative methods highlight the difficulties in identifying a reasonable model for taxonomic or phylogenetic diversity. Not only must there be a decision as to what is to be predicted or represented by a particular definition, but also there must be a decision as to how phylogenetic information relates to this goal.

Node-counting and divergence information

The use of divergence (branch length) information, interpreted in a cladistic (or phylogenetic, *sensu* Farris 1979) framework, appears to be both a strength and a weakness of phylogenetic diversity. The divergence information helps to link the phylogeny to predicted feature diversity (Faith 1992*b*), but information about relative amounts of divergence can be difficult to obtain.

One alternative strategy is to disregard the potential role of divergence information, by using phylogenetic diversity in its nodes-only form (Faith 1992*a*), or by adopting one of the nodes-based methods of Williams *et al.* (1991). All of these nodes-based methods have an information-availability advantage; all that is required is an estimate of the hierarchical topology, derived either through cladistic analysis or from an existing classification.

However, PD in the node-counting form, and the node-counting methods of Williams *et al.* (1991), are not as effective (relative to PD with branch-length information) as predictors of feature-diversity (Faith 1992*b*). An example (Fig. 18.1) highlights the practical limitations of any measure, including PD, when restricted to its node-counting form (Faith 1992*b*). For the phylogeny in Fig. 18.1, the species that is the sole member of 'genus 1' should make a strong contribution to phylogenetic diversity, based on the length of the corresponding branch (in fact, only the branch length is of importance; it is not necessary for purposes of this initial example for the species to be in its own genus). However, such a branch will not have any intervening branching points (nodes) that could be counted-up as an indication of its degree of divergence from other taxa; consequently none of the nodes-based measures of Williams *et al.* (1991) can consistently reflect the contribution to overall diversity of such a species (but see Williams and Humphries, this volume).

Phylogenetic diversity and classification information

Another information-availability problem relates to possible absence of phylogeny-related topological information (branching pattern), possibly in combination with absence of divergence information. In such cases, the

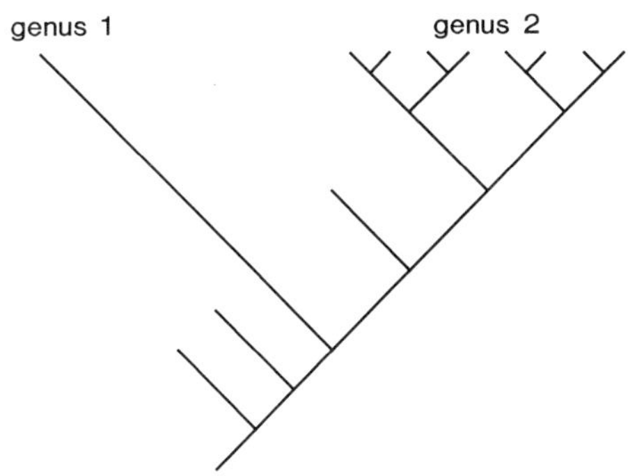

Fig. 18.1 A hypothetical phylogeny in which one species is at the end of a long branch and is the sole member of genus 1. Node-counting cannot reflect the anagenetic divergence represented by this branch.

existing classification of the species forms the only usable information about the possible phylogenetic relationships among taxa.

Even in cases where additional divergence information is not directly associated with the classification, interpreting the different taxonomic groups as separated along phylogenetic paths by nominated amounts of change, means that the basic classification can be interpreted in the PD framework (Faith, in press). For example, phylogenetic information meant that the species representing 'genus 1' (Fig. 18.1) would make a strong contribution to phylogenetic diversity, based on the length of the corresponding branch. But even in the absence of the phylogenetic information that includes this branch length, the recognition of the species as forming a unique genus provides information compatible with phylogenetic diversity calculations (Faith, in press). If the average degree of divergence (i.e. the branch length) to the common ancestor of a genus is 'dg', and the average divergence represented by a species is 'ds', then the increase in phylogenetic diversity I is approximated by:

$$I = G\ d_g + N\ d_s, \qquad (18.2)$$

where G is the number of previously unrepresented genera and N is the number of new species present. This equation is analogous to equation (18.1) both record the amount of additional branch length gained with the addition of the taxa in question.

Figure 18.2 presents another example. The classification is drawn as a hierarchy with different levels corresponding to each taxonomic rank. The

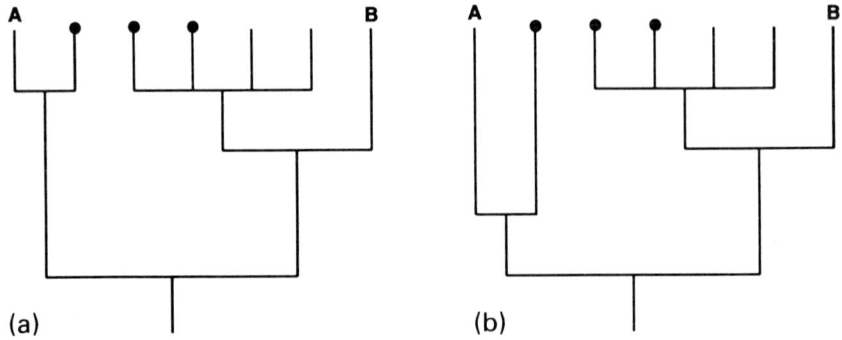

Fig. 18.2 Two hierarchies recording the classification levels (species, genera, families, tribes) as different levels in the hierarchy. The species represented by solid circles are taken to be protected, and a choice is to be made between species A or B. In (a), species A's sister is in another family, while in (b), it is in another genus. Node counting methods cannot distinguish between these two scenarios.

lengths of the resulting branches can then be used in PD calculations. Suppose that the species represented by solid circles are taken as already protected, and there is a choice between species A and B. For the case in Fig. 18.2a, it is clear that choosing B would maximise the gain in PD, while in Fig. 18.2b, species A would be chosen, because it adds a new family, associated with a large amount of divergence.

Node-counting and classifications

This same hierarchical representation of classification information has been used as input to the nodes-based measures of Williams *et al.* (1991) and Williams *et al.* (1993. While the large divergence associated with the genus 1 (in the phylogeny of Fig. 18.1) would not be detected by these measures, it might appear that in cases where the classification hierarchy is represented as in Fig. 18.2 that such divergence would be detected. For example, in Fig. 18.2, species B (in its own genus) would be chosen before any of the species within the 4-species genus, because the classification hierarchy shows an extra node 'gained' with the addition of B. However, the contrast between Fig. 18.2a and 18.2b shows that the nodes-based methods can still fail. Note that node-counting cannot detect any difference between the case where species A is just another species in an already-protected genus (Fig. 18.2a) versus the case where it is a species in a new family (Fig. 18.2b). For this reason methods based on nodes are not recommended, even when only classification information is available.

Another strategy for using basic classification information

Mares (1992) used existing classification information in his comparison of the biodiversity contributions of lowland Amazon rainforest and drylands of South America. Based on tabulations of the numbers of species, genera, and families in the two areas, Mares showed that the drylands in fact make the greater contribution to diversity, and used this study to emphasize the need to take what Mares referred to as 'hierarchical diversity' into account. While comparisons based on the taxonomic hierarchy were easily achieved for his study (the drylands clearly had many more endemic genera), Mares also recommended a particular quantitative measure of hierarchical diversity devised by Pielou (1969). This same measure has been recommended, following an IUBS workshop (Solbrig 1991), as an approach to taking hierarchical diversity into account through the apparent partitioning of diversity into hierarchical components.

However, as shown below, the partitioning of diversity in Pielou's measure is not useful as a general measure of the contribution of both species and genera to overall diversity.

The measure, $H'_G(S)$ (indicating the measure H' applied to genera and species), is equal to:

$$= -\Sigma_i\Sigma_j q_i p_{ij} \log (q_i p_{ij})$$

$$= -\Sigma_i q_i \log (q_i)\ \Sigma_i q_i\ [-\Sigma_j p_{ij} \log (p_{ij})]$$

$$= A + B,$$

where there are G genera, N species, M_i species in genus i, q_i is the proportion of individuals belonging to genus i, and p_{ij} is the proportion of the genus i members belonging to species j (in this discussion all species effectively have equal abundance as species are recorded as present or absent).

First, note that the overall value of $H'_G(S)$ cannot be used as a taxonomic diversity summary. When all species are simply recorded as present or absent, the measure reduces to:

$$A + B = -\Sigma_i q_i \log (q_i) + \Sigma_i q_i) + \Sigma_i q_i\ [-\Sigma_j\ p_{ij} \log (p_{ij})]$$

$$= -1/N\Sigma_i M_i \log M_i + \log N + 1/N\Sigma_i M_i \log M_i$$

$$= \log N.$$

Thus, the overall quantity (A + B) is not influenced in any way by the number of different genera that are present. Consequently, two areas (e.g. with taxa as in I versus II in Fig. 18.3) having the same number of species but a different number of endemic genera (I > II), would not differ in 'hierarchical diversity' (the $H'_G(S)$ value is log4 in both cases; Table 18.1).

An alternative might be some index that contrasts the two components, A and B. Pielou (1969: p. 231) points out that, for two areas having the same number of species, the area with the greater number of genera will have a *higher* A term and a lower B term. In Fig. 18.3, the area with the taxa shown in I would follow this pattern relative to the area with the taxa shown in II. This suggests that some measure contrasting these components, such as A/(A + B), or A − B, might be used. However, examination of the additional area shown in III shows that these measures are also misleading. Area III has the same number of genera but twice the number of species compared to area I. The diversity value (Table 18.1) for either of these alternative measures actually is smaller for the case (III versus I) where there are more species for the same number of genera! The hierarchical diversity components, A and B, therefore do not appear to be combinable in any way that consistently allows useful comparisons of the diversity represented by different areas.

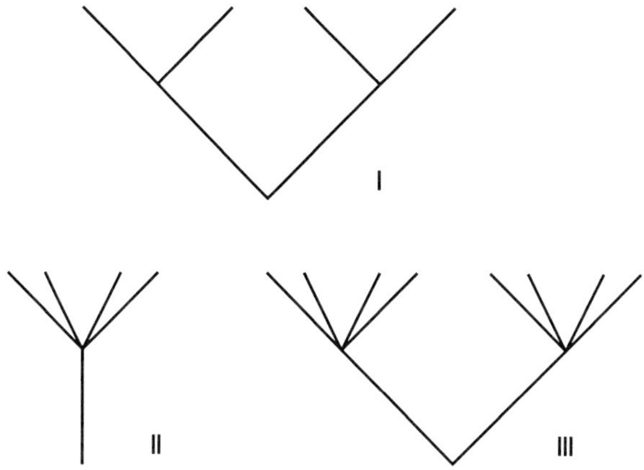

Fig. 18.3 Three different areas, I, II, and III, each contributing new species represented by the corresponding phylogenies. Area I has 2 new genera each with 2 species, area II has 1 new genus with 4 species, and area III has 2 new genera each with 4 species.

Table 18.1 Diversity values for different measures under the three scenarios from Figure 18.13.

	Diversity values		
Measure	I	II	III
A + B	log4	log4	log8
A / (A + B)	1/2	0	1/3
A − B	0	−log4	−log2
PD	6	5	10

The PD values (using equation 18.2) for the three areas are shown in Table 18.1, under the assumption that $d_g = d_s = 1$. Area III shows the greatest phylogenetic diversity, with the diversity for area I slightly greater than that for II. The calculations implied by equation (18.2) represent nothing more than a more formal version of common weighting schemes that take taxonomy into account (e.g. Ahern *et al.* 1985, and references within). Here, all genera and species are implicitly treated as equal in their contributions to diversity. This approximation represents the best estimate available in the absence of any other divergence information, and uses the classification-based information about hierarchy and divergence in a way consistent with the phylogenetic diversity model.

It should be emphasized that the results can differ dramatically from those that might be found from the corresponding cladogram for the same species; both topology and amounts of divergence can differ between the two. The two case studies presented below further explore the utility of classification information in phylogenetic diversity assessment. In the first of these, an existing classification is used in the absence of a phylogeny. The second example below (rainforest amphipods), shows one way in which a classification can provide information different from that obtained from the phylogeny. A cladogram of existing genera is presented, in which one genus makes a relatively large contribution to diversity, in representing a large degree of divergence from the root of the tree.

Two case studies

1. *Australian orchids*

Because 80 per cent of Australian vascular plants are endemic to the country, Australia is uniquely responsible for the conservation of a large number of plant species. Beginning in 1975, a working group representing all states and territories began compiling a list of rare or threatened Australian plants (ROTAP). This listing highlights those plants arguably most deserving of special conservation consideration. Approximately 17 per cent of the vascular flora of Australia is on the ROTAP list. Only about 11 per cent of the more than 3000 ROTAP plant species are regarded currently as adequately preserved.

The ROTAP list provides one form of 'weighting' of different plant species for conservation, but does not include other factors that might contribute to conservation interest in particular species. The ROTAP list has been criticized as under-valuing species that are widespread to the extent that they may not be conventionally regarded as 'rare', but nevertheless are vulnerable (McIntyre 1991). Complementary to the ROTAP listing is information on phylogenetic diversity; plant species that are phylogenetically unique may also deserve special conservation attention that would not be guaranteed by the ROTAP criteria.

The Orchidaceae illustrate how phylogenetic diversity can yield different values for areas relative to normal species-based approaches, and how this framework is applicable in cases of limited information. Phylogenetic diversity criteria can be applied to help determine which species, and which regions, deserve special attention for conservation. While Australia is very rich in orchid species (Clements 1989), a large number of the species are classified as rare or threatened (endangered or vulnerable). For purposes of

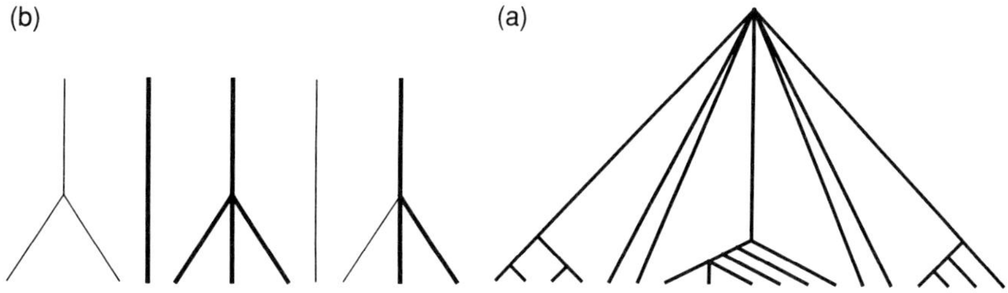

Fig. 18.4 A diagram representing the poorly known phylogenetic relationships for orchids. The thick lines represent taxa that are not endemic to Australia. (a) At the higher taxonomic level (tribes), there are no taxa endemic to Australia. (b) At the level of genera and species, some whole genera and some species are endemic to Australia (thin lines).

this study, species found outside Australia will be regarded as adequately protected. A total of 106 of the ROTAP orchid species (Briggs and Leigh 1988) are endemic to Australia, and so can be regarded as in need of protection within Australia.

Figure 18.4a shows a diagram of the poorly known phylogenetic relationships among orchids (Dressler 1981). At this coarse level, none of the terminal taxa are endemic to Australia. In fact, application of the phylogenetic diversity criterion for Australian orchids needs only to address the phylogeny at the generic level and below (Fig. 18.4b); some of these genera can be regarded as protected elsewhere because they are not endemic to Australia (the thick lines in Fig. 18.4b), while other genera and species are Australian endemics. Clements (pers. comm.) is now developing a cladogram for Australian orchids, but the existing classification information can nevertheless be incorporated into phylogenetic diversity framework.

The areas used for summary in this example (Fig. 18.5) will be those defined by the usual Australian botanic regions (e.g. Clements 1989). Table 18.2 lists those endemic genera and (one or more) component species whose members are all on the ROTAP list.

The Cape York region (40) contributes the largest amount to phylogenetic diversity, implying protection of 32 species and two of the endemic genera (*Trichoglottis* and *Aphyllorchis*). The two regions next in importance are very similar to each other in the number of additional species represented (note that complementarity is applied here in that only those species not found in region 40 are counted; in this case, however, there is no overlap among these regions). Region 1 in the far south-west has 17 endemic species from the ROTAP list and region 63 has 16. This information on its own would suggest that region 1 would make the greater contribution to overall diversity, but region 63 contains five more endemic genera, while region 1 would imply protection of only one additional genus. Thus, even without any knowledge

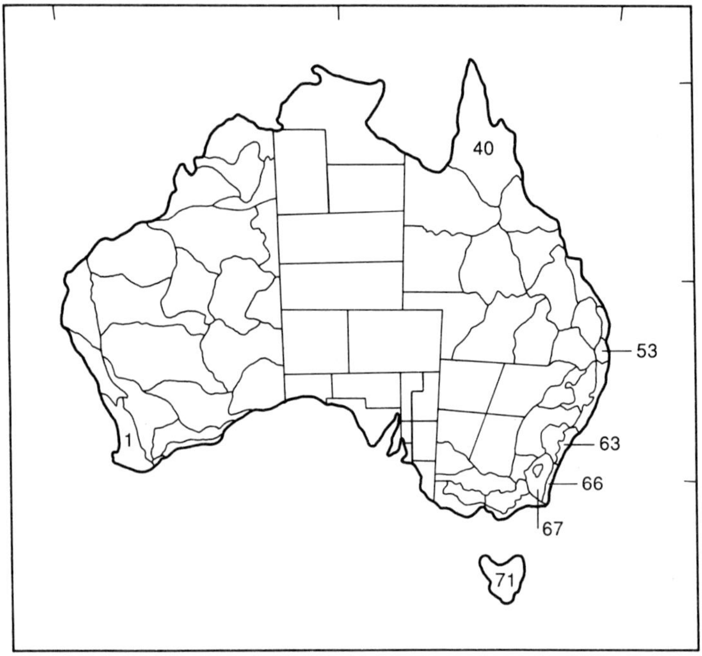

Fig. 18.5 A map of Australia showing botanic regions. Numbered regions are those making high contributions to phylogenetic diversity.

Table 18.2 Endemic genera and component species on the ROTAP list.

Adenochilus	*nortonii*
Aphyllorchis	*anomala*
	queenslandica
Burnettia	*cuneata*
Drakaea	*jeanensis*
	(plus 3 other species)
Genoplesium	*baueri*
Papillilabrium	*beckleri*
Rhizanthela	*gardneri*
	slateri
Schistotylus	*purpuratus*
Trichoglottis	*australiensis*

of the relative divergence among genera versus among species (i.e. the relative lengths of branches in Fig. 18.4), it appears that region 63 is somewhat more important in its contribution to total phylogenetic diversity.

The map (Fig. 18.5) also shows the two regions that would subsequently be chosen to complement the first three. Region 53 would add 10 more species, and then region 71 would add nine more. Overall, these five regions represent 84 of the 106 endemic ROTAP species of orchids, and an even higher proportion of the total phylogenetic diversity to be protected, given that these regions contain eight out of nine of the unprotected genera.

The orchids exemplify a case where areas receive different weights using phylogenetic diversity versus species-counting. The cladogram of orchids (Clements, in prep.), when available, may change the result found here; for example, a given genus may be highly divergent and so contribute proportionally more to diversity. The result reported here is also subject to other limitations on available information. In fact, a number of new species are reported as endemic to region 1 and its neighbours (Clements, pers. comm.). Also, one of the genera (*Genoplesium*) previously represented by a single species, and found only in regions 63 and 66, now has several newly described species, some relatively well-protected (Jones 1992). The importance of region 40 might also be adjusted by additional information; a number of species found only in this region in Australia are found outside Australia only in Papua New Guinea. If any of these species are determined to be poorly protected there, then the conservation importance of region 40 will increase, through its greater potential contribution to the protection of the phylogenetic diversity of this group.

The list of endemic species and genera provides a different conservation perspective relative to the list of endangered orchids (Briggs and Leigh 1988); many of the endangered species are found outside Australia, are only endangered in part of Australia, or are members of genera with many species. Also, the endangered species give a different priority ordering on the areas; for example, region 67 easily exceeds the other regions in having four of the endangered species, but is not among the highest contributors in the ordered list for phylogenetic diversity.

2. *Tasmanian rainforest*

The cool temperate rainforest of Tasmania (Fig. 18.6) has been a conservation focus in Australia, facing various threats in different areas from clearing for agriculture and plantation forest, inundation for hydroelectric schemes, and invasion of alien plants (Jarman *et al.* 1987). Over three-quarters of the Tasmanian rainforest is either reserved, classed as 'recommended for protection', or classed as 'deferred'. These areas presently are not available for logging. The Working Group for Forest Conservation has recognized the need to evaluate the levels of protection required for various

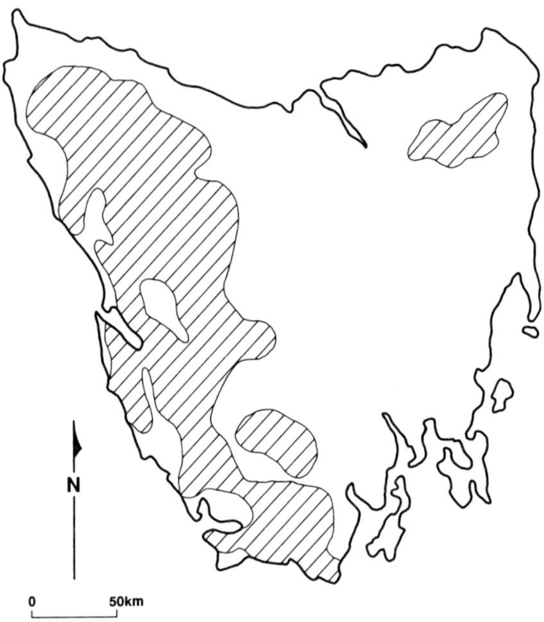

Fig. 18.6 A map of Tasmania showing approximate distribution of rainforest.

areas. While vascular angiosperms and gymnosperms are well-protected, current work focuses on evaluating non-vascular plants (Jarman *et al.* 1991) and beginning assessments based on invertebrate diversity (Greenslade, pers. comm.). Patterns in invertebrate diversity may provide important information; vascular plants are already well-reserved and, while non-vascular plants may not be as well-reserved, they do not have a high level of endemicity in Tasmania (Jarman *et al.* 1991). Additionally, a focus on the conservation of vegetation alone may not imply adequate protection for invertebrates (Taylor 1991).

The goal of our study (Greenslade and Faith, unpublished data) will be to assess material for a number of different invertebrate groups, in order to identify sites or areas that are complementary to other areas in their represented phylogenetic diversity. Greenslade (unpublished data) describes the vegetation at representative sample sites, their land tenure, and sampling methods. Phylogenetic diversity will be calculated using cladistic information for the study groups, with comparisons made to corresponding results using classification information. As an example, the preliminary results for one group, Amphipoda, are presented here.

Friend (1987) reports that approximately 93 per cent of the Tasmanian terrestrial amphipod (Amphipoda; 'landhoppers') species are endemic to the state; three of seven genera are also endemic. Overall, the distribution information over the Greenslade sites corroborates and extends distribution information found in Friend (1987). Preliminary results from the Greenslade sites suggest that the two north-western sites are distinguished in having species in two of the endemic genera.

An initial estimate has been made of the phylogenetic relationships among many of the amphipod species found in Tasmania (Friend, unpublished data). Figure 18.7 presents a diagram of the basic form of this cladogram using letter-codes for taxon names (the complete cladogram with taxon names is contained in Friend, unpublished). Using the divergence information from the cladogram (the heavier line represents diversity that is assumed to be protected elsewhere, given the distribution of certain genera outside Tasmania), the two sample sites in the north-west (in uniquely having species from endemic genera 'O' and 'T' on the cladogram) contribute 19 diversity (branch length) units. Another sample site in the south-east (uniquely having species A1, A2, and A3) contributes 16 or more units. A final assessment of phylogenetic diversity patterns awaits the sorting of material from other sample sites (Greenslade, unpublished).

A tentative conclusion from these preliminary data is that the area defined by north-west sites will be a major contributor to phylogenetic diversity, over amphipods and several other invertebrate groups. Unlike many of the other sites which are in reserves, the sample sites in the north-west are part of an area that is state forest, administered by the Forestry Commission

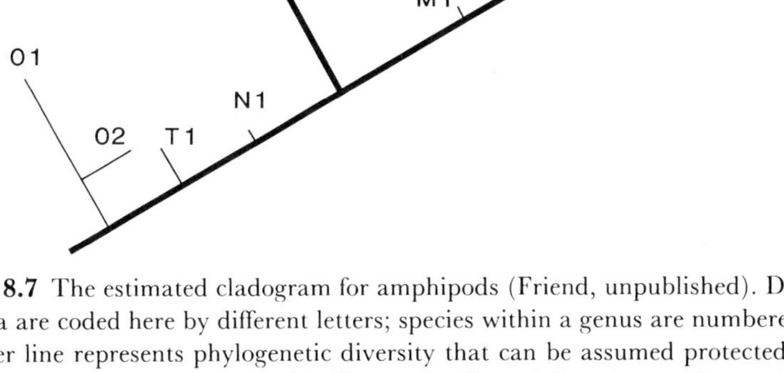

Fig. 18.7 The estimated cladogram for amphipods (Friend, unpublished). Different genera are coded here by different letters; species within a genus are numbered. The heavier line represents phylogenetic diversity that can be assumed protected as the spanning taxa are present outside Tasmania. Branch lengths are drawn in units proportional to the estimated relative number of character changes in each branch (for example, the length of the branch to species O2 is 4 units).

(Greenslade, pers. comm.); if this area proves to be highly complementary to other areas in its phylogenetic diversity, then sites in this area may be candidates for increased conservation efforts.

The amphipod cladogram also serves to illustrate the potential additional information available from a cladogram over that contained only in the classification. The classification can only indicate all genera as equal in their contribution to diversity, but here the cladogram shows that the hypothetical loss of the genus M might represent a smaller loss in underlying feature diversity, relative to the loss of the genus K, which is highly divergent.

Conclusion

Phylogenetic diversity as a framework for systematics-based conservation evaluation has several useful properties. One is that its optimization goal (maximizing represented feature diversity) is clearly defined. This also means that the predictive success of the measure is testable; a subset of taxa with high phylogenetic diversity should have high feature diversity for a new set of characters. Another important property of phylogenetic diversity is that this same framework is applicable over a realistic range of information availability—as summarized in the sequenced list shown in Table 18.3. The

Table 18.3 Available information for phylogenetic diversity.

	Information available	Version of phylogenetic diversity
1	existing classification information that may not reflect phylogeny	equation (18.2), d_g, d_s subjectively defined, or equal weighting for all species
2	classification information concordant with phylogeny	equation (18.2), d_g, d_s subjectively defined
3	classification information with some additional divergence information	equation (18.2), d_g, d_s vary among taxa, as defined by divergence information
4	well corroborated phylogenetic estimates, topology only	equation (18.1), branch lengths defined by some model, or defined as all equal to some constant
5	well-corroborated phylogenetic estimates, including branch lengths	equation (18.1)

information that is available determines which of two different formulae is used; however, these formulae are different expressions for the same quantity—the sum of branch lengths along the corresponding minimum spanning path. Further, the goal under all these information scenarios is therefore the same; to identify the subset of species that can be expected to maximize the represented feature diversity.

Acknowledgements

I thank the organizers of the 'Systematics and Conservation Evaluation Symposium' for the invitation to attend the meeting. The work of Vane-Wright, Humphries, and Williams has stimulated much of my recent work on this topic. I am grateful to P. Greenslade, A. Friend, and M. Clements for discussion and collaboration, and to F. Sheard for figure preparations. C. Margules and P. Williams provided helpful reviews of the manuscript.

References

Ahern, L. D., Brown, P. R., Robertson, P., and Seebeck, J. H. (1985). Application of a taxon priority system to some Victorian vertebrate fauna. *Technical Report Series.* **32**. Arthur Rylah Institute for Environmental Research, Victoria.

Altschul, S. F. and Lipman, D. J. (1991). Equal animals. *Nature*, **348**, 493–4.

Briggs, J. D. and Leigh, J. H. (1988). *Rare or threatened Australian plants.* ANPWS, Australia.

Clements, M. A. (1989). Catalogue of Australian Orchidae. *Australian Orchid Research*, Vol. 1. Australian Orchid Foundation, Victoria.

Dressler, R. L. (1981). *The orchids: natural history and classification.* Harvard University Press, Cambridge, MA.

Faith, D. P. (1992*a*). Conservation evaluation and phylogenetic diversity. *Biological Conservation*, **61**, 1–10.

Faith, D. P. (1992*b*). Systematics and conservation: on predicting the feature diversity of subsets of taxa. *Cladistics*, **8**, 361–73.

Faith, D. P. and Cawsey, E. M. (1990). PHYLOREP, software for phylogenetic representativeness. Distributed privately.

Farris, J. S. (1979). The information content of the phylogenetic system. *Systematic Zoology*, **28**, 483–519.

Friend, J. A. (1987). Terrestrial Amphipods (Amphipoda: Talitridae) of Tasmania: systematics and zoogeography. *Records of the Australian Museum*, Suppl. 7.

Hopper, S. D. and Coates, D. J. (1990). Conservation of genetic resources in Australia's flora and fauna. *Proceedings of the Ecological Society of Australia*, **16**, 567–77.

IUCN (1980). *World conservation strategy: living resource conservation for sustainable development.* IUCN–UNEP–WWF, Gland, Switzerland.

Jarman, S. J., Brown, M. J., and Kantvilas, G. (1987). The classification, distribution and conservation status of Tasmanian rainforests. In *The rainforest legacy*, AGPS, Canberra.

Jarman, S. J., Kantvilas, G., and Brown, M. J. (1991) Floristic and ecological studies in Tasmanian rainforest. *Tasmanian NRCP Report*, **3**.

Jones, D. L. (1991). New taxa of Australian Orchidaceae. *Australian Orchid Research*, **2**, Australian Orchid Foundation, Victoria.

McIntyre, S. (1991). Risks associated with the setting of conservation priorities from rare plant species lists. *Biological Conservation*, **60**, 31–7.

McNeely, J. A., Miller, K. R., Reid, W. V., Mittermeier, R. A., and Werner, T. B. (1990). *Conserving the world's biological diversity*. IUCN, WRI, CI, WWF–US, World Bank [International Union for Conservation of Nature and Natural Resources, World Resources Institute, Conservation International, World Wildlife Fund—US], Gland, Switzerland.

Mares, M. A. (1992). Neotropical mammals and the myth of Amazonian biodiversity. *Science*, **255**, 976–9.

Pielou, E. C. (1969). *An introduction to mathematical ecology*. John Wiley, New York.

Solbrig, O. T. (1991). *From genes to ecosystems: a research agenda for biodiversity*. IUBS, France.

Taylor, R. J. (1991). *Fauna conservation in production forests in Tasmania*. Forestry Commission of Tasmania, Hobart.

Vane-Wright, R. I., Humphries, C. J., and Williams, P. H. (1991). What to protect—Systematics and the agony of choice. *Biological Conservation*, **55**, 235–54.

Weitzman, M. L. (1992). On diversity. *Quarterly Journal of Economics*, **62**, 363–405.

Williams, P. H., Humphries, C. J., and Vane-Wright, R. I. (1991). Measuring biodiversity: taxonomic relatedness for conservation priorities. *Australian Systematic Botany*, **4**, 665–79.

Williams, P. H., Vane-Wright, R. I., and Humphries, C. J. (1993). Measuring biodiversity for choosing conservation areas. In *Hymenoptera and biodiversity* (ed. J. LaSalle and I. D. Gauld), pp. 309–28. CAB International, Wallingford, UK.

19. Biodiversity, taxonomic relatedness, and endemism in conservation

P. H. WILLIAMS and C. J. HUMPHRIES
Biogeography and Conservation Laboratory, Department of Botany, The Natural History Museum, London SW7 5BD, UK.

Abstract

Practical approaches to measuring biodiversity are described. Within any defined geographical area biodiversity is considered to be a measure of two main variables: number of taxa and degrees of difference between taxa. Three approaches for the measurement of biodiversity are described, phenetic, phylogenetic, and cladistic, the latter being considered the most appropriate. Using the computer program WORLDMAP comparative trials of four different measures of cladistic diversity, root-weight, higher taxon richness, spanning-tree length, and cladistic dispersion, were undertaken. Even representation of sub-groups within taxa using the cladistic dispersion measure agrees best with the popular, intuitive ideal of what consititutes the most diverse fauna or flora. The taxonomic measures were used together with a measure of endemism in conjunction with complementarity of floras and faunas to determine priority sequences of areas for conservation. It was concluded that the greatest contribution to the conservation and representation of biodiversity is from complementarity, but differences between organisms can be most readily represented by using available information of group membership from taxonomic classifications. Endemism is not a measure of biodiversity but of range size, and was shown to be ineffective as a means for selecting areas for conservation if not all selected areas could be included.

Introduction

If we are to conserve as much as possible of biodiversity with limited resources, we shall need to know where most of the biodiversity is found (McNeely *et al.* 1990; WRI/IUCN/UNEP 1992; Groombridge 1992), and

Systematics and Conservation Evaluation (ed. P. L. Forey, C. J. Humphries, and R. I. Vane-Wright), Systematics Association Special Volume No. 50, pp. 269–87, Clarendon Press, Oxford, 1994.

therefore we shall need to be able to measure it (Vane-Wright *et al.* 1991; Pressey *et al.*, in press). Here we review practical approaches to measuring biodiversity within areas, and we discuss the suitability of these approaches in different situations.

Concepts of biodiversity

Biodiversity is a 'pseudocognate' term (Salt 1979), users of which assume that everyone shares the same intuitive definition. Most people do agree on which assemblages of organisms represent small amounts of biodiversity, such as crop monocultures and bowling greens. But at the other extreme, lack of a common, explicit, and clear definition hinders discussion and precise agreement on what constitutes the greatest amount of biodiversity.

One of the most widely cited definitions of biodiversity is that of McNeely *et al.* (1990): 'It is an umbrella term for the degree of nature's variety, including both the number and frequency of ecosystems, species, or genes in a given assemblage.' To date, this broad definition has defied rigorous, practical measurement. Most measurements have fallen short of the ideal because they have consisted of simple counts of the numbers of species (species richness).

Factors other than species richness have sometimes been included in diversity measures. Ecologists interested in 'community structure' have developed diversity measures that combine species richness with information on the evenness of abundance among species (Peet 1974; May 1975; Southwood 1978; Taylor 1978; Magurran 1988; Cousins 1991), whereas conservationists interested in preventing extinction have used measures that combine species richness with information on the perceived vulnerabilities of those species (Usher 1986). But workers in both fields have also identified the need to include information on taxonomic distinctness (Pielou 1975; IUCN/UNEP/WWF 1980; Groombridge 1992; WRI/IUCN/UNEP 1992).

The *Shorter Oxford English Dictionary* (1990) defines diversity more generally as difference or unlikeness. This definition has the consequence that, when applied to species, not all should be considered as equal. For example, a flora consisting of a dandelion and the unique *Welwitschia mirabilis* would be considered more diverse than a flora consisting of two species of dandelions. One suggestion is that a general measure of biodiversity ought to include not only a measure of the number of taxa in a sample, but also some weight that measures the *degree of difference* (in the sense of dissimilarity) among them (Vane-Wright *et al.* in May 1990; Vane-Wright *et al.* 1991; Williams *et al.* 1991, 1993). Thus, we concentrate on measuring present-day patterns, and seek to minimize assumptions about possible diversification processes (cf. Erwin 1991; Brooks *et al.* 1992) because of the problems of foretelling future patterns of diversification (Anderson 1974; Gould 1989; Green 1991) in patchy and changing environments.

Approaches to measuring biological differences

Three approaches could be used to measure degrees of difference among species (Table 19.1).

1. *Phenetic approach*

This approach depends on measuring the difference among species directly using dissimilarity in their characteristics. The data must be accepted as truly representative of the overall differences among species (alternatively they are the only characters of interest). Furthermore, for the recognition of predictive (natural) monophyletic groups data must also be accepted as free of homoplasy (homoplasy is the phenomenon of patterns of character states that are irreconcilable with a single tree classification). This direct approach

Table 19.1 Summary of approaches to measuring biological difference among taxa to assess biodiversity.

1. PHENETIC—direct attribute dissimilarity

from a pairwise matrix of attribute dissimilarity assumes:

(a) data are truly representative of overall differences among taxa or are representative of the only attributes of interest,

(b) no homoplasy.

2. PHYLOGENETIC—anagenetic branch-length estimates

from distance estimates across n-trees assumes:

(a) unbiassed sampling of attributes within and between branches,

(b) attribute state transitions show constant relative rates of change between sampled and unsampled attributes.

3. TAXONOMIC—group membership

from topology (shape) of any (preferably cladistic) classification assumes:

(a) attribute state differences are inherited,

(b) attribute state differences accumulate along branches.

could, in principle, be applied to dissimilarity measures based on ecological or genetic attributes (Williams *et al.* 1991). However, homoplasy is now universally considered to have invalidated phenetics as an approach to estimating genealogies (e.g. Farris 1977), so a phenetic approach to predicting unsampled character differences cannot be supported.

2. *Phylogenetic approach*

The phylogenetic approach recognizes the problem of homoplasy outlined for the phenetic approach, but retains quantitative measures of character differences as a representation of overall anagenetic change (i.e. the degree of evolutionary change without speciation) along the branches of a proposed phylogenetic tree (Altschul and Lipman 1990; Crozier 1992; Faith 1992, in press, submitted, this volume). Consequently, the phylogenetic approach has been used when character data are believed to be truly representative samples of the unstudied differences among species. Indeed, characters, or 'features' as they have been called, may be seen as the appropriate foci of conservation rather than species (Faith 1992). The phylogenetic approach has to assume unbiased sampling of characters, both within and between branches, and assumes that character changes show constant relative rates of change with time between sampled and unsampled characters (Williams *et al.* 1993). An even more restrictive assumption of regular, clock-like character change is made by Weitzman's (1992) method.

In the absence of discrete character data, information from classifications in the form of nomenclatural rank has been interpreted as an estimate of anagenetic character changes (Faith, this volume). We have avoided this interpretation because the assignment of nomenclatural rank, other than for sister-groups, is essentially arbitrary (e.g. Gauthier *et al.* 1988).

The phylogenetic approach could be applied to all data, but we expect it to work best with data from molecular sequences (see Farris 1985). We suspect that for morphological studies both of the assumptions above are likely to be violated severely. Crozier (1992) argued that if maximum parsimony is applied to morphological data, then there are several reasons why it is hard to estimate branch lengths. One is that there are usually a number of character state changes which cannot be assigned to particular branches. Another is that, in the case of morphological data, the degree of independence between characters is particularly hard to determine.

3. *Taxonomic approach*

In practice, usually the only reliable information available for most groups of organisms is some taxonomic pattern of group membership. Therefore we have chosen to investigate the taxonomic approach, which uses only the information in patterns of groups, from the nodes or branching points in these classifications (Vane-Wright *et al.* 1991; Williams *et al.* 1991, 1993). The assumptions of the taxonomic approach are less demanding, because group membership in genealogical classifications should predict overall differences among species (Farris 1983) if character differences are inherited, and if differences accumulate along branches.

The best results from the taxonomic approach should be obtained using well-supported and fully-resolved cladistic classifications. However, in principle the approach could also be applied to any hierarchical classification, and useful information may be gleaned from even partially resolved pre-cladistic classifications.

WORLDMAP computer software

We have used the WORLDMAP program (version 3.04) as a test platform for trials of measures of taxonomic diversity, endemism, and complementarity (e.g. Vane-Wright *et al.* 1991; Williams 1992). This software is not intended to be yet another fully-flexible, geographic information system (GIS), but rather to give rapid results for its specialized purpose.

Measures of taxonomic diversity

Figure 19.1 summarizes the criteria behind the four taxonomic diversity measures that we have examined (along with unweighted species richness and endemism-weighted species richness). The challenge presented in Fig. 19.1 is to decide which is the most diverse biota of three species to be drawn from among the total group of eight. Once this has been identified, the preferred criterion can be applied to more complex cases.

1. *Root weight*

The first measure, root weight, was developed in response to a need to weight groups such as the tuatara, *Sphenodon punctatus* and *S. guntheri*, more highly than any two species from the sister-group squamates, so as to express the greater value that has been attached to these 'relic' species (Vane-Wright *et al.* in May 1990; Vane-Wright *et al.* 1991). The two surviving species of *Sphenodon* branch off close to the root of a classification, with no surviving close relatives, whereas the sister-group of lizards and snakes has 6000 or so species. The root weight measure weights species individually by measuring how close-to-root a species diverges by comparison to the remainder of species in the classification (Vane-Wright *et al.* 1991). The method proceeds by counting the number of nodes (branching points) between the root of the classification and each species in turn. In order to produce higher weights for the most close-to-root species, the total node count for the classification is

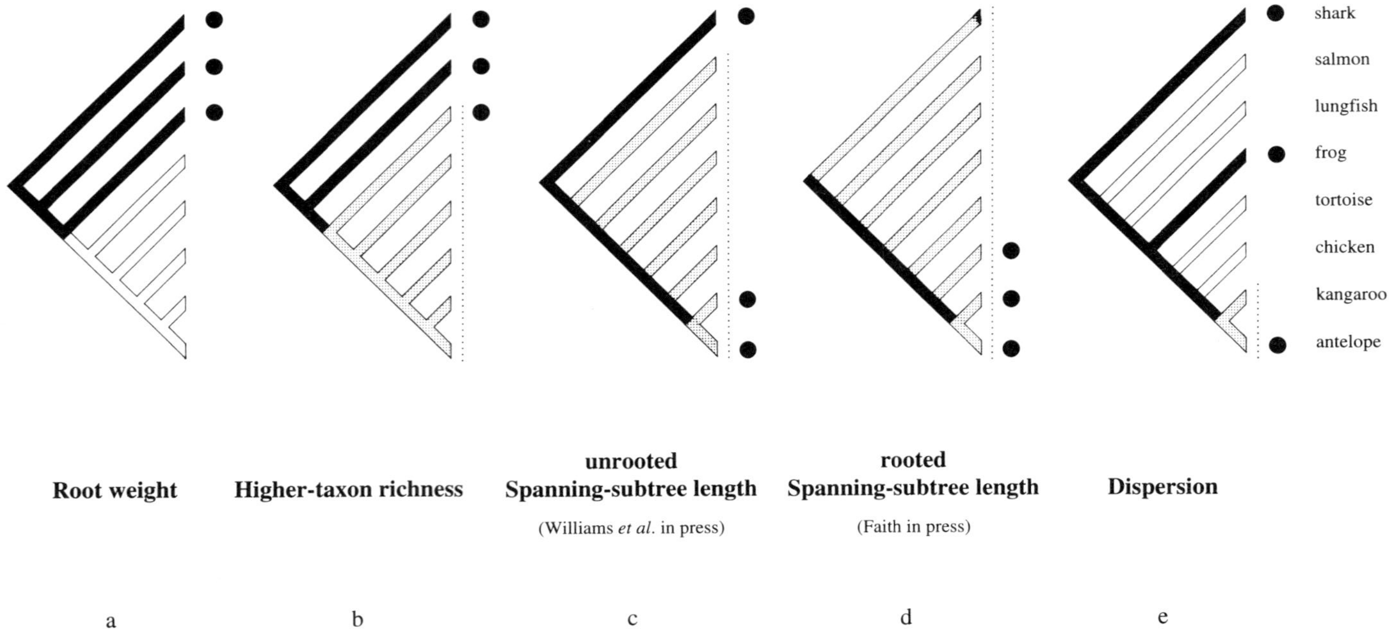

Fig. 19.1 Classification of eight surviving species showing which biotas of just three species would score most highly when using each of the four measures of taxonomic diversity. The selected species are shown by black spots, with any equally-highly scoring alternatives for each choice shown bracketed by dotted lines. The groups within the classification that are essential to represent high diversity value using each measure are shown in black, groups which remain alternative choices are shown stippled, and groups which are of low diversity value are shown in white outlined in black.

divided by the node count for each species. These individual species scores can be expressed as a percentage of the total node count for the classification and can be simply added together to give the diversity score for a biota.

'Relic' status may be a valid criterion, but its implementation in the root weight measure has some undesirable properties. With incompletely-resolved classifications, underestimated divergences may grossly overweight some biotas (May 1990; Crozier 1992). But a more fundamental problem is that with more symmetrical classifications it fails to give greater weight to those biotas with the most divergent species (Williams *et al.* 1991).

2. *Higher-taxon richness*

The second measure, higher-taxon richness, is a development of root weight that values higher taxa irrespective of the numbers of species within each (Williams *et al.* 1991, 1993).

One way of measuring the number of higher taxa in a biota is to use pairwise measures of taxonomic divergence between species. For a biota in which more of the higher taxa are represented, the species will tend to diverge closer to the root of the classification. Consequently, species will have a lower average number of shared nodes from the root of the classification. Therefore, the higher-taxon diversity of a biota can be measured as the product of its richness of species with the mean inverse number of shared nodes from the root of the classification (i.e. number of species × mean $1/S_{ij}$, for all pairwise comparisons between species i and j; Williams *et al.* 1991). Note that because species are weighted from pairwise comparisons, only biotas with at least two species will receive a meaningful diversity score. For our computer implementation of these measures, biotas with single species are given an arbitrary score of zero, although their presence is still registered on the map. Just as for root weight, biota scores can be expressed as a percentage of the total score for the entire classification. However, unlike root weight scores, the scores for higher-taxon richness are no longer simply additive, because the cumulative score must take into account the cross comparisons between those species not shared by the separate biotas. So the diversity increment, when one biota is added to another, is calculated as the difference between the score for the pooled biota and the score for the original biota.

Higher-taxon richness solves the fundamental problem of symmetric trees identified for root weight, and unresolved 'bushes' of species (terminal polytomies) are less of a problem. However, like root weight it still does not yield a good representation of the variety of species in highly asymmetric trees, as shown in Fig. 19.1. For instance, using a highly simplified hypothetical example of say, just eight surviving vertebrates (shark, salmon, lungfish, frog, tortoise, lizard, kangaroo, antelope), higher-taxon richness would give a higher score to a fauna consisting of the shark, the salmon, and lungfish (Fig. 19.1b), rather than to a fauna consisting of the shark, the frog, and antelope. Faunal and floral representativeness can only be improved by using measures of taxonomic representativeness.

3. *Spanning-subtree length*

Spanning-subtree length is the most obvious measure of taxonomic representativeness. It simply measures how much of a classification is represented in a biota (this is numerically equivalent to the node-counting version of 'phylogenetic diversity', with the addition of +1 (Faith 1992, in presss, this volume; Williams *et al.* 1993)).

The measure of spanning-subtree length (or cladistic 'path' length) of Williams *et al.* (1993) measures the taxonomic divergence among the species of a biota by adding up the number of intervening nodes of the classification between all of the species included in the biota, counting each node once. It has to count each species as an additional terminal node. Note that because species are weighted from numbers of nodes between species, only biotas with at least two species will receive a meaningful diversity score (just as for higher-taxon richness, for our computer implementation, biotas with single species are given a score of zero, although their presence is still registered on the map). Biota scores can be expressed as a percentage of the total score for the entire classification. Like higher-taxon richness scores, the scores for spanning-subtree length are not simply additive, because the cumulative score must take into account only those nodes between the new species and the previous spanning subtree. So the diversity increment when one biota is added to another is calculated as the difference between the score for the pooled biota and the score for the original biota.

Spanning-subtree length may work well with branch length data as advocated by Faith (1992, in press, this volume) if the data are good enough, but without this extra information, the 'node-counting' version of 'phylogenetic diversity' may be equivocal, with no guarantee of discrimination against taxonomic clumping that is required of a diversity measure. Williams *et al.* (in press) did not include any nodes between the root of the classification (which need not be identified) and the most basal node of the spanning subtree itself. Faith (in press) in challenging their dispersion measure, advocated treating an outgroup of a classification as an already protected part of the biota subtree, with the consequence that all nodes between the root and the spanning subtree should be included in the measure of subtree length. This is reasonable in so far as some organisms of other groups will always persist to provide 'secure' outgroups. However, it does greatly reduce discrimination of differences in the shape of biota subtrees that have similar numbers of species.

Using the simplified vertebrate example (Fig. 19.1), spanning-subtree length would give the same score to a fauna consisting of the shark, kangaroo, and antelope (Fig. 19.1c), as to a fauna consisting of the shark, frog, and antelope. The latter choice is more evenly representative of the variety of subgroups (more cladistically dispersed). Rooted spanning-subtree length is even less successful at representing the variety of subgroups, because it would give the same score to a fauna consisting of the chicken, kangaroo, and antelope (Fig. 19.1d).

4. *Cladistic dispersion*

The fourth measure, taxonomic dispersion, is a refinement of the representativeness criterion that lies behind spanning-subtree length. It values the representation of not only a broad range of subgroups from across the classification, but also a particularly *even representation* (Williams *et al* 1991, 1993). The logical basis of this criterion is that more evenly spread choices maximize the degree to which subtrees represent the overall shape of a classification, and thereby are more representative of the variety of subgroups with their different character *combinations*.

Cladistic evenness can be measured using the relationship between the mean and the standard deviation among species. An appropriate measure of taxonomic divergence is provided by a count of the number of nodes of the classification between two species (i.e. $U_{ij} + U_{ji} + 1$, the sum of the nodes unique to species i in relation to species j, and unique to j in relation to i, plus one). For a biota with species that are distantly and evenly spread across a classification (i.e. highly dispersed), the average divergence among the species will be high and the distribution of divergences around this average will be narrow.

The first implementation of a cladistic dispersion measure had two undesirable properties arising from the balance of the three terms in its calculation. Williams *et al.* (1991) realized that under certain circumstances it could yield negative diversity increments when species were added to a biota. Faith (in press) showed that under certain other circumstances the balance of preference between adding one of two species to a biota could change as more species were added across the cladogram.

Here we propose a revised measure of cladistic dispersion that overcomes the implementation problems identified by Williams *et al.* (1991) and by Faith (in press). It is based on the measure of unrooted spanning-subtree length in order to take account of the number of species in a biota and the cladistic divergences between them. This is a simple function that always increases (monotonically) with species richness (see section on spanning-subtree length). But in order improve its discrimination among different subtree shapes in favour of the more representative, cladistically-dispersed biotas, a dispersion term is added. This term is scaled in the range zero to one, so that it will only take effect among biotas of the same subtree length (if required, its effect could be increased by multiplying the dispersion term by a factor for the number of nodes of subtree length that may be sacrificed for greater eveness of subtree shape):

$$D = \begin{matrix}\text{unrooted} \\ \text{spanning} \\ \text{subtree} \\ \text{length}\end{matrix} + e^{\{-SD(U_{ij}+U_{ji}+1)/\text{mean}(U_{ij}+U_{ji}+1)\}}$$

where e is the base of the natural logarithm and SD is the standard deviation of the pairwise divergences between species. Note that because species are weighted from the eveness of pairwise comparisons, only biotas with at least three species will receive a meaningful dispersion score (for our computer implementation, biotas of one or two species are given an arbitrary score of zero, so their diversity score reverts to the spanning-subtree length score). Biota scores can be expressed as a percentage of the total score for the entire classification. However, as for higher-taxon richness scores, the scores for dispersion are not simply additive, because the cumulative score must take into account the cross comparisons between those species not shared by the separate biotas. So the diversity increment, when one biota is added to another, is calculated as the difference between the score for the pooled biota and the score for the original biota.

When cladistic dispersion is applied to the simplified example of vertebrates, it discriminates a fauna consisting of shark, frog, and either kangaroo or antelope as the most diverse (Fig. 19.1e). This even representation of subgroups agrees well with the popular, intuitive ideal of what constitutes the most diverse fauna.

Priority-areas analysis for conservation

All of the measures of taxonomic diversity can be applied to compare biotas of areas at any spatial scale. These measures are easily identified with the idea of local diversity within homogeneous habitat patches (α-diversity). This has been distinguished from differences along local gradients (β-diversity) or between regions at increasing spatial scales (γ-diversity) through partial species turnover (e.g. Whittaker 1960; Magurran 1988). However, species turnover can be combined with measures of taxonomic difference to increase the efficiency of conservation planning by using the simple idea of complementarity.

The complementary portion of a biota comprises those species unrepresented in another biota with which it is being compared. For our sequential priority-areas analyses, once a first high-diversity biota has been chosen, the second priority biota is selected for its highest incremental diversity score, based solely on its biotic complement. Once the second choice has been established, the third choice is based on considering the complement of the first two, and so on, ideally until the complement is reduced to zero (i.e. all species have been represented at least once; Vane-Wright *et al.* 1991). Alternatively, complementarity can be applied to choosing sets of areas at each step as described by Margules *et al.* (1988, this volume) and Pressey *et al.* (this volume). This approach is more complicated to implement, but can give more efficient solutions (Pressey and Nicholls 1989). The idea of complementarity is also fundamental to the method of Kirkpatrick (1983), to *Critical Faunas Analysis* (Ackery and Vane-Wright 1984) and to *Gap Analysis* (e.g. Burley 1988).

Assessment of complementarity, biodiversity measures and endemism

It could be argued that specialists should already know where the primary 'hotspots' for particular groups are found (e.g. Myers 1988, 1990), so why bother with biodiversity measures?

1. *Complementarity*

The most powerful principle is complementarity, which can only be applied to measures of biodiversity that retain information on the identity of species in biotas. Once a primary biodiversity 'hot spot' has been identified, the most efficient way to capture maximum additional biodiversity in a second chosen area is by complementarity. This also provides the flexibility to remove from consideration the biotas of areas that are already adequately protected before the start of an analysis, so that maximum additional biodiversity can be located for further protection. Because all of the remaining areas can be scored for complementary diversity, optimal choices can still be made even when constrained by socio-economic costs.

The effect of complementarity can be seen on representation of diversity for bumble bees of the *sibiricus*-group (Fig. 19.2). The stippled area shows the range of scores for priority-area sequences by species richness and by the taxonomic diversity measures (all re-scored by taxonomic dispersion). These diversity scores are generally significantly higher than would be expected from choosing areas at random. Indeed, all of the direct diversity measure sequences give fairly similar diversity scores, including unweighted species richness. This shows that the largest contribution to performance in accumulating biodiversity comes from using any direct measure of diversity coupled with complementarity.

2. *Taxon richness and distinctness*

The best measures of biodiversity should be those that require and receive the most complete information about biotas. Consequently, diversity

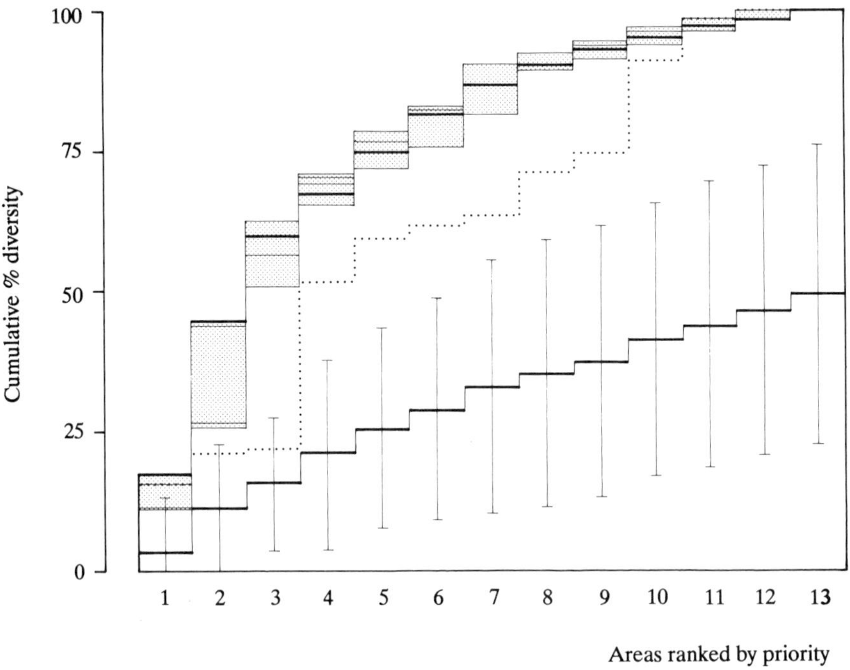

Fig. 19.2 Superimposed cumulative histograms of taxonomic dispersion scores for priority-area sequences for bumble bees of the sibiricus-group. The sequence shown by the lower solid line represents selection of areas at random, using the mean scores from 1000 simulated sequences, with vertical bars extending two standard deviations on either side of the mean. The sequence shown by the middle dotted line represents selection of areas by an absolute endemism criterion, preferentially choosing areas that contain the most narrowly-distributed species without regard to species richness. The stippled part of the figure covers the range of scores obtained by selecting areas by species richness and the four taxonomic diversity measures.

measures based on the phylogenetic approach, given perfect information, might be expected to perform especially well. However, in practice, *reliably representative* data for overall differences are rarely available, so this approach will not be considered.

Our preference among taxonomic diversity measures has been for the criterion of evenness of subgroup representation, despite the mathematical difficulties of its implementation in the taxonomic dispersion measure. Spanning-subtree length may also be useful, because it provides a simple measure of representativeness that is easier to calculate (Williams *et al.* in press).

How useful are taxonomic measures of biodiversity? At present, 25 species-based datasets have been compiled and analysed (Table 19.2). Of these, 22

Table 19.2 List of the species datasets analysed using the WORLDMAP computer program. Abbreviations: D = R, + if the area with maximum taxonomic dispersion score is among the areas with maximum species richness, otherwise − if areas are different; D = E, + if the area with maximum taxonomic dispersion is among the areas with maximum endemism score, otherwise − if areas are different (only for the Australian *Fraus* data does the same area score maximally for species richness and endemism but not taxonomic dispersion).

Area	Taxon	No. spp	D=R	D=E	Source
GLOBAL					
	Bombini (bumble bees inc. *sibiricus*-group)	240	+	−	P. Williams
	Strigiformes (owls)	207	−	−	Vane-Wright *et al.* this volume
	Danainae (milkweed butterflies)	161	+	−	R. Vane-Wright pers. comm.
	Pteropodidae (fruit bats)	159	+	+	Mickleburgh *et al.* in press
	Rhyssinae (wood wasp parasitoids)	62	+	+	P. Eggleton pers. comm.
	sibiricus-group (bumble bees)	43	−	−	P. Williams
REGIONAL					
Australia	*Monocalyptus* (eucalypt trees)	102	+	+	Humphries *et al.* 1988
Africa	Bovidae (antelopes)	68	+	−	East 1988, 1989, 1990
Africa	*Francolinus* (partridge-like birds)	36	+	−	T. Crowe pers. comm.
Australia	*Fraus* (ghost moths)	25	−	−	Nielsen and Kristensen 1989

Table 19.2 (*cont.*)

Area	Taxon	No. spp	D=R	D=E	Source
LOCAL					
Thailand	Sphingidae (hawk moths)	177	+	−	Vane-Wright *et al.* this volume
Philippines	Nymphalinae (brush-footed butterflies)	148	+	+	R. Vane-Wright and C. Treadaway pers. comm.
Philippines	Hesperiidae (skipper butterflies)	133	+	+	R. Vane-Wright, C. Treadaway and R. de Jong pers. comm.
Thailand	Cicindelidae (tiger beetles)	103	+	−	see Vane-Wright *et al.* this volume
Philippines	Polyommatini (blue butterflies)	81	+	−	R. Vane-Wright and C. Treadaway pers. comm.
Philippines	Satyrinae (brown butterflies)	63	+	−	R. Vane-Wright and C. Treadaway pers. comm.
Philippines	Arhopalini (blue butterflies)	62	+	−	R. Vane-Wright and C. Treadaway pers. comm.
Philippines	Pieridae (white butterflies)	52	+	+	R. Vane-Wright and C. Treadaway pers. comm.
Philippines	Papilionidae (swallowtail butterflies)	51	+	+	R. Vane-Wright and C. Treadaway pers. comm.
Philippines	Danainae (milkweed butterflies)	34	+	−	R. Vane-Wright and C. Treadaway pers. comm.
Philippines	Iolaini (blue butterflies)	34	+	+	R. Vane-Wright and C. Treadaway pers. comm.

Table 19.2 (*cont.*)

Area	Taxon	No. spp	D=R	D=E	Source
Philippines	Miletinae (blue butterflies)	29	+	−	R. Vane-Wright and C. Treadaway pers. comm.
Philippines	Poritiinae (blue butterflies)	20	+	+	R. Vane-Wright and C. Treadaway pers. comm.
Thailand	Strigiformes (owls)	17	+	−	I. Kitching pers. comm.
Philippines	Amathusiinae (butterflies)	15	+	−	R. Vane-Wright and C. Treadaway pers. comm.

datasets (88 per cent) show the highest-scoring biota by taxonomic dispersion to be among the highest-scoring biotas by unweighted species richness. So for the majority of these groups, the most species-rich biotas are not taxonomically clumped but are already fairly well-dispersed across the classification. Consequently the refinement of biodiversity appraisal added by using a complex measure is distinct, but not always great.

The divergence between simple species richness and more complex diversity measures should be greatest for datasets in which taxonomic resolution is good and the taxonomic clumping of biotas is more extreme, as shown by the bumble bees of the *sibiricus*-group (Vane-Wright *et al.* 1991; Williams *et al.* 1991, 1993). But even for this example, the diversity scores from the taxonomic diversity measures and species richness cover fairly narrow ranges in the stippled area of Fig. 19.2, showing that even unweighted species richness provides a reasonable first approximation as a measure of diversity.

The effect of cladistic weighting is likely to decline for datasets with larger numbers of species, because within larger biotas, species are less likely to be highly taxonomically clumped. The contribution of taxonomic weighting would then be to provide better discrimination among biotas with near ties in species richness. However, discrimination for taxonomic differences can be enhanced for these cases by manipulating the exponents on the taxon richness or mean taxonomic difference weighting terms of the diversity measures (Williams *et al.* 1991, implemented in WORLDMAP version 3.2). The disadvantage of this procedure is that, because the degree of taxonomic clumping within the biotas clearly remains fixed, increased taxonomic discrimination can only be achieved at the expense of discrimination for species richness, so priority-area sequences will tend to become longer.

3. *Endemism*

Measures of endemism (defined as rarity in the form of restricted range size) have been used as surrogates for measures of biodiversity (Bibby *et al.* 1992; Thirgood and Heath this volume). Indicator or surrogate variables have great potential for reducing the substantial costs of sampling and identification that severely constrain biodiversity inventory projects. However, the use of surrogates must always be supported with demonstrations that the chosen variable has a close and reliable relationship with the target variable, in this case biodiversity (Williams and Gaston, in press).

Convergence among taxonomic diversity scores does not necessarily mean that there will be convergence with endemism scores, particularly at larger spatial scales. In only nine of the 25 species-based datasets in Table 19.2 (i.e. 36 per cent) does endemism identify the same first priority area as taxonomic dispersion. For some of these datasets, such as the sub-Saharan antelopes and the African francolins, four out of five of the diversity measures (including species richness) agree that the same top biotas within each dataset are the most diverse (e.g. for antelope: Tanzania-Serengeti and Zambia-Kafue). In these cases the species in the most species-rich areas are also particularly broadly dispersed across the classification, and have particularly broad geographical distributions. Yet in the same data sets the endemism scores are highest for very different, relatively low diversity areas (e.g. for antelope: South Africa-De Hoop). This result is obtained irrespective of whether endemism is measured as absolute (choosing first the area with the most restricted species without regard to species richness), or as endemism-weighted species richness as implemented in the WORLDMAP program (for which the area weight is the sum of each species' inverse frequency of occurrence among all grid-squares, e.g. Usher 1986; Howard 1991). This divergence of endemism from biodiversity scores should not be surprising, because endemism is, after all, a measure of distribution range, not of diversity.

The poor performance of endemism as a surrogate for biodiversity measures in priority-areas analysis can also be seen for the bumble bee example in Fig. 19.2. The dotted line represents the taxonomic dispersion score for the priority-area sequence selected using absolute endemism. The position of the endemism histogram below almost all of the diversity measure sequence scores confirms the poor diversity scores for endemism on its own. Some improvement can be achieved by combining endemism with species richness as a combined measure with diversity. However, the resulting scores still remain generally lower than for any of the other direct diversity measures, including species richness on its own, if not all selected areas could be included.

Conclusions

1. The greatest contribution to the conservation and representation of biodiversity is from complementarity.
2. The differences between organisms can be represented in biodiversity measures by using readily available information on group membership from taxonomic classifications.
3. Empirically, in the absence of all other information, unweighted species richness provides a good first approximation to more demanding taxonomic diversity measures.
4. In contrast, because endemism is not a measure of biodiversity but of range size, its use can reduce the effectiveness of procedures for conservation of biodiversity.

Acknowledgements

We thank Dan Faith, Clive Moncrieff, Laurence Packer, and Dick Vane-Wright for helpful discussion. We are also grateful to Tim Crowe, Paul Eggleton, Ian Kitching, Ebbe Nielsen, and Dick Vane-Wright, who have shared the data in Table 19.2 with us. The work is supported by NERC research grant GR3/8335.

References

Ackery, P. R. and Vane-Wright, R. I. (1984). *Milkweed butterflies, their cladistics and biology, being an account of the natural history of the Danainae, a subfamily of the Lepidoptera.* British Museum (Natural History), London and Cornell University Press, Ithaca, NY.

Altschul, S. F. and Lipman, D. J. (1990). Equal animals. *Nature*, **348**, 493–4.

Anderson, S. (1974). Patterns of faunal evolution. *Quarterly Review of Biology*, **49**, 311–32.

Bibby, C. J., Collar, N. J., Crosby, M. J., Heath, M. F., Imboden, C., Johnson, T. H. *et al.* (1992). *Putting biodiversity on the map: priority areas for global conservation.* ICBP, Cambridge.

Brooks, D. R., Mayden, R. L., and McLennan, D. A. (1992). Phylogeny and biodiversity: conserving our evolutionary legacy. *Trends in Ecology and Evolution*, **7**, 55–9.

Burley, F. W. (1988). Monitoring biological diversity for setting priorities in conservation. In *Biodiversity* (ed. E. O. Wilson), pp. 227–30. National Academy Press, Washington, DC.

Cousins, S. H. (1991). Species diversity measurement: choosing the right index. *Trends in Ecology and Evolution*, **6**, 190–2.

Crozier, R. H. (1992). Genetic diversity and the agony of choice. *Biological Conservation*, **61**, 11–15.

East, R. (ed.) (1988). *Antelopes. Global survey and regional action plans. Part 1. East and Northeast Africa.* IUCN, Gland, Switzerland.

East, R. (ed.) (1989). *Antelopes. Global survey and regional action plans. Part 2. South and South-Central Africa.* IUCN, Gland, Switzerland.

East, R. (ed.) (1990). *Antelopes. Global survey and regional action plans. Part 3. West and Central Africa.* IUCN, Gland, Switzerland.

Erwin, T. L. (1991). An evolutionary basis for conservation strategies. *Science*, **253**, 750–2.

Faith, D. P. (1992). Conservation evaluation and phylogenetic diversity. *Biological Conservation*, **61**, 1–10.

Faith, D. P. (in press). Systematics and conservation: on predicting the feature diversity of subsets of taxa. *Cladistics.*

Faith, D. P. (submitted). Genetic diversity: no agony of choice among hierarchical measures.

Farris, J. S. (1977). On the phenetic approach to vertebrate classification. In *Patterns of vertebrate evolution* (ed. M. K. Hecht, P. C. Goody, and B. M. Hecht), pp. 823–50. Plenum, New York.

Farris, J. S. (1983). The logical basis of phylogenetic analysis. In *Advances in cladistics II* (ed. N. I. Platnick and V. I. Funk), pp. 7–36. Columbia University Press, New York.

Farris, J. S. (1985). Distance data revisited. *Cladistics*, **1**, 67–85.

Gauthier, J., Kluge, A. G., and Rowe, T. (1988). Amniote phylogeny and the importance of fossils. *Cladistics*, **4**, 105–209.

Gould, S. J. (1989). *Wonderful life.* Penguin Books, London.

Green, D. M. (1991). Chaos, fractals and nonlinear dynamics in evolution and phylogeny. *Trends in Ecology and Evolution*, **6**, 333–7.

Groombridge, B. (ed.) (1992). *Global biodiversity, status of the Earth's living resources.* Chapman and Hall, London.

Howard, P. C. (1991). *Nature conservation in Uganda's tropical forest reserves.* IUCN, Gland, Switzerland.

Humphries, C. J., Ladiges, P. Y., Roos, M., and Zandee, M. (1988). Cladistic biogeography. In *Analytical biogeography: an integrated approach to the study of animal and plant distributions* (ed. A. A. Myers and P. S. Giller), pp. 371–404. Chapman and Hall, London.

IUCN, UNEP, and WWF (1980). *World conservation strategy, living resource conservation for sustainable development.* IUCN, UNEP and WWF, Gland, Switzerland.

Kirkpatrick, J. B. (1983). An iterative method for establishing priorities for the selection of nature reserves: an example from Tasmania. *Biological Conservation*, **25**, 127–34.

McNeely, J. A., Miller, K. R., Reid, W. V., Mittermeier, R. A., and Werner, T. B. (1990). *Conserving the world's biological diversity.* IUCN, WRI, CI, WWF and World Bank, Washington, DC.

Magurran, A. E. (1988). *Ecological diversity and its measurement.* Croom Helm, London.

Margules, C. R., Nicholls, A. O., and Pressey, R. L. (1988). Selecting networks of reserves to maximise biological diversity. *Biological Conservation*, **43**, 63–76.

May, R. M. (1975). Patterns of species abundance and diversity. In *Ecology and evolution of communities* (ed. M. L. Cody and J. M. Diamond), pp. 81–120. Harvard University Press, Cambridge, MA.

May, R. M. (1990). Taxonomy as destiny. *Nature*, **347**, 129–30.

Mickleburgh, S. P., Hutson, A. M., and Racey, P. A. (in press). *Old World fruit bats, an action plan for their conservation.* IUCN, Gland, Switzerland.

Myers, N. (1988). Threatened biotas: 'hot spots' in tropical forests. *The Environmentalist*, **8**, 187–208.

Myers, N. (1990). The biodiversity challenge: expanded hot-spots analysis. *The Environmentalist*, **10**, 243–56.

Nielsen, E. S. and Kristensen, N. P. (1989). Primitive ghost moths: morphology and taxonomy of the Australian genus *Fraus* Walker (Lepidoptera: Hepialidae s. lat.). *Monographs on Australian Lepidoptera*, **1**. CSIRO, Melbourne.

Oxford English Dictionary (1990). *The Shorter Oxford English Dictionary.* (3rd edn). Clarendon Press, Oxford.

Peet, P. K. (1974). The measurement of species diversity. *Annual Review of Ecology and Systematics*, **5**, 285–307.

Pielou, E. C. (1975). *Ecological diversity.* John Wiley, New York.

Pressey, R. L. and Nicholls, A. O. (1989). Efficiency in conservation evaluation: scoring versus iterative approaches. *Biological Conservation*, **50**, 199–218.

Pressey, R. L., Humphries, C. J., Margules, C. R., Vane-Wright, R. I., and Williams, P. H. (in press). Beyond opportunism: key principles for systematic reserve selection. *Trends in Ecology and Evolution.*

Salt, G. (1979). A comment on the use of the term emergent properties. *American Naturalist*, **113**, 145–8.

Southwood, T. R. E. (1978). *Ecological methods with particular reference to the study of insect populations.* (2nd edn). Chapman and Hall, London.

Taylor, L. R. (1978). Bates, Williams, Hutchinson—a variety of diversities. In *Diversity of insect faunas* (ed. L. A. Mound and N. Waloff), pp. 1–18. Blackwell, Oxford.

Usher, M. B. (ed.) (1986). *Wildlife conservation evaluation.* Chapman and Hall, London.

Vane-Wright, R. I., Humphries, C. J., and Williams, P. H. (1991). What to protect?—Systematics and the agony of choice. *Biological Conservation*, **55**, 235–54.

Weitzman, M. L. (1992). On diversity. *Quarterly Journal of Economics*, **107**, 363–405.

Whittaker, R. M. (1960). Vegetation of the Siskiyou Mountains, Oregon and California. *Ecological Monographs*, **30**, 279–338.

Williams, P. H. (1992). *WORLDMAP priority areas for biodiversity. Using version 3.* Privately distributed, London.

Williams, P. H. and Gaston, K. J. (in press). Measuring more of biodiversity: can higher-taxon richness predict wholesale species richness?

Williams, P. H., Humphries, C. J., and Vane-Wright, R. I. (1991). Measuring biodiversity: taxonomic relatedness for conservation priorities. *Australian Systematic Botany*, **4**, 665–79.

Williams, P. H., Vane-Wright, R. I., and Humphries, C. J. (1993). Measuring biodiversity for choosing conservation areas. In *Hymenoptera and biodiversity* (ed. J. LaSalle and I. D. Gauld), pp. 309–28. CAB International, Wallingford, UK.

WRI, IUCN, and UNEP (1992). *Global diversity strategy: guidelines for action to save study and use Earth's biotic wealth sustainably and equitably.* WRI, New York.

20. Biodiversity of the Mediterranean region

PJOTR OOSTERBROEK
Institute of Taxonomic Zoology, University of Amsterdam, Plantage Middenlaan 64, 1018 DH Amsterdam, The Netherlands

Abstract

A review of the present status of biodiversity and conservation throughout six terrestrial areas in the Mediterranean region is presented. Measures of biodiversity, species richness, and taxonomic diversity are compared. Species richness is considered for four vertebrate (birds, mammals, reptiles, and amphibians), and three insect (lacewings, butterflies, and crane flies) groups. Taxonomic measures were applied to cladograms of nine taxa of amphibians (frogs and newts), scorpions, moths, and crane flies. The six regions included north-west Africa, the Iberian Peninsula, Italy, the Balkans, west Asia Minor, and east Asia Minor. A priority list of the six areas using species richness and complementarity targets east Asia Minor as the richest area of biodiversity. The taxonomic indices give different results. In all measures Italy, Iberia and east Asia Minor contain the relictual faunal elements, with phylogenetically more recent taxa distributed throughout the Mediterranean but predominantly Balkanic or of west Asia Minor in origin. Although this is a preliminary analysis there are clear implications for conservation. Most nature reserves occur in northern European countries rather than the Mediterranean. Among the six Mediterranean areas there is a reasonable number of nature reserves in Iberia and Italy, but both are less than the European average. Conservation is a neglected issue for much of the Mediterranean and, as indicated by the figures on species richness and taxonomic diversity, the Balkans and Asia Minor ought to have conservation action plans as a high priority.

Introduction

In this paper information is presented on the biodiversity and the present status of conservation throughout the Mediterranean region. The Mediterranean region is of interest to conservation biologists for a number of reasons

Systematics and Conservation Evaluation (ed. P. L. Forey, C. J. Humphries, and R. I. Vane-Wright), Systematics Association Special Volume No. 50, pp. 289–307, Clarendon Press, Oxford, 1994.

(Cody 1986) but, as pointed out by Balletto and Casale (1991), conservation is still a long-neglected issue in most Mediterranean countries.

Simply put, three means of measuring biodiversity can be recognized. Species richness (the number of species in a particular country or region), ecological diversity (species richness in terms of standard sample sites and habitat differentiation), and taxonomic diversity (the amount of biodiversity as revealed from taxonomic information) (Bond 1989; May 1990; Erwin 1991; Vane-Wright *et al.* 1991; Williams *et al.* 1991, 1993; Faith 1992*a*,*b*; Vane-Wright 1992; Nixon and Wheeler 1992).

Information on biodiversity is presented here for terrestrial animal groups and six large areas covering most of the Mediterranean region (Fig. 20.1). Species richness is illustrated by the number of species in four vertebrate groups (birds, mammals, reptiles, and amphibians), and three insect groups (lacewings, butterflies, and crane flies). Taxonomic information is derived from nine cladograms for species groups of amphibians (frogs and newts), scorpions, moths, and crane flies.

The Mediterranean region

Two regions of the Palaearctic are inhabited by relatively large numbers of species, the Far East (Japan and China), and the Mediterranean region. Mediterranean-climate regions, characterized by cool wet winters and dry summers, are found between latitudes 32° and 40° north and south of the equator in California, Chile, South Africa, Australia, and the whole Mediterranean basin, extending as far east as the Persian Gulf and the Caspian Sea (Aschmann 1973; Thrower and Bradbury 1973; Dell 1986).

The six areas dealt with here largely cover the major subregions of the Mediterranean biogeographic realm (Fig. 20.1). Not included are the areas

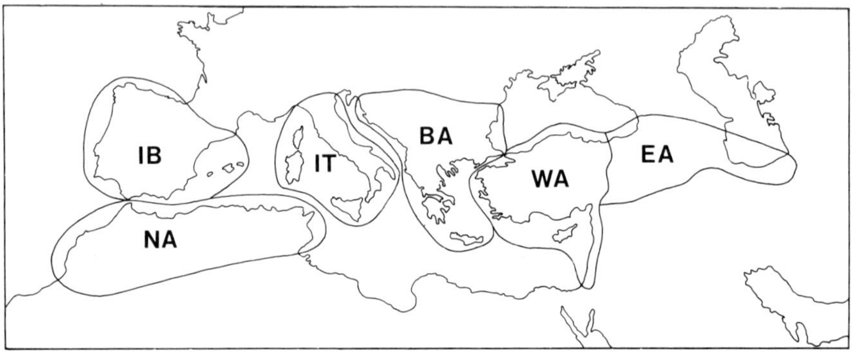

Fig. 20.1 The six areas in the Mediterranean region for which biodiversity measures are discussed (see text for details).

with Mediterranean type ecosystems along the north-east coast of Africa and towards the Persian Gulf, as none of the taxa used for measuring taxonomic diversity occurs here. The six areas are:

NA: North-west Africa, as far south as the southern slopes of the High Atlas and the Atlas Saharien, to the Gulf of Gabès in Tunisia.

IB: The Iberian Peninsula, south of the Pyrenees.

IT: Italy, south of the river Po and including Corsica, Sardinia, and Sicily.

BA: The Balkans, with former Yugoslavia (except Slovenia), Albania, southern half of Romania, Bulgaria, European part of Turkey, Greece.

WA: West Asia Minor, including the Greek islands along the coast of Turkey, Cyprus, Turkey as far east as the eastern borderlines of the provinces Hatay, Adana, Maras, Sivas, and Ordu, Syria north of Lebanon, Lebanon, and Israel except the Negev desert.

EA: East Asia Minor, with eastern Turkey, Armenia, southern halves of Georgia and Azerbaydzhan, north-west Iran.

The largest of the six areas is NA (*ca* 815 000 km^2), the smallest IT (*ca* 250 000 km^2), the other areas are of intermediate size (from *ca* 480 000 to *ca* 610 000 km^2; Table 20.1). The six areas together cover about 3 300 000 km, some 3 per cent of the world's non-frosted area.

Species richness

In order to evaluate differences in species richness among the six areas, species numbers are estimated for four vertebrate groups (breeding birds, mammals, reptiles, and amphibians), and species and subspecies numbers for three insect groups (lacewings [Neuroptera], butterflies [Rhopalocera], and crane flies [Tipulidae s. str.]) (Table 20.1). Subspecies of the vertebrate groups are not accounted for because of the different opinions in the literature on the status of many subspecies, and because of the sometimes large number of subspecies recognized within narrowly distributed species (extreme examples are *Podarcis lilfordi* (Günther) from the Balearics and *P. pityusensis* (Boscá) from the Pityusen with 27 and 23 subspecies respectively (Böhme 1986)).

Of the four vertebrate groups, the species found in the region represent 70–80 per cent of the western Palaearctic fauna. For the entire order of insects the estimate is 75 per cent (Balletto and Casale 1991). So, in a regional perspective, species richness can be considered high. On a global scale, however, other and even much smaller areas are inhabited by several times as many species, as illustrated in Fig. 20.2 for vertebrates. Most of the

Table 20.1 Estimated number of species (vertebrate groups) or species and subspecies (insect groups) occurring in the areas exemplified on Fig. 20.1, NR-values, and percentage of endemics

Areas: sq. 100 km:	NA 81.50	IB 58.50	IT 25.00	BA 61.00	WA 56.00	EA 48.00	Total 330.00	NR-value	% end
Breeding birds (1)	215	221	203	268	294	282	399	3.72	
Endemics	22	16	11	13	16	15	45	2.16	11
Mammals (2)	75	79	74	104	99	103	195	2.74	
Endemics	9	9	3	12	14	10	40	1.43	21
Reptiles (3)	48	39	34	55	63	82	163	1.97	
Endemics	19	16	10	23	17	25	77	1.43	47
Amphibians (3)	13	22	24	26	16	14	65	1.77	
Endemics	5	7	11	6	3	2	27	1.26	42
Neuroptera (4)	118	162	124	201	193	191	461	2.15	
Endemics	41	59	29	70	77	59	230	1.46	50
Rhopalocera (5)	137	205	179	265	275	372	677	2.12	
Endemics	85	72	41	71	109	192	416	1.37	61
Tipulidae (6)	42	104	110	218	111	116	498	1.41	
Endemics	34	56	58	102	90	77	361	1.16	72

selected countries are in the tropics where biodiversity is in general much higher than in other parts of the world. However, the lower diversity in the Mediterranean region might be due in part to anthropogenic impoverishment of the biota all round the Mediterranean basin. The extent to which this is the case cannot be quantified at present, but human influences have been very extensive over a long period of time, and much of the Mediterranean basin has been completely transformed from its native state (Mooney 1988). Faunistic studies comparing Mediterranean type ecosystems with different histories of human disturbance are therefore of major interest to evaluate the effects of this disturbance on biotic diversity (see also Mylonas 1984; Cody 1986; Mooney 1988; Holloway and Stork 1991).

The four vertebrate groups differ in the number of species ranging from 65 (amphibians) to 399 (breeding birds); the percentage of endemics from 11 per cent (birds) to 47 per cent (reptiles); and the average number of regions (NR value) in which each species occurs, from 1.77 (amphibians) to 3.92 (birds) (see Table 20.1 and Fig. 20.3). These differences are of some interest for nature conservation. Birds and mammals are seen to be characterized by low percentage endemism and high NR-values (Fig. 20.3). In other words,

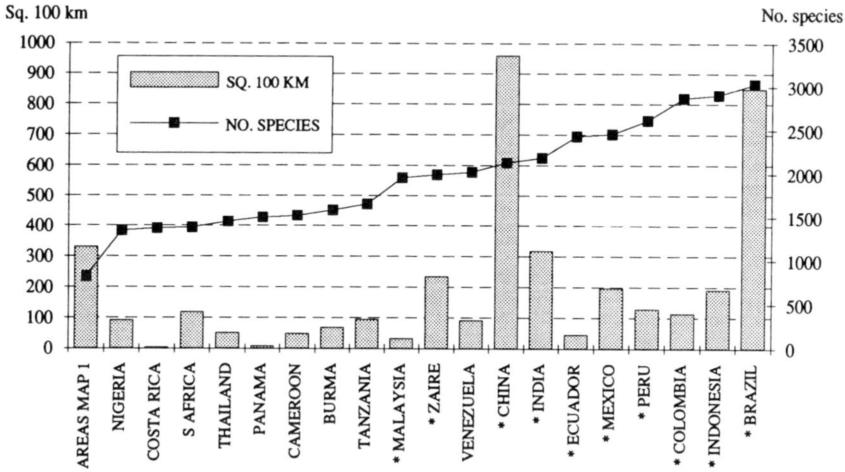

Fig. 20.2 Total number of birds, mammals, reptiles, and amphibians as estimated for the area under study and as presented for other selected countries by McNeely *et al.* (1990), with the size of these countries in comparison to the area under study (* = megadiversity countries (60–70 per cent of the World's total)).

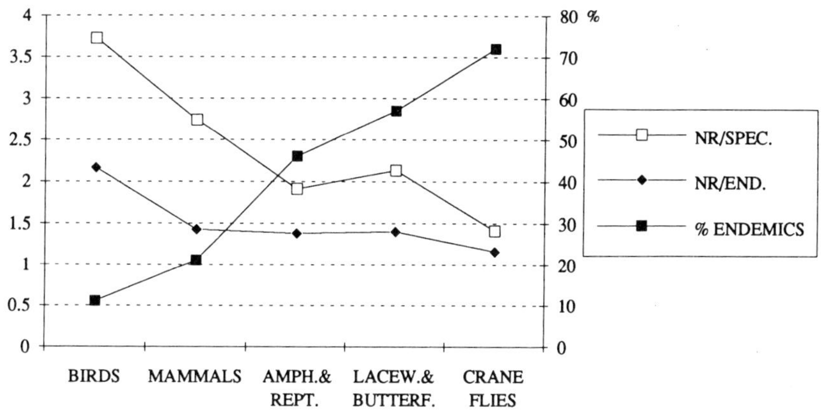

Fig. 20.3 NR-values for the species (or species and subspecies) and for the endemics, and percentages of endemics.

most bird and mammal species are not confined to the Mediterranean region, and within the region they occur in several of the six areas. Reptiles and amphibians show much higher percentages of endemics and lower NR-values. These figures indicate that the emphasis in site selection on the presence of large-bodied and attractive species with a high amenity value, such as birds and mammals (Moore 1991; Morowitz 1991; Soulé 1991), is in fact undesirable. Reptiles and amphibians show more endemism and are known to be very sensitive to environmental changes, which makes them especially useful as indicator groups in local situations. The same applies in general for insects (see also Collins and Thomas 1991; Holloway and Stork 1991). The percentages of endemics among the three insect groups considered are notably higher than the corresponding values in the vertebate groups (Table 20.1, Fig. 20.3), which is due to the fact that species and subspecies are taken into account. The NR-values demonstrate that the distributions of species and subspecies of lacewings and butterflies cover about as many areas as do species of amphibians and reptiles, and that on average distributions are more limited in crane flies.

The distribution of the four vertebrate groups and the three insect groups over the six areas is rather similar (Figs 20.4–7). The richest areas are EA, BA, and WA, together accounting for 60 per cent of the (sub)species, followed by IB. Fewer species or subspecies are found in IT and NA. Endemicity follows the same pattern (62 per cent in EA + WA + BA), but with relatively more endemics in NA among birds and mammals (Fig. 20.4), and with fewer endemics in Asia Minor among the amphibians and reptiles (Fig. 20.5). Given the distribution of the four vertebrate and three insect groups, the order of overall species richness is:

	Percentage of all non-endemics	Percentage of all endemics	Percentage of all sp./spps
1. EA	62	32	47
2. BA	67	25	46
3. WA	57	27	43
4. IB	47	20	34
5. IT	46	14	30
6. NA	34	18	26

The sequence of areas given above does not account for the size of the respective areas. Alpha diversity (species richness expressed in standard surface size) is highest in the smallest area, IT, with the other areas in the same sequence as above:

	100 km^2	non-endemics/ 100 km^2	endemics/ 100 km^2	sp. spp/ 100 km^2
1. IT	25	23	7	30
2. EA	48	16	8	24
3. BA	61	13	6	19
4. WA	56	13	5	18
5. IB	59	10	4	14
6. NA	82	5	3	8

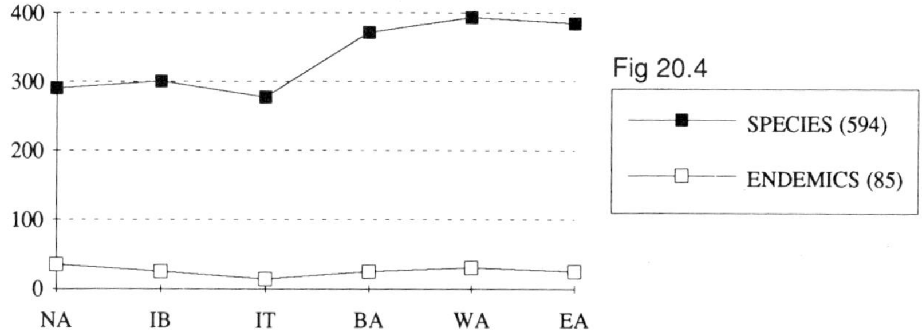

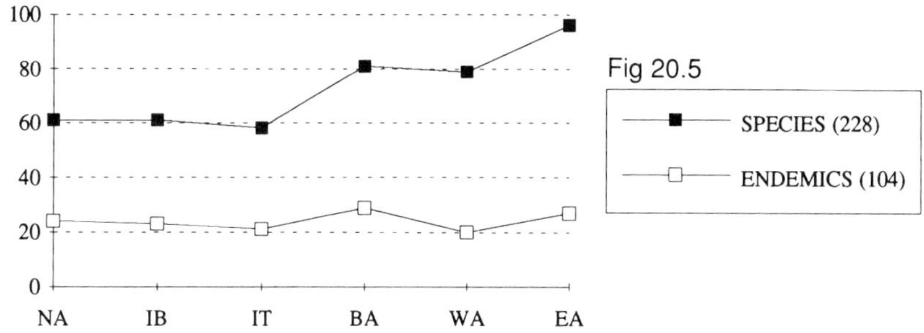

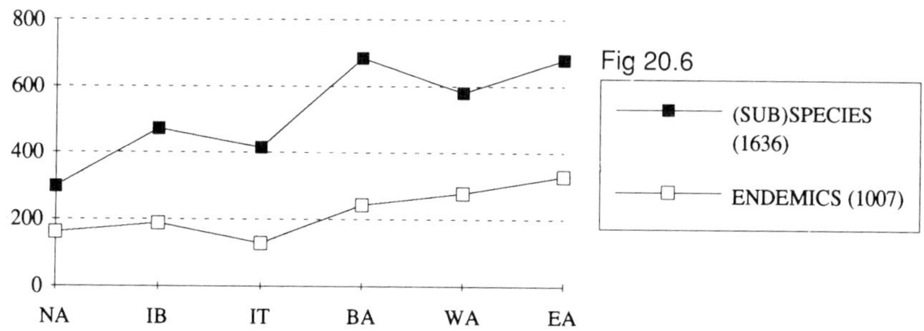

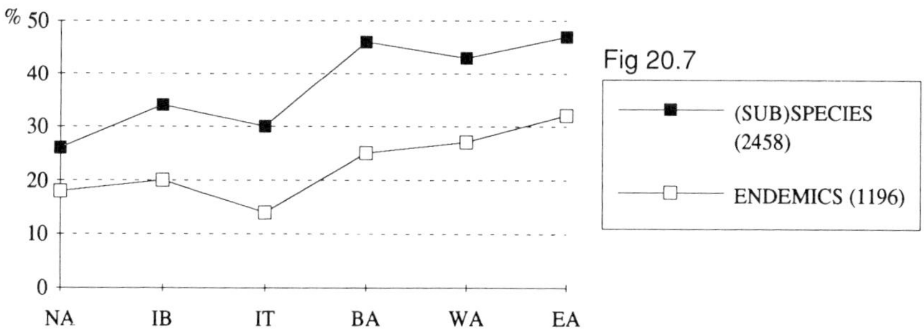

Figs. 20.4–7 Number of species (or species and subspecies) and endemics in breeding birds and mammals (Fig. 20.4), reptiles and amphibians (Fig. 20.5), the three insects groups (Fig. 20.6), and the seven groups combined (Fig. 20.7).

A priority list of the six areas, based on species richness, can be generated by making use of the principle of complementarity. According to the first list above, EA is the richest area and is given first priority. This area includes non-endemic species that also occur in other areas. In the next steps of the analysis, area EA and its endemics as well as the non-endemics occurring in EA are eliminated. The second area is thus drawn from the complement of the first, viz. among the remaining areas the one with the highest species richness. Once the first two areas have been selected, the reduced complement is searched for the third area, etc.

Complementarity applied to the distributions of all 2458 species (or subspecies) mentioned in Table 20.1 results in the following priority list of areas:

1. EA (47%) 2. BA (21%) 3. IB (12%) 4. NA (8%) 5. WA (7%)
6. IT (5%).

The first two areas account for 68 per cent of the species (or subspecies), the first three for 80 per cent. As shown in the first list, 43 per cent of the species (or subspecies) occur in WA. Once EA and BA have been selected, the remaining complement of WA is reduced to only seven per cent. Higher priorities of adjacent regions are also responsible for IT, the area with the highest alpha diversity, having the lowest priority.

Taxonomic diversity

The recognition that unique or distinct species deserve special attention in conservation has stimulated the development of various indices for measuring biodiversity on the basis of taxonomic information (see references in Faith, in press). At present the indices for taxonomic diversity presented by Williams *et al.* (1993, WORLDMAP 2.5; also in part as 'taxic diversity' by Vane-Wright *et al.* 1991; Vane-Wright 1992) are the most helpful because they provide the necessary link between taxonomic information (in the form of cladograms), and the selection of important areas for conservation (by priority analysis of areas).

The taxonomic diversity indices of Williams *et al.* (1993; see also Williams and Humphries, this volume) have been developed to facilitate the efficient recognition of global or regional priority lists of areas. They make use of the principle of complementarity and take advantage of our knowledge about cladistic relationships of well studied plant and animal groups. For the Mediterranean region only a limited number of such studies is available. In their paper on the correspondence between biogeographic patterns and Mediterranean palaeogeography, Oosterbroek and Arntzen (1992) used nine cladograms of taxa distributed throughout the Mediterranean region. These

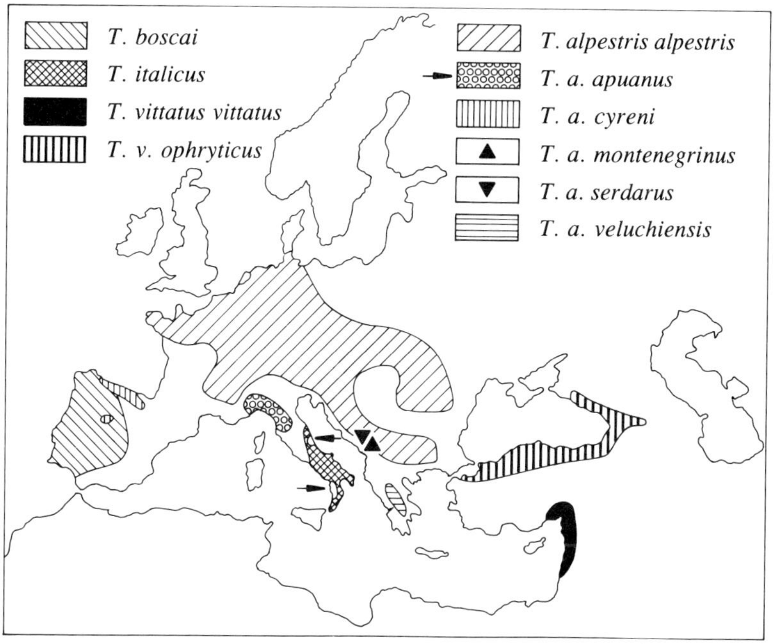

Fig. 20.8 Distribution of part of the European species of the genus *Triturus* (Salamandridae, Newts), after MacGregor *et al.* (1990).

nine cladograms are employed here as well. They include 94 species or subspecies, belonging to one group of moths, three groups of crane flies, one group of scorpions, two groups of frogs, and two groups of newts (see for details Table 20.2). The majority of the species or subspecies are endemic to one of the six areas recognized (as exemplified in Fig. 20.8 for some of the newts), but north-west African endemics are found only among the crane flies and frogs. Species richness, in percentages of complementarity of the 94 distributions, results in the following priority list of the areas:

1. BA (37%) 2. IB (18%) 3. EA (14%) 4. IT (13%) 5. NA (10%) 6. WA (8%).

The four cladistically based indices of Williams *et al.* are briefly outlined here (Fig. 20.9; see for details Williams *et al.* 1993).

Close-to-root species. Searches for areas with distinctive 'relic' species. Originally developed by Vane-Wright *et al.* (1991) as 'taxic diversity'. Slight modification was proposed by May (1990). Selects area A of Fig. 20.9 as first priority.

Table 20.2 Taxonomic lists of the nine groups (after Oosterbroek and Arntzen 1992) with the actual distribution of (sub)species (western borderlines of WA and EA are somewhat more toward the east than in Oosterbroek and Arntzen 1992, so that Crete is part of BA and Cyprus part of WA). For cladograms see Oosterbroek and Arntzen 1992.

Procis (Procis) statices group (Zygaenidae, Lepidoptera)

	NA	IB	IT	BA	WA	EA
schmidti		1				
italica			1			
alpina				1		
storaiae						1
drenowskii				1	1	
statices				1		1

Tipula (Acutipula) maxima group (Tipulidae, crane flies)

	NA	IB	IT	BA	WA	EA
luctuosa			1			
anormalipennis	1					
repentina	1					
triangulifera		1				
repanda		1				

Nephrotoma flavescens group (Tipulidae, crane flies)

	NA	IB	IT	BA	WA	EA
saccai			1			
nasuta					1	1
schaeuffelei						1
beckeri					1	1
cretensis				1		
guestfalica		1	1	1	1	
hartigiana			1			
surcoufi	1					
quadrifaria			1	1		1
farsidica						1
fontana	1					
minuscula					1	
theowaldi				1	1	
appendiculata			1	1	1	1
pertenua	1	1	1			
sullingtonensis	1	1				
exastigma	1					
flavescens		1		1		

Rana temporaria group (Ranidae, frogs)

	NA	IB	IT	BA	WA	EA
temporaria		1	1	1		
iberica		1				
holtzi					1	
cameranoi						1
macrocnemis					1	1
graeca			1	1		
latastei			1			
dalmatina			1	1	1	1

Triturus (Triturus) group (Salamandridae, newts)

	NA	IB	IT	BA	WA	EA
vittatus					1	
ophryticus					1	1
marmoratus		1				
pygmaeus		1				

citae	1					
macra						1
isparta					1	
cypriensis					1	
doriae	1	1	1			
cretensis				1		
corsica			1			
rifensis	1					
maxima			1			
balcanica				1		
pseudogigantea		1				
libanica					1	
transcaucasica						1
latifurca				1	1	1

Nephrotoma cornicina group (Tipulidae, crane flies)

	NA	IB	IT	BA	WA	EA
forcipata		1				
eugeniae						1
tenuipes				1		1
aculeata				1	1	1
moravica				1		
cornicina		1	1	1	1	1
sardiniensis			1			

astigma	1					
submaculosa	1	1	1	1		
lempkei		1				

Chactidae (Scorpionida, scorpions)

	NA	IB	IT	BA	WA	EA
nordmanni						1
xambeui		1				
flavicaudis	1	1	1			
italicus			1	1		1
carpathicus			1	1	1	
germanicus				1	1	
mesotrichus			1	1	1	

Rana ridibunda group (Ranidae, frogs)

	NA	IB	IT	BA	WA	EA
saharica	1					
perezi		1				
ridibunda				1	1	1
lessonae				1		
italian species			1			
shqiperica				1		
epeirotica				1		

dobrogicus				1		
karelini				1	1	1
cristatus				1		
carnifex			1	1		

Triturus (Palaeotriton) group (Salamandridae, newts)

	NA	IB	IT	BA	WA	EA
boscai		1				
italicus			1			
helveticus		1				
montandoni				1		
vulgaris				1		
ampelensis				1		
graecus				1		
kosswigi					1	
lantzi						1
meridionalis			1			
schmidlerorum					1	

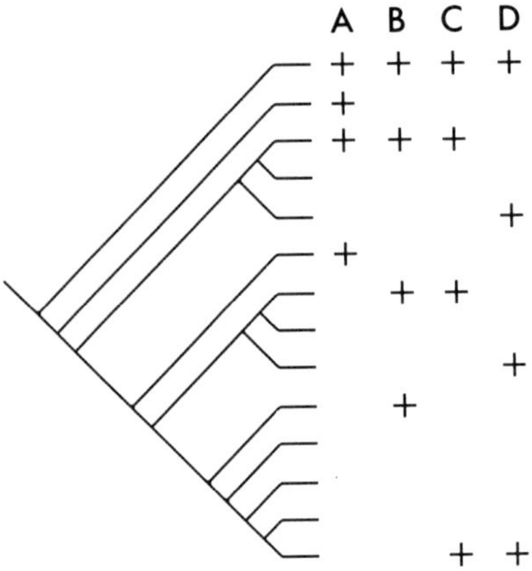

Fig. 20.9 Hypothetical cladogram for 14 species, out of which different selections of four species occur in the areas indicated by A, B, C, and D.

Higher-taxon richness. Searches for areas with as many higher taxa as possible. Selects area B of Fig. 20.9 as first priority.

Spanning-subtree length. Searches for areas with the largest proportion of cladistic divergence. Selects area C of Fig. 20.9 as first priority.

Taxonomic dispersion. Searches for areas of which the species are cladistically very divergent and most evenly spread across the cladograms. Selects area D of Fig. 20.9 as first priority.

These four indices have been calculated separately for each of the nine cladograms by using WORLDMAP (version 2.5). Subsequently, the relative contribution of each group within the total of 94 species (or subspecies) was taken into account. For the nine cladograms combined the results with complementary diversity increments are as follows :

Close-to-root
1. IT (31%) 2. EA (22%) 3. IB (19%) 4. BA (13%) 5. NA (8%) 6. WA (7%).

Higher-taxon richness
1. IT (29%) 2. IB (20%) 3. EA (16%) 4. BA (14%) 5. WA (11%) 6. NA (10%).

Spanning-subtree length
1. BA (30%) 2. IT (24%) 3. EA (16%) 4. IB (12%) 5. WA (11%) 6. NA (7%).

Taxonomic dispersion
1. BA (24%) 2. IT (21%) 3. EA (19%) 4. IB (17%) 5. WA (10%) 6. NA (9%).

None of the indices follows the priorities 1–4 as based on species richness among the nine cladograms. All the indices, species richness included, have WA and NA as the last two priorities. Close-to-root and higher-taxon richness produce similar outcomes, as do spanning-subtree length and taxonomic dispersion. The two first-mentioned indices do select areas with a larger share of basal taxa, with IT as first priority, and EA and IB as second or third. These results are in line with the conclusion of Oosterbroek and Arntzen (1992), that the oldest faunal elements among the nine cladograms are found in Iberia and Italy, the second oldest in east Asia Minor.

According to Oosterbroek and Arntzen (1992), phylogenetically more recent groups became distributed over the entire Mediterranean, but are predominantly of Balkanic and west Asia Minor origin. Furthermore, the Balkans and Transcaucasus are enriched by temperate widespread European and Palaearctic species (Theowald and Oosterbroek 1983, 1985, 1986, 1990). It is apparently largely for these reasons that BA, WA, and EA possess such a high overall species richness (above).

Williams *et al.* (1993) mention that their 'present preference among all the diversity measures described [by them and among others in Willams *et al.* 1991] is for taxonomic dispersion' because it 'appears to be the closest to seeking out the pattern of difference valued in the popular concept of diversity'. The comparison made here illustrates that taxonomic dispersion does indeed generate results which are intermediate between close-to-root and overall species richness. The first priority is species-rich BA, the second is 'close-to-root' IT, and the third is 'close-to-root' and species-rich EA. It should be noted, however, that spanning-subtree length produces the same priority list.

Measuring biodiversity by way of species richness or taxonomic diversity leads to different results. These differences are nevertheless comprehensible if placed in a historical perspective, as indicated by the comparison with the historical biogeographic study of Oosterbroek and Arntzen (1992). The implications for understanding biodiversity historically are often neglected. The emphasis has been on species richness, providing very valuable information on species representation and richness in relation to endemism, area size, habitat differentiation, and ecosystems, and playing a dominant role in the second round of biodiversity analysis (see below). The historical aspects of biodiversity, however, must be considered as important as the more general or ecological approaches toward biodiversity. It does explain how biodiversity evolved in concordance with the evolving earth and ecosystems, what kind of relationships exist between centres of biodiversity, and how differences in biodiversity are to be evaluated. Especially spanning-subtree length and taxonomic dispersion apparently take full account of the historical backgrounds of biodiversity and thus might have significant implications with respect to conservation strategies.

Status of conservation

Vane-Wright *et al.* (1991) conclude that practical planning of centres for preservation requires two rounds of analysis. First, recognition of priority areas by taxonomic diversity techniques and secondly, within any such area, analysis without taxonomic weighting to identify a network of reserves to contain all local taxa and ecosystems.

For the Mediterranean region both rounds of analysis are still in the initial phase. The nine cladograms, used here to illustrate what kind of results can be obtained from various taxonomic diversity indices, only represent a minor fraction of the Mediterranean fauna. They nevertheless demonstrate that information from different terrestrial taxa can be meaningfully combined (see also Vane-Wright *et al.*, this volume). In the near future several more cladistic analyses for circum-Mediterranean crane fly taxa will become available. It is hoped that the outcome of these studies can be integrated with results from many more other groups.

The second round of analysis requires, among other things, the recognition of areas of endemism on a much smaller scale than the six large areas dealt with in the present study. Enough detailed biogeographic information is available, especially in the entomological literature, to identify these smaller areas of endemism. An important next step will be to compare these results with the location of existing nature reserves.

Throughout the Mediterranean region many reserves already exist, but the percentage of land set aside as protected areas is relatively low. Comparison with other parts of the world (Fig. 20.10) shows that in almost all major regions a distinctly higher percentage of land is protected. Regarding strict nature reserves and national parks (Fig. 20.10: left columns) the Mediterranean region even scores the lowest percentage.

Furthermore, wildlife preservation is not at all in line with the observation that 70–80 per cent of all western Palaearctic species occur in the Mediterranean region. In several European countries between 9–20 per cent of land is protected (Austria, Czechoslovakia, Denmark, France, Germany, The Netherlands, Norway, United Kingdom). Among the six Mediterranean areas, a reasonable percentage of protected reserves is found only in IB and IT (Fig. 20.11), both remaining below the average of 7.6 per cent for the total of Western and Central Europe. Strict nature reserves and national parks are very limited throughout the region (Fig. 20.11: left columns). These figures show distinctly that conservation is indeed a much neglected issue in most

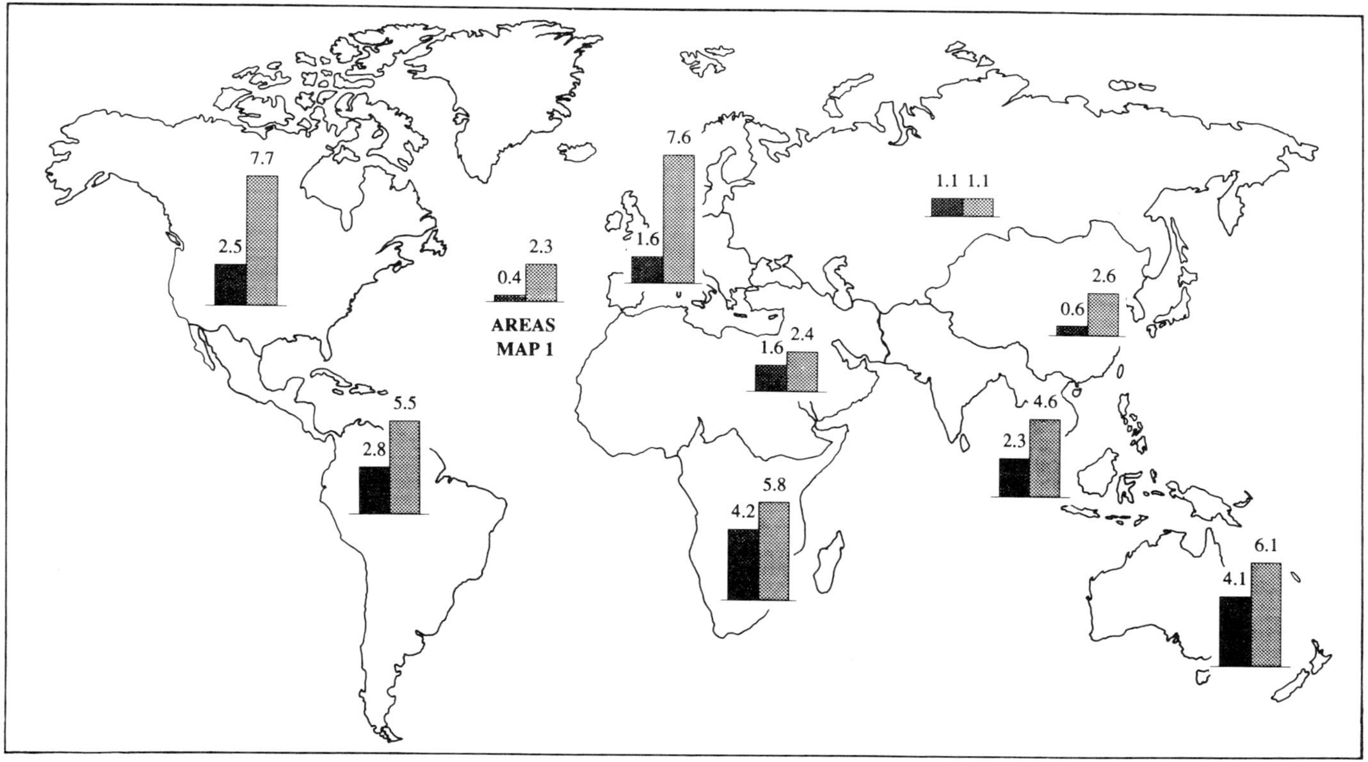

Fig. 20.10 Percentage of allocated as protected reserves in North America (except Greenland); Central and South America, total for NA, IB, IT, BA, WA, and EA; Western and Central Europe (except former USSR, including IB, IT, and BA); former USSR; northern part of Africa and the Middle East (including NA, WA, and EA); southern part of Africa; East Asia (including Japan); South-east Asia (including Taiwan and as far east as Papua New Guinea); Australia and New Zealand.
Left column: Strict nature reserves/Scientific reserves and National parks (IUCN categories I and II). *Right column*: Strict nature reserves/Scientific reserves, National parks, Natural monuments/Natural landmarks, Nature conservation reserves/Managed nature reserves/Wildlife sanctuaries and Protected landscapes or seascapes (IUCN categories I, II, III, IV and V). Source: IUCN (1990*a*).

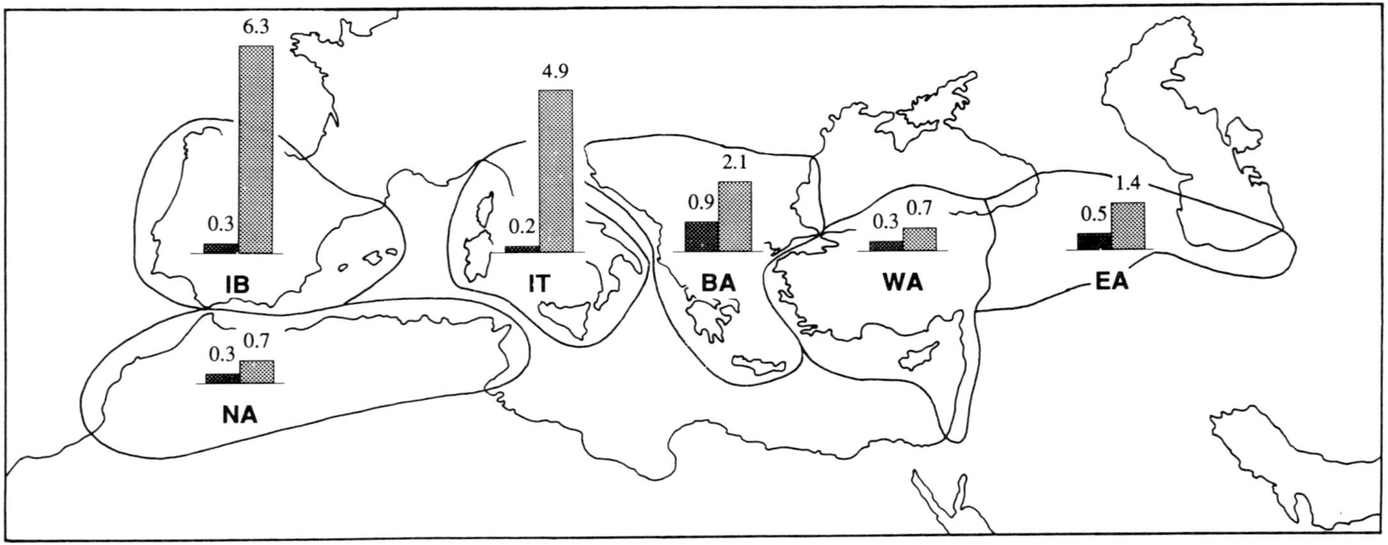

Fig. 20.11. Percentage of land set aside as protected reserves in the Mediterranean region. Left and right columns as in Fig. 20.10. The percentages given for EA is an estimate, most of the Iranian reserves listed by the IUCN (1990*a*) could not be located. Sources: Bibelreither and Schreiber (1989), IUCN (1990*a*, 1990*b*, 1991*a*, 1991*b*), Kasparak and Kasparak (1990).

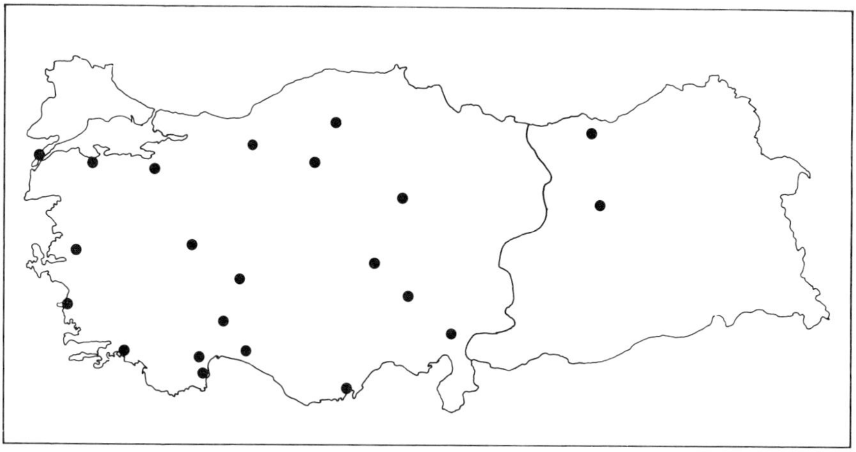

Fig. 20.12 Distribution of protected reserves in Turkey. Sources: Bibelreither and Schreiber (1989), IUCN (1990*a*), Kasparak and Kasparak (1990).

Mediterranean countries. As indicated by species richness and taxonomic diversity, the Balkans and Asia Minor are very important regions. Establishing more reserves in these parts of the Mediterranean ought to have particularly high priority. Of special interest are the high index values for EA, coupled with the fact that most of the existing Turkish nature reserves are in WA (Fig. 20.12).

Acknowledgements

Special thanks go to Chris Humphries and Paul Williams for their help with WORLDMAP (version 2.5) and to Dan Faith, for providing copies of manuscripts in press; to Peter van Bree, Annemarie Zuiderwijk, Harry van Oorschot, Willy de Prins, and Siegbert Wagener for their assistance in estimating species numbers in the Mediterranean for mammals, reptiles, amphibians, and butterflies, to Hans Duffels, Ronald Sluys, and Dick Vane-Wright, who greatly improved an earlier version of the manuscript, and to Dick Langerak for his help with some of the maps. I would like to thank Peter Forey for the opportunity to participate in the very stimulating 'Systematics and Conservation Evaluation' meeting.

References

Aschmann, H. (1973). Distribution and peculiarity of Mediterranean ecosystems. In *Mediterranean type ecosystems: origin and structure* (ed. F. di Castri and H. A. Mooney), pp 11–19. Springer-Verlag, Berlin.

Balletto, E. and Casale, A. (1991). Mediterranean insect conservation. In *The conservation of insects and their habitats* (ed. N. M. Collins and J. A. Thomas), pp. 121–42, Academic Press, London.

Bibelriether H. and Schreiber, R. L. (1989). *Die Nationalparke Europas*. Pro Natur GMBH, Frankfurt/M.

Böhme, W. (ed.) (1986). *Handbuch der Reptilien und Amphibien Europas*. Band 2/II. Aula Verlag, Wiesbaden.

Bond, W. J. (1989). Describing and conserving biotic diversity. In *Biotic diversity in Southern Africa: concepts and conservation* (ed. B. J. Huntley), pp. 2–18, Oxford University Press, Cape Town.

Cody, M. L. (1986). Diversity, rarity, and conservation in Mediterranean-climate regions. In *Conservation biology. The science of scarcity and diversity* (ed. M. E. Soulé), pp. 123–52, Sinauer, Sunderland, MA.

Collins N. M. and Thomas, J. A. (ed.) (1991). *The conservation of insects and their habitats*. Academic Press, London.

Dell, B., Hopkins, A. J. M., and Lamont, B. B. (1986). Introduction. In *Resilience in Mediterranean-type ecosystems* (ed. B. Dell, A. J. M. Hopkins, and B. B. Lamont), pp. 7–12. Junk, Dordrecht.

Erwin, T. L. (1991). An evolutionary basis for conservation strategies. *Science*, **253**, 750–2.

Faith D. P. (1992*a*). Conservation evaluation and phylogenetic diversity. *Biological Conservation*, **60**, 1–10.

Faith, D. P. (1992*b*). Systematics and conservation: on predicting the feature diversity of subsets of taxa. *Cladistics*, **8**, 361–73.

Holloway, J. D. and Stork, N. E. (1991). The dimensions of biodiversity: the use of invertebrates as indicators of human impact. In *The biodiversity of microorganisms and invertebrates: its role in sustainable agriculture* (ed. D.L. Hakwsworth), pp. 37–61. CAB International, Wallingford, UK.

IUCN (1990*a*). 1990 *United Nations list of national parks and protected areas*. IUCN, Gland, Switzerland.

IUCN (1990*b*). Protected areas in Eastern and Central Europe and the USSR (An interim review). *Environmental Research Series*, **1**, 1–99.

IUCN (1991*a*). *Environmental status reports: 1990. Vol. 2: Albania, Bulgaria, Romania, Yugoslavia*. IUCN East European Programme, Cambridge.

IUCN (1991*b*). *Environmental status reports: 1990. Vol. 3: USSR*. IUCN East European Programme, Cambridge.

Kasparak, M. and Kasparak, A. (1990). *Reiseführer Natur Türkei*. BLV Verlag, München.

Macgregor, H. C., Sessions, S. K., and Arntzen, J. W. (1990). An integrative analysis of phylogenetic relationships among newts of the genus *Triturus* (family Salamandridae), using comparative biochemistry, cytogenetics and reproductive interactions. *Journal of Evolutionary Biology*, **3**, 329–73.

May, R. M. (1990). Taxonomy as destiny. *Nature*, **347**, 129–30.

McNeely, J. A., Miller, K. R., Reid, W. V., Mittermeier, R. A., and Werner, T. B. (1990). *Conserving the world's biological diversity*. IUCN, WRI, CI, WWF-US, The World Bank, Gland, Switzerland.

Mittermeier, R. A. (1988). Primate diversity and the tropical forest: case studies from Brazil and Madagascar and the importance of the megadiversity countries. In *Biodiversity* (ed. E. O. Wilson), pp. 145–54, National Academic Press, Washington, DC.

Mooney, H. A. (1988). Lessons from Mediterranean-climate regions. In *Biodiversity* (ed. E. O. Wilson), pp. 157–65, National Academic Press, Washington, DC.

Moore, N. W. (1991). Observe extinction or conserve diversity? In *The conservation of insects and their habitats* (ed. N. M. Collins and J. A. Thomas), pp. 1–8. Academic Press, London.

Morowitz, H. J. (1991). Balancing species preservation and economic considerations. *Science*, **253**, 752–4.

Mylonas, M. (1984). The influence of man: a special problem in the study of the zoogeography of terrestrial molluscs on the Aegean islands. In *World-wide snails* (ed. A. Solem and A.C. van Bruggen), pp. 249–59. Brill, Leiden.

Nixon, K. C. and Wheeler, Q. D. (1992). Measures of phyletic diversity. In *Extinction and phylogeny* (ed. M. J. Novacek and Q. D. Wheeler), pp. 216–34. Columbia University Press, New York.

Oosterbroek, P. and Arntzen, J. W. (1992). Area-cladograms of circum-Mediterranean taxa in relation to Mediterranean palaeogeography. *Journal of Biogeography*, **19**, 3–20.

Oosterbroek, P. and Theowald, Br. (1992). Family Tipulidae. In *Catalogue of Palaearctic Diptera, Vol. 1* (ed. A. Soos, L., Papp, and P. Oosterbroek), pp. 56–178. Budapest.

Soulé, M. E. (1991). Conservation: tactics for a constant crisis. *Science*, **253**, 744–50.

Theowald, Br. and Oosterbroek, P. (1983). Zur Zoogeographie der west-palaearktischen Tipuliden. III. Die Tipuliden der europäischen Tiefebenen (Diptera, Tipulidae). *Bonner Zoologisches Beiträge*, **34**, 371–94.

Theowald, Br. and Oosterbroek, P. (1985). Zur Zoogeographie der west-paläarktischen Tipuliden. VI. Die Tipuliden der montanen, alpinen und borealen Gebiete (Insecta, Diptera, Tipulidae). *Bonner Zoologisches Beiträge*, **36**, 185–220.

Theowald, Br. and Oosterbroek, P. (1986). Zur Zoogeographie der west-palaearktischen Tipuliden. VII. Die Tipuliden der Balkanhalbinsel (Diptera, Tipulidae). *Tijdschrift voor Entomologie*, **129**, 1–13.

Theowald, Br. and Oosterbroek, P. (1990). Zur Zoogeographie der west-paläarktischen Tipuliden. IX. Die Tipuliden des Vorderen Orients. *Tijdschrift voor Entomologie*, **133**, 85–95.

Thrower, N. J. W. and Bradbury, D. E. (1973). The physiography of the Mediterranean lands with special emphasis on California and Chile. In *Mediterranean type ecosystems: origin and structure* (ed. F. di Castri and H. A. Mooney), pp. 37–52. Springer-Verlag, Berlin.

Vane-Wright, R. I. (1992). Systematics and the global biodiversity strategy. *Antenna*, **16**, 49–56.

Vane-Wright, R. I., Humphries, C. J., and Williams, P. H. (1991). What to protect?—systematics and the agony of choice. *Biological Conservation*, **55**, 235–54.

Williams, P. H., Humphries, C. J., and Vane-Wright, R. I. (1991). Measuring biodiversity: taxonomic relatedness for conservation priorities. *Australian Systematic Botany*, **4**, 665–79.

Williams, P. H., Humphries, C. J., and Vane-Wright, R. I. (1993). Measuring biodiversity for choosing conservation areas. In *Hymenoptera and biodiversity* (ed. J. LaSalle and I. D. Gauld), pp. 309–28. CAB International, Wallingford, UK.

21. Systematic assessment of taxic diversity by summation

R. I. VANE-WRIGHT, C. R. SMITH, and I. J. KITCHING
Biogeography and Conservation Laboratory, Department of Entomology, The Natural History Museum, Cromwell Road, London SW7 5BD, UK

Abstract

Combining data from many different taxonomic groups (up to 20 000 taxa has been suggested) will be necessary if meaningful priority areas analysis for conservation is to be achieved. Rapid analysis of data sets of more than about 1000 taxa using measures more complex than spanning-subtree weighting becomes prohibitively expensive, and coping with 20 000 is beyond present capability. Therefore, a method that combines the results of separate analyses in order to approximate the result of an all-inclusive analysis is required. Using data derived from owls, hawkmoths, milkweed butterflies, and tiger-beetles, analysed with the WORLDMAP program at regional (SE Asia) and national (Thailand) levels, we investigate methods for performing such combinations. An exact summation method is described for taxic richness, but for taxic difference measures (e.g. dispersion) we conclude that no similar exact solution exists. A heuristic measure, based upon re-ranking priority sequences, which can approximate the result of a combined analysis, is described and its limitations are discussed.

Introduction

Systematic conservation evaluation is a relatively new discipline. It acknowledges that, if protection of biological diversity is now accepted as a major conservation goal (Reid *et al.* 1992), use of the limited resources available for practical action must be optimized through the application of three operational principles: efficiency, flexibility, and viability (Vane-Wright, in press; cf. Pressey *et al.* 1993). These principles are seen as basic to developing what May (1990*a*) has called 'the calculus of biodiversity'. This

Systematics and Conservation Evaluation (ed. P. L. Forey, C. J. Humphries, and R. I. Vane-Wright), Systematics Association Special Volume No. 50, pp. 309–26, Clarendon Press, Oxford, 1994.

paper is primarily concerned with the question of efficiency (Pressey and Nicholls 1989) in the selection of conservation sites.

Systematic conservation evaluation depends on basic information about the existence of taxa, their distributions, and their taxonomic relationships (Vane-Wright *et al.* 1991; Williams and Humphries, this volume). These apparently simple requirements can only be met by good taxonomy and detailed, systematic survey data. But there is an unresolved taxonomic problem: which groups of organisms should be used in systematic conservation evaluation for biodiversity, and at what taxonomic rank? For example, however desirable it might be to include groups like bacteria and fungi (Hawksworth and Colwell 1992), or nematodes, which between them might conceivably account for 50 per cent or more of all biodiversity, our knowledge of the taxonomy of these groups is so poor that they cannot be used for detailed systematic evaluations now, or in the foreseeable future. Instead, we have to hope that decisions based on groups which, for historical reasons, are better known taxonomically and geographically (e.g. birds: ICBP 1992) will accurately reflect, or act as good surrogates for, the whole of biodiversity. This predictive ability would need to work not only for plants and vertebrates, but for all species—including the vast diversity of life represented by myriad bacteria, protists, algae, fungi, and invertebrate animals (Fernholm *et al.* 1989).

Which surrogate?

As a result, at this stage of the development of systematic conservation evaluation, it is inevitable that we will enter a debate on the problem of 'which surrogate'? Currently, five broad approaches to this problem can be identified: energy-flux, ecosystem, higher-taxon, target-taxon, and taxon summation.

The energy-flux (Currie and Paquin 1987; Currie 1991; Turner *et al.* 1988) and ecosystem approaches (e.g. McNaughton 1989; Walker 1992), although very valuable, do not rely on taxic measures for their primary input, and thus lie outside the scope of this paper. Furthermore, the value of energy-flux methods can only be assessed, ultimately, by use of the direct, biotic measures, and so the problem of choosing a taxonomic surrogate remains. This difficulty also applies, with only slightly less force, to ecosystem methods.

1. *Higher-taxon surrogates*

Gaston and Williams (1993) report a strong correlation between species richness and higher taxon richness for a number of regions and taxonomic groups. Based on this finding, they suggest that higher taxa (e.g. flowering plant families) are likely to prove a particularly tractable and all-encompassing surrogate for total biodiversity. While this may be true at global levels, especially if tempered by knowledge of the interrelationships of the higher taxa, it remains uncertain whether this approach will provide sufficient resolution at finer scales, to permit identification of maximally representative regional networks (Margules *et al.* 1988, 1991).

2. *Target-taxon surrogates*

The conservation literature has long embodied the concept of 'indicator species'—particular animals or plants that can be monitored to provide an effective measure of local ecosystem type, quality, richness, or change. More recently this concept has been extended to the idea of using indicator groups to identify priority areas for biodiversity as a whole (e.g. Brown 1991; Pearson and Cassola 1992). The work on endemic bird areas as a guide to global conservation priorities (ICBP 1992) is the largest published analysis that relies on this concept, in so far as it is argued that what is good for birds will be good for a wide variety of organisms (Thirgood and Heath, this volume). The most refined development of this general idea of choosing specific groups to act as surrogates is Kremen's (1992) 'target taxon' approach. A variety of techniques are applied in an effort to identify particular, notably highly endemic taxa, sensitive to ecological boundaries, and which correlate strongly with the diversity of other organisms in differing local ecosystems (Kremen, in press).

3. *Taxon summation*

Humphries and Vane-Wright (1992) proposed the global analysis of up to 100 or more groups (genera, tribes, or families) of widely differing organisms, each group to include, on average, roughly 200 species. In this case the emphasis would be on well-known groups, chosen to provide high quality data, rather than seeking 'target taxa' to act as particularly sensitive indicators. Integration, or summation of the data from many well-known groups would, it was suggested, offer a good subsample or surrogate for total biodiversity.

No convincing evidence for the validity of this idea was provided, and progress towards this goal has remained slow. Summation can be justified, however, as a logical method in theory. At the limit, summation of all taxa, from kingdoms down to subspecies, would be the Holy Grail for systematic conservation evaluation and priority area analysis. But in practice, both ignorance of total diversity of life (e.g. May 1990*b*; Hammond 1992) and, assuming we had data, the sheer scale of the analytical problems involved, means that only a relatively limited attempt, such as the '100 groups' approach, can be contemplated. Here, we explore technical problems associated with summation of data sets of different taxonomic groups.

Goals and methods

At local levels, our purpose corresponds closely to a primary goal of Margules *et al.* (1988, 1991), and of Rebelo (this volume). We seek maximally efficient solutions for fully representative reserve networks, either as the minimum number of areas, or smallest configuration of sites, capable of providing sustainable protection (living space) for all species found in a given region.

At global or continental levels, our objective is subtly different. We seek to recognize a network of networks, the minimal global set of regional networks (Margules *et al.* 1991) capable of maintaining all species. Such a set is likely to involve virtually all countries. If all countries were to implement fully effective national networks immediately, our trouble at this level would not be necessary, but this ideal is far from being realized (IUCN 1992; Oosterbroek, this volume). So we seek to differentiate amongst areas at global, continental or national levels, by providing sufficient data and rational methods, so that relative priorities can be set for international aid for biodiversity conservation. Thus we need not only to recognize an optimal configuration of areas, but also to set priorities for action within that configuration.

The work presented here utilizes taxic richness, with or without taxic dispersion (Williams *et al.* 1991), and complementarity, as discussed by Williams *et al.* (1993) and Williams and Humphries (this volume). In the examples, these measures have been implemented using Williams' (1992) WORLDMAP computer program.

Some problems of summation

Consider the global pattern for milkweed butterflies (Ackery and Vane-Wright 1984): rich in the old world tropics, but poor in South America and absent from high latitudes. Now consider pteropodid fruit bats, a group restricted to the old-world (Mickleburgh *et al.* 1992). Analysed by the dispersion measure, it takes six optimized areas to reach 50 per cent of fruit bat diversity, but only four areas to reach this level for milkweed butterflies (Table 21.1). Two of the areas are the same, one with even the same priority rank, but the other two areas differ. Such data immediately raise a series of general questions. How can we sum data for groups of different size and rank that do not even occur together, have only a partial range overlap, or exhibit quite different priorities for the same areas? How do we combine data at all?

The most obvious, and we believe ultimately the best way, is to do it directly, by making joint data sets and assembling the classifications according to known relationships. This procedure has been adopted by Oosterbroek (this volume) to combine data for nine taxonomic groups in

Table 21.1 The minimal configuration of countries, regions, or islands, in priority sequence, sustaining 50 per cent taxic diversity for old world fruit bats (Pteropodidae) and milkweed butterflies (Nymphalidae: Danainae). Six areas are required for the bats, but only four (partially congruent) areas for the butterflies (see text).

	bats	danaines
Sulawesi	***1***	2
Cameroon	2	
Mindanao	3	
Irian Jaya	4	4
Sumatra	5	
Choiseul (Solomons)	6	
Java		***1***
Ituri (Zaïre)		3

each of six subregions of the Mediterranean, and is applied below to various taxa in South East Asia.

Combining groups for analysis is a simple process. Williams' (1992) WORLDMAP makes use of a hexadecimal clade or taxon notation, based on Hennig's (1981) numerical sequencing system, to specify precisely any given hierarchical relationships for a set of taxa. Thus the mutual relationships of a rat, a robin, and a rook can be written as 1, 21, and 22 respectively. Likewise, the relationships of a moth together with three butterflies of unknown mutual affinity can be specified as 1, 21, 22, and 23, respectively. To add groups together, this system is simply extended according to mutual relationships. Thus, in the example, we can add all seven terminal taxa together by coding them as 11 (moth), 121, 122, 123 (three butterflies), 21 (rat), 221 (robin), and 222 (rook), where the initial (1) specifies Lepidoptera and the alternative (2) specifies Vertebrata.

The balance between taxonomic rank and species richness embodied within the taxic measures of Williams *et al.* (1993) automatically gives powerful correction for inequalities in group size. However, if we are combining, say, two groups of vertebrates with one group of insects, the addition procedure will inevitably overestimate the taxonomic difference between the two groups of vertebrates in relation to the invertebrate group. Despite this weakness, the taxic weighting will still have the right sign, and be better than making no allowance for hierarchical position at all (as when relying upon species richness alone).

Yet, even with this very simple approach, we encounter technical limitations. The computational load in rapid handling of the more sophisticated taxic diversity measures is enormous, so much so that the current PC-DOS version of WORLDMAP can only accept data sets of up to 1000 or so species. As already discussed, Humphries and Vane-Wright (1992) proposed analysing 20 000 species, or more. The number of mathematical operations required to calculate taxic dispersion for 20 000 species currently far exceeds the capacity of any practical desktop (personal computer) system. So just what can be achieved?

A regional analysis

In order to test combinatorial methods for different groups we looked at 14 adjacent large-scale areas in South East Asia, using data from three families that we are working on individually: owls (Smith, in prep.), hawkmoths (Kitching, in prep.), and milkweed butterflies (Vane-Wright, in prep.). We do not propose to publish this particular test data set, as the components have been abstracted from larger data sets still under development, but it can be obtained from the authors, on request, as a WORLDMAP file.

According to the data assembled for this investigation (Table 21.2), the 14 areas between them support 38 species of owls (Strigidae), 218 hawkmoths (Sphingidae), and 56 milkweed butterflies (Danainae). The total of 312 species is well within the capability of the current WORLDMAP program. Thus we can use analyses of the combined data set, including a combined three-taxon classification (milkweed butterflies plus hawkmoths represent the Lepidoptera, to the exclusion of owls), as a test standard, to see if indirect ways of combining data or results from the individual analyses can reproduce the results derived from analysis of this combined data set.

For taxic richness, we are confident there is an exact, if somewhat tedious solution. At the first step the species score for every site is listed separately for each group. These separate values are then summed for each site and the joint top-scoring area identified. In the present example, this procedure selects Thailand, with a grand total for the three groups of 226 species (Table 21.2).

The groups are then analysed separately on WORLDMAP, excluding the top-scoring site (Thailand), and implementing complementarity. For each group the incremental number of additional species represented by the other 13 areas is tabulated (Table 21.3). Summing across the three incremental scores for each area then identifies the optimized next best step. In the example, this indicates Sumatra as the best second choice, adding 29 species (3 owls, 19 hawkmoths, and 7 danaines) to the total of 226 for Thailand (5

Table 21.2 Species richness for owls (Strigidae), hawkmoths (Sphingidae), and milkweed butterflies (Danainae) in 14 contiguous areas of South East Asia. Even though the richest area for owls is Burma, and for danaines is Java, by summation it is evident that the highest aggregate score is provided by Thailand, with a combined total of 226 owl, moth, and butterfly species (see also text).

	owls	moths	danaines	sum
Thailand	18	***175***	33	***226***
Sumatra	12	124	33	169
Java	11	100	***35***	146
North Borneo	11	106	27	144
Malay Peninsula	14	96	33	143
Burma	***19***	69	32	120
North Indo-China	16	62	28	106
Palawan	5	42	22	69
South Indo-China	18	21	29	68
Mentawi Is	6	33	26	65
W Lesser Sundas	7	28	28	63
Andaman Is	5	35	10	50
South Borneo	10	3	27	40
Nicobar Is	3	4	14	21
Total different species	38	218	56	312

more than Java, its nearest 'rival' at this step—Table 21.3). At the third step, incremental complements are then calculated for the 12 remaining areas in relation to the top two sites combined, and so on. Note how, at the third step, because of the effects of complementarity, Java again takes second place, with the Western Lesser Sunda Islands, having already selected Thailand and Sumatra, offering the best third choice (Table 21.4). This whole process is iterated until full representation is achieved. This gives a result (Table 21.5) identical to analysing the whole data set as a single WORLDMAP analysis using taxic richness. Again, because of the sequential effects of complementarity, note how this priority order compares to the sequence in Table 21.2, where the 14 areas are simply listed by decreasing species richness.

In contrast, we do not believe that an exact solution exists for the taxic measures (higher taxon richness, spanning-subtree weighting, or dispersion) without data (tree) combination. We examined several methods for

Table 21.3 Species richness for owls (Strigidae), hawkmoths (Sphingidae), and milkweed butterflies (Danainae) in 14 contiguous areas of South East Asia. Having accepted Thailand as the first priority (Table 21.2), the number of additional (complementary) species for each group for each of the 13 remaining areas has been tabulated, using Williams' WORLDMAP. Even though the richest additional complement for owls is W Lesser Sundas, and for danaines is Java, by summation it is evident that the highest aggregate additional complement to Thailand would be provided by Sumatra, with a combined total of 29 additional owl, moth, and butterfly species (see also text).

	owls	moths	danaines	sum
(***Thailand***) +				
Sumatra	3	***19***	7	***29***
Java	2	11	***11***	24
North Borneo	2	16	4	22
W Lesser Sundas	***6***	2	***11***	19
Palawan	3	6	4	13
Malay Peninsula	0	11	1	12
Burma	3	6	1	10
Mentawi Is	3	2	4	9
North Indo-China	2	7	0	9
Andaman Is	2	1	2	5
South Borneo	1	0	4	5
Nicobar Is	1	0	2	3
South Indo-China	2	1	0	3

combining the results of separate analyses to try to reproduce the exact rank-order recognized by using the combined WORLDMAP analysis as standard. For example, a simple approach sums the priority rank scores of the groups for each area. Any group for which an area is unrepresented is allocated a score of zero. Each total is then divided by the number of non-zero scores of which it is comprised, to give a mean rank. We investigated alternative methods of scoring unrepresented areas, but all were less satisfactory than a zero score when compared to the combined analysis standard. Thus, for three taxa, if an area ranks 1 for taxon A, 3 for taxon B, and taxon C does not have a score, the mean rank of the area is 2 ((1 + 3)/2). Finally, the areas that have mean rank values are then re-ranked in priority order, with the lowest score representing the highest priority.

Table 21.4 Species richness for owls (Strigidae), hawkmoths (Sphingidae), and milkweed butterflies (Danainae) in 14 contiguous areas of South East Asia. Having accepted Thailand followed by Sumatra as the first two priority areas (Table 21.3), the number of additional (complementary) species for each group for each of the 12 remaining areas has been tabulated, using Williams' WORLDMAP. Even though the richest additional complement for hawkmoths is North Indo-China, by summation it is evident that the highest aggregate additional complement to (Thailand + Sumatra) would be provided by W Lesser Sundas, with a combined total of 16 additional owl, moth, and butterfly species (see also text).

	owls	moths	danaines	sum
(***Thailand + Sumatra***) +				
W Lesser Sundas	***6***	2	***8***	***16***
Java	2	3	6	11
Burma	3	6	1	10
North Indo-China	2	***7***	0	9
North Borneo	1	5	1	7
Palawan	3	2	2	7
Andaman Is	2	0	2	4
Mentawi Is	3	1	0	4
South Indo-China	2	1	0	3
Nicobar Is	1	0	1	2
Malay Peninsula	0	1	0	1
South Borneo	0	0	1	1

When applied to the species richness scores in Table 21.2, this approach allocates the correct rank to only 3 out of 12 areas (as judged by the combined data standard), with modest deviations for the rest (Table 21.5). This compares unfavourably with the exact method (which correctly ranked all 12 areas), but is not necessarily a poor result considering that there are over 468 million (i.e. 12 factorial) possible priority sequences for 12 sites. A similar result is found when the procedure is applied to rank sequences derived from taxic dispersion (Table 21.6). While the correct configuration of 12 sites is found, again, the correct rank, as determined by analysis of the combined data, is allocated to only 3 areas.

Furthermore, this particular example masks a fundamental failing, which we shall refer to as the '4–3–1' problem. Imagine three sites, I–III, of which you will only be able to keep one for conservation purposes. Mammals rank I

Table 21.5 Priority area sequences based on species richness for owls (Strigidae), hawkmoths (Sphingidae), and milkweed butterflies (Danainae) in 14 contiguous areas of South east Asia. Separate rank sequences are given for the three groups (owls, moths, dan = danaines), followed by the mean rank (mr), and re-ranked scores (rr). This last sequence should be compared with that obtained by the exact method (ex) and by direct analysis of the combined data set (comb). The areas are placed in the priority sequence determined by analysis of the combined data set. See also text.

	owls	moths	dan	mr	rr	ex	comb
Thailand	–	1	–	1	1	1	1
Sumatra	2	2	4	2.7	2=	2	2
W Lesser Sundas	3	–	3	3	4=	3	3
Burma	1	5	2	2.7	2=	4	4
North Borneo	–	4	6	5	8	5	5
Java	5	6	1	4	6=	6	6
North Indo-China	–	3	–	3	4=	7	7
Palawan	7	7	5	6.3	9	8	8
Mentawi Is	4	–	–	4	6=	9	9
Andaman Is	6	–	8	7	10=	10	10
South Indo-China	–	8	–	8	12	11	11
Nicobar Is	–	–	7	7	10=	12	12
South Borneo	–	–	–	–	–	–	–
Malay Peninsula	–	–	–	–	–	–	–

top, with 4 species, complementarity places III second, with a single endemic, and II drops out, having just a subset of species in area I. Suppose that birds are an exact reverse—III top, with 4 species, I second with a single endemic, with II again dropping out. Using re-ranked scores, areas I and III, both of which support a total of 5 species, would be tied, with area II discounted. However, should it so happen that the subsets in area II consist of 3 of the 4 mammals in area I, and 3 of the 4 birds in area III, on summation it becomes immediately apparent that the optimum biodiversity choice, based on species richness alone and other factors being equal, would be area II, supporting a total of 6 species.

Combining the results of separate analyses is fraught with dangers of this '4–3–1' type. Because of the potential for '4–3–1' problems in most data sets, we consider that, where possible, the raw data should be combined and a single analysis carried out, rather than being tempted by the easier option of combining individual group results.

Table 21.6 Priority area sequences based on taxic dispersion for owls (Strigidae), hawkmoths (Sphingidae), and milkweed butterflies (Danainae) in 14 contiguous areas of South east Asia. Separate rank sequences are given for the three groups (owls, moths, dan = danaines), followed by the mean rank (mr) and re-ranked scores (rr). This last sequence should be compared with that by direct analysis of the combined data set (comb). See also text.

	owls	moths	dan	mr	rr	comb
Thailand	–	1	–	1	1	1
Sumatra	2	2	4	2.7	3	2
W Lesser Sundas	3	8	3	4.7	5	3
Burma	1	3	2	2	2	4
Palawan	6	7	5	6	9	5
Java	4	6	1	3.7	4	6
North Borneo	–	4	6	5	6=	7
North Indo-China	–	5	–	5	6=	8
Mentawi Is	5	–	–	5	6=	9
Andaman Is	7	–	7=	7	10=	10
Nicobar Is	–	–	7=	7	10=	11
South Indo-China	–	9	–	9	12=	12
South Borneo	–	–	–	–	–	–
Malay Peninsula	–	–	–	–	–	–

A local analysis

1. *Hawkmoth species-richness pattern*

Thailand has by far the most diverse hawkmoth fauna in South East Asia. Its 175 species represent one-sixth of all world species, and only Indonesia is known to hold more. Fieldwork by IJK during the last five years, museum material, and information gathered by numerous other entomologists, has provided the basic data. Despite this effort, and although we anticipate much better coverage in the future, we currently have records for only 90 distinct sites (national parks, wildlife sanctuaries, and comparable non-gazetted areas: Fig. 21.1).

The richest area is represented by Doi Suthep-Pui National Park (N.P.), with 121 (68.75 per cent) species. Applying complementarity, the second choice would be Doi Inthanon N.P. which, despite being only 50 km away, would add 17 species (9.66 per cent), mainly due to its high number of endemics. The third area in the optimised sequence is Sri Sakhon, an unprotected, more or less secondary site that would add many southern,

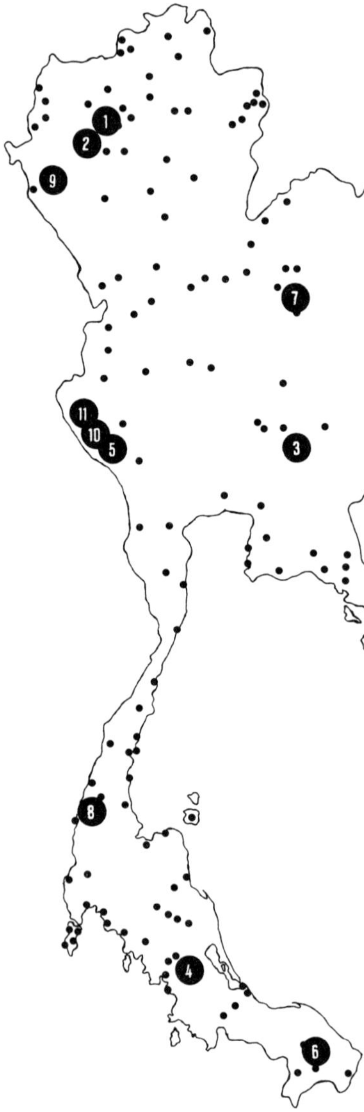

Figure 21.1 Outline map of Thailand showing distribution of hawkmoth and tiger-beetle survey sites. Individual sites referred to in text, and in Tables 21.7 and 21.8, are as follows: 1, Doi Suthep-Pui N.P.; 2, Doi Inthanon N.P.; 3, Khao Yai N.P.; 4, Khao Banthad W.S.; 5, Sai Yok N.P.; 6, Sri Sakhon; 7, Phu Khieo W.S.; 8, Khlong Nakha W.S.; 9, Mae Sariang; 10, Kroeng Krawia N.H.A.; 11, Sangkhlaburi; 12, Phu Phan N.P.

Table 21.7 Priority sequences for the top five areas in Thailand for hawkmoth (Sphingidae) and tiger-beetle (Cicindelidae) species richness, selected from 133 areas in Thailand. Separate sequences are given for the two groups (moths, beetles), mean ranks (mr), re-ranked sequences (rr), and by direct analysis of the combined data set (comb). Note that Khao Banthad W.S. and Sai Yok N.P. do not form part of the top five configuration for either group, and therefore do not appear in the configuration derived by combining the results of the separate analyses. They do, however, appear in the single analysis of the combined raw data set, due to effects of the '4–3–1' type. See also text.

	moths	beetles	mr	rr	comb
Doi Suthep-Pui N.P.	1	2	1.5	2	1
Doi Inthanon N.P.	2	–	2	3	2
Khao Yai N.P.	–	1	1	1	3
Khao Banthad W.S.	–	–	–	–	4
Sai Yok N.P.	–	–	–	–	5
Sri Sakhon	3	–	3	4=	–
Phu Khieo W.S.	4	–	4	–	–
Khlong Nakha W.S.	5	–	5	–	-
Mae Sariang	–	3	3	4=	–
Kroeng Krawia N.H.A.	–	4	4	–	–
Sangkhlaburi	–	5	5	–	–

Malaysian species. The fourth area is Phu Khieo Wildlife Sanctuary (W.S.). On adding the fifth area, Khao Banthad W.S., we would reach 90 per cent Thai hawkmoth species diversity. This sequence of top five sites is tabulated in Table 21.7. To represent all Thai hawkmoths at least once, using the existing database, 14 of the 90 sites are required.

2. *Tiger-beetle species richness pattern*

We do not yet have owl or milkweed butterfly data for Thailand on a scale similar to the hawkmoth information. However, a recent revision of Thai tiger-beetles by Naviaux (1991) provides comparable data. Naviaux's records have been divided amongst 75 areas, of which 32 are in common with hawkmoths (giving a total of 133 areas in Thailand for which we have records for one or both groups: Fig. 21.1).

On present evidence, the species of tiger-beetles are much more restricted in distribution than are the species of hawkmoths. Khao Yai N.P. has the highest species richness, with 24.27 per cent of the total of 103 species. Doi

Suthep-Pui comes second, with a 14.56 per cent increment—so this national park is very valuable for both tiger-beetles and hawkmoths. Mae Sariang is third—a locality for which we have no hawkmoth records (equally, there are no tiger-beetle records for Sri Sakhon, the third hawkmoth choice). By area 5 (Sangkhlaburi), about 60 per cent of the tiger-beetles would be taken into account. This sequence of the top five areas is given in Table 21.7. Twenty areas would be needed for 90 per cent, and a total of 30 to achieve full representation.

3. *Summed hawkmoth/tiger-beetle species richness pattern*

When the hawkmoth and tiger-beetle are analysed together, Doi Suthep-Pui N.P. comes top (Table 21.7), with 51.61 per cent. This is not surprising, as this was the richest site for the more speciose hawkmoths, and second for tiger-beetles. Doi Inthanon N.P. is next, then Khao Yai N.P. These results are not surprising either, as these sites cover the top two for both groups. However, one might expect that the next step, judging by the separate results, would involve Sri Sakhon versus Mae Sariang, but this is not so. At this step a '4–3–1' type situation arises, to give Khao Banthad W.S. as the fourth step, and Sai Yok N.P. in fifth place (even though neither appears in the separate top-five lists). So these summed results are not intuitively obvious from examination of the results for the individual groups. In particular, note that calculation of mean ranks and re-ranking cannot identify the optimum configuration of sites (Table 21.7).

These 5 sites (Table 21.7) account for 75 per cent of all species in the combined set; 90 per cent is reached by step 17, and 38 areas are needed in all, for complete representation. Implementing measures sensitive to taxic distinctness, including dispersion, gives broadly comparable results (Table 21.8). The sequence for hawkmoths alone is the same, while for tiger-beetles Kroeng Krawia Non-hunting Area (N.H.A.) is replaced by Phu Phan N.P. Analysis of the combined data reverses the order of Khao Yai N.P. and Doi Inthanon N.P., and substitutes Mae Sariang for Sai Yok N.P.

It should be emphasized that these results are based on incomplete data for only two families of insects. Far more work would have to be done, on a much wider range of taxa, before the priority areas sequence could be considered to be informative.

The next steps

If we take a global analysis of owls, milkweed butterflies or, notably, hawkmoths, Thailand is seen to have very high diversity. Analysed at a regional scale, within a portion of South East Asia, the combined data for these three groups identifies the fauna of Thailand as the most speciose, and thus likely to be a particularly valuable country within any global

Table 21.8 Priority sequences for the top five areas in Thailand for hawkmoth (Sphingidae) and tiger-beetle (Cicindelidae) taxic dispersion, selected from 133 areas in Thailand. Separate sequences are given for the two groups (moths, beetles), mean ranks (mr), re-ranked sequences (rr), and by direct analysis of the combined data set (comb). For comparison with Table 21.7; see also text.

	moths	beetles	mr	rr	comb
Doi Suthep-Pui N.P.	1	2	1.5	2	1
Khao Yai N.P.	–	1	1	1	2
Doi Inthanon N.P.	2	–	2	3	3
Khao Banthad W.S.	–	–	–	–	4
Mae Sariang	–	3	3	4=	5
Phu Phan N.P.	–	5	5	–	–
Sri Sakhon	3	–	3	4=	–
Phu Khieo W.S.	4	–	4	–	–
Khlong Nakha W.S.	5	–	5	–	–
Sangkhlaburi	–	4	4	–	–

conservation network. Within the country itself, limited as we undoubtedly are by a lack of intensive survey data, the same methods can still be applied, in theory, to ranking parks and other parcels of land in terms of their potential importance in contributing to an efficient area network for conserving national biodiversity.

As yet it is too early to generalize from the few cases examined here. We need to explore the problems of summation through examination of a far larger number of data sets. One obvious priority is to determine if the raw data summation method described, when applied using taxic measures, usually identifies the minimum configuration, even if it is unreliable in such circumstances with respect to placing the elements of the configuration in the 'correct' optimised priority sequence.

There is also an evident need to test both individual and large combined analyses against a much wider sample of biodiversity. Even if we do succeed in analysing 100 groups averaging 200 species each, this would still only be 20 000 species—little more than 0.1 per cent of the number now believed to exist (Hammond 1992). One approach, already briefly alluded to, being vigorously followed up by Paul Williams, Kevin Gaston, Chris Humphries, and others, is to map the global distribution of plant and various insect families on WORLDMAP. This will give us some grasp of a major portion of known terrestrial biodiversity.

However, as we have discussed above, the summation of species data may prove to be of greatest value at national level. For the UK, widely regarded as having the best recorded biota in the world (cf. Lawton *et al.*, this volume), the number of species mapped on the 10 km grid has yet to exceed 20 000. Development of the summation procedures outlined here could play a particularly important role at the key national level of analysis.

Thus we are now in a position to begin to pursue two of the most basic approaches to estimating global patterns and total biodiversity—analysing genera, tribes, or families of major groups such as flowering plants or beetles on the one hand, and through summation of data for a variety of groups recorded at species level on the other. What we lack is the information in suitable form. But one hundred groups of specialists could surely assemble global summary versions of their own data in a few months, if there was a real will to do it!

Acknowledgments

We thank our colleagues Chris Humphries, Paul Williams, and Peter Forey for help in various ways with preparation of this paper; IJK thanks Adam Cotton, Bob Kennett, Philip Round, and many others for help in Thailand; CRS thanks Tim Crowe, Alan Kemp, and Storrs Olsen for advice about owls.

References

Ackery, P. R. and Vane-Wright, R. I. (1984). *Milkweed butterflies: their cladistics and biology*. British Museum (Natural History), London and Cornell University Press, Ithaca, NY.

Brown, K. S. (1991). Conservation of neotropical environments: insects as indicators. In *The conservation of insects and their habitats* (ed. N. M. Collins and J. A. Thomas), pp. 349–404. Academic Press, London.

Currie, D. J. (1991). Energy and large-scale patterns of animal- and plant-species richness. *American Naturalist*, **137**, 27–49.

Currie, D. J. and Paquin, V. (1987). Large-scale biogeographical patterns of species richness of trees. *Nature*, **329**, 326–7.

Fernholm, B., Bremer, K., and Jornvall, H. (ed.) (1989). *The hierarchy of life*. Excerpta Medica, Amsterdam.

Gaston, K. J. and Williams, P. H. (1993). Mapping the world's species—the higher taxon approach. *Biodiversity Letters*, **1**, 2–8.

Hammond, P. M. (1992). Species inventory. In *Global biodiversity* (ed. B. Groombridge), pp. 17–39. Chapman and Hall, London.

Hawksworth, D. and Colwell, R. R. (ed.) (1992). Biodiversity amongst micro-organisms and its relevance. *Biodiversity and Conservation*, **1**, 221–345.

Hennig, W. (1981). *Insect phylogeny* (ed. A. C. Pont). John Wiley, Chichester.

Humphries, C. J. and Vane-Wright, R. I. (1992). *Systematic evaluation of the global network of protected areas: objectives, alternatives, prospects and proposals*. Abstract and draft papers presented at IVth World Congress on National Parks and Protected Areas, Caracas, February 1992.

ICBP (1992). *Putting biodiversity on the map: priority areas for global conservation.* International Council for Bird Preservation, Cambridge.

IUCN (1992). *Regional reviews* (IVth World Congress on National Parks and Protected Areas, Caracas, Venezuela). IUCN, Gland, Switzerland.

Kremen, C. (1992). Assessing the indicator properties of species assemblages for natural areas monitoring. *Ecological Applications*, **2**, 203–17.

Kremen, C. (in press). Biological inventory using target taxa: a case study of the butterflies of Madagascar. *Ecological Applications.*

Margules, C. R., Nicholls, A. O., and Pressey, R. L. (1988). Selecting networks of reserves to maximise biological diversity. *Biological Conservation*, **43**, 63–76.

Margules, C. R., Pressey, R. L., and Nicholls, A. O. (1991). Selecting nature reserves. In *Nature conservation: cost effective biological surveys and data analysis* (ed. C. R. Margules and M. P. Austin), pp. 90–7. CSIRO, Australia.

May, R. M. (1990*a*). Taxonomy as destiny. *Nature*, **347**, 129–30.

May, R. M. (1990*b*). How many species? *Philosophical Transactions of the Royal Society (B)*, **330**, 293–304.

McNaughton, S. J. (1989). Ecosystems and conservation in the twenty-first century. In *Conservation for the twenty-first century* (ed. D. Western and M. Pearl), pp. 109–20. Oxford University Press, New York.

Mickleburgh, S. P., Hutson. A. M., and Racey, P. A. (1992). *Old World fruit bats: an action plan for their conservation.* IUCN, Gland, Switzerland.

Naviaux, R. (1991). Les cicindeles de Thailande, étude faunistique (Coleoptera: Cicindelidae). *Bulletin Mensuel de la Société linnéenne de Lyon*, **60**, 209–88.

Pearson, D. L. and Cassola, F. (1992). World-wide species richness patterns of tiger beetles (Coleoptera: Cicindelidae): indicator taxon for biodiversity and conservation studies. *Conservation Biology*, **6**, 376–91.

Pressey, R. L., Humphries, C. J., Margules, C. R., Vane-Wright, R. I., and Williams, P. H. (1993). Beyond opportunism: key principles for systematic reserve selection. *Trends in Ecology and Evolution*, **8**, 124–8.

Pressey, R. L. and Nicholls, A. O. (1989). Efficiency in conservation evaluation: scoring versus iterative approaches. *Biological Conservation*, **50**, 199–218.

Reid, W. V., Barber, C., and Miller, K. R. (1992) *Global biodiversity strategy: guidelines for action to save, study and use Earth's biotic wealth sustainably and equitably.* World Resources Institute, New York.

Turner, J. R. G., Lennon, J. J., and Lawrenson, J. A. (1988). British bird species distributions and the energy theory. *Nature*, **335**, 539–41.

Vane-Wright, R. I. (in press). Insect systematics and the conservation of biodiversity: global, national and local perspectives. In *Perspectives on insect conservation* (ed. K. J. Gaston, T. R. New, and M. Samways). Intercept, Andover, UK.

Vane-Wright, R. I., Humphries C. J., and Williams, P. H. (1991). What to protect?—systematics and the agony of choice. *Biological Conservation*, **55**, 235–54.

Walker, B. H. (1992). Biodiversity and ecological redundancy. *Conservation Biology*, **6**, 18–23.

Williams, P. H. (1992). *WORLDMAP, priority areas for biodiversity (using version 3)*, 10 pp. Privately distributed, London.

Williams, P. H., Humphries, C. J., and Vane-Wright, R. I. (1991). Measuring biodiversity: taxonomic relatedness for conservation priorities. *Australian Systematic Botany*, **4**, 665–79.

Williams, P. H., Vane-Wright, R. I., and Humphries, C. J. (1993). Measuring biodiversity for choosing conservation areas. In *Hymenoptera and biodiversity* (ed. J. LaSalle and I. D. Gauld), pp. 309–28. CAB International, Wallingford, UK.

22. A scientific basis for establishing networks of protected areas

C. R. MARGULES*, I. D. CRESSWELL[†],
and A. O. NICHOLLS*
**CSIRO Division of Wildlife and Ecology, P.O. Box 84, Lyneham, ACT 2602, Australia*
[†]ERIN (Environmental Resources Information Network), GPO Box 636, Canberra, ACT 2601, Australia

Abstract

A method is proposed for identifying networks of protected areas to maximize the likelihood that they sample biological diversity. Environmental regionalization is followed by the use of iterative heuristic algorithms to determine where protected areas should be located within regions so that all known species, communities, or environments are represented. While there is no single best solution, iterative procedures incorporate the principle of complementarity and offer the advantages of flexibility, efficiency, and explicitness. The most urgent scientific challenge is to improve the quality of the data bases on which these algorithms operate.

Introduction

The scientific purpose of a system of protected areas, or a reserve network, the term used here, is to sample and thus help sustain the biological diversity of the region in which it is situated. There are other valid reasons for protecting nature, including beauty, inspirational value, and the sustained use of natural resources, but the primary scientific purpose of a reserve network is to help protect biological diversity; the variety of life.

Biological diversity is defined for this purpose as variation both within and between species. A sample is defined statistically as a subset of a population with the same mean and variance as the population. A sample of biological diversity then would be a sub-population of all species with the same mean

Systematics and Conservation Evaluation (ed. P. L. Forey, C. J. Humphries, and R. I. Vane-Wright), Systematics Association Special Volume No. 50, pp. 327–50, Clarendon Press, Oxford, 1994.

and variance as the wider population; the same demographic mean and variance, the same morphological mean and variance, and the same genetic mean and variance. Reserve networks alone will not be sufficient to sustain the variety of life, but it is a valid scientific goal to maintain a sample in reserves.

Unfortunately, we lack the knowledge to judge the success or failure to achieve this goal. There probably will never be a complete enumeration of biological diversity. Describing the spatial and temporal patterns of species distributions, and the demographic, morphological, and genetic means and variances, seems an impossible task. On the practical side, decisions are being made now whether, for example, to log a forest or set it aside for nature conservation. Even if the scientific tools for enumerating biological diversity were available, time might preclude their effective use. Yet it is worth stating this goal explicitly because it gives rise to the following working definition of conservation value. The conservation value of a site in a region is the contribution it makes to sampling regional biological diversity (Margules and Nicholls, in press).

In applying this definition, the lack of appropriate ecological knowledge and an urgency in the face of changing land use means compromises are necessary. Biological diversity may have to be represented by subsets of species, communities, or even environments. The goal of maintaining a sample in reserve networks may have to be reduced to encompassing as many species, communities, or environments as possible. However, accepting these compromises, it is still possible to derive a measure of success or failure. The question can be asked, 'does the reserve network encompass the known subset of species, communities, or environments present in this region?' As an aside, it is worth noting that according to this definition, sites need not necessarily be exceptionally diverse or rich in species to have high conservation value. If a site, even a species-poor site, contributes a sample of regional biological diversity not contributed by any other sites, then it has a very high conservation value.

If reserves are to be located so that they complement one another in this way, then two things are required; a suitable database and a procedure for using those data to identify complementary sites. It is useful to distinguish these two requirements because the second is often neglected in favour of a quest for the best database, or, cynically, a quest for the technical capacity to display a database at its best.

Databases consist of objects and attributes. For conservation planning objects are defined geographical areas of land or water, such as sample sites, grid cells, catchments, or classes of tenure. Attributes are the features which characterize those objects; species, habitats types, communities, or environments.

A realistic database might contain information on the distribution patterns of some species. Some species may be adequate indicators of a broader range of species, but in the absence of proven correlations between so-called indicator species and others, it might be prudent for a database to contain maps of habitat types or environments as a sample of each habitat or environment may stand a higher chance of sampling biological diversity than a sample of some species. Recent methodological developments in the design of biological surveys, the analysis of survey data and the description of species, community, habitat, or environmental distribution patterns will help in the development of more comprehensive databases (e.g. Austin and Heyligers 1989, 1991; Faith and Norris 1989; Mackey *et al.* 1989; Nicholls 1989, 1991; Austin *et al.* 1990; Austin 1991; Belbin 1991, in press; Margules and Austin 1991). In the meantime, decisions on land allocation for conservation or exploitation are being made and will continue to be made regardless of the quality of databases. This paper tries to show, by example, how those decisions can be better informed in the short term so as to maximize the likelihood that systems of protected areas set aside now, or in the near future, have a high probability of sampling biological diversity. The proposed procedure has two steps. The first is an environmental or biogeographic regionalization. The second is the use of explicit iterative algorithms to search for a set of objects within each region which together sample all attributes; a set of sites (grid cells, units of tenure, remnant habitat patches, etc.) which represent all species (or habitats, communities, environments, etc.). The result is a nominal core reserve network, which could serve as the base on which to build regional conservation management plans.

Environmental and biogeographic regionalization

The idea of an environmental regionalization is that components of the landscape, climate, geological substrate, landforms, and soils co-occur in unique associations which can be described and mapped. Vegetation is often mapped as the surface expression of these determining factors to provide a biological component to the regionalization. Regionalization is a tool for describing broad scale patterns of co-occurring variables.

Environmental and biogeographic regionalization has a long history in Europe, particularly eastern Europe (see Isachenko 1973 for an historical review). By the mid 1960s many parts of Africa (e.g. Milne 1936; Trapnell 1953), North America (e.g. Hills 1961), and Australia (e.g. Christian and Stewart 1968) had been mapped into biogeographic regions and sub-regions variously called land systems, ecoregions, land units, catenas, etc. Two problems arose. The first is that, using the traditional techniques of aerial photograph interpretation, experience, and common sense, the degree of variance of the attributes within a region could not be quantified (Spence

1968). Modern computers and numerical classification methods such as the one described below have overcome this problem.

The second problem was that the location of boundaries between regions was a difficult and contentious issue because it was a subjective decision. Numerical classification techniques have not resolved this problem because no precise methodology has been uniformly accepted that optimizes boundary location. The choice of objects (e.g. grid cells vs irregular polygons), the choice of variables (e.g. enviornmental parameters), and the decision to stop a classification at a given number of classes, determines the membership of each class and hence the resultant boundaries. However, the precise location of boundaries should not be a major issue. Because there are so many solutions, and because regionalizations are a way of pigeon-holing information and providing a spatial framework for planning and management, it does not matter so much where a boundary is precisely, as how it was chosen. Numerical classifications are explicit so it is clear how boundaries are determined.

Figure 22.1a is an environmental regionalization of Australia (Thackway and Cresswell 1992). Thirty regions have been identified by the numerical classification of 281,177 1/20th by 1/20th degree grid cells based on the similarities they share in the composition of the 12 environmental attributes listed in Table 22.1. The decision to derive 30 regions is arbitrary, but the method for deriving them is explicit. If less within-region heterogeneity is

Table 22.1 The 12 environmental attributes used to produce the regionalization in Fig. 22.1.

Major attribute	Sub-attribute
Terrain	Mean elevation Range of elevation Roughness (maximum/mean)
Temperature	Annual mean temperature Maximum temperature warmest month Minimum temperature coldest month
Rainfall	Annual mean precipitation Precipitation warmest quarter Precipitation coldest quarter
Soils	Soil profile (mainly texture) Available water holding capacity Moist state permeability (potential)

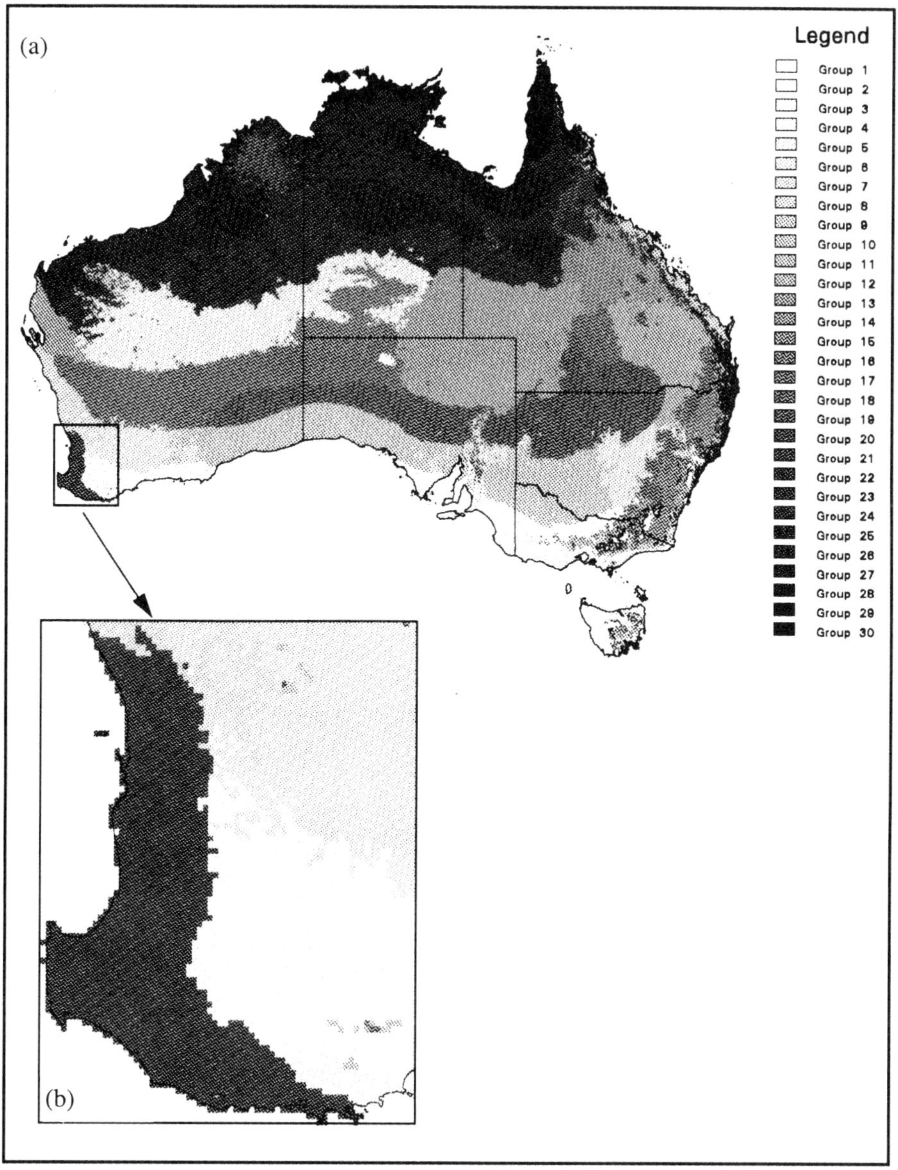

Fig. 22.1 A map of 30 environmental regions of Australia derived from a numerical non-hierarchical classification (see text for details of the method).

required, then it is possible to specify a greater number of regions, or to divide them further into sub-regions.

A non-hierarchical method was used for the classification. Non-hierarchical methods begin with a set of seed groups, which in this case was created by arbitrarily splitting the initial matrix into 30 equal parts. Then, the distance in 12 dimensional (12 variables) space from each grid cell (object) in each group to every other grid cell in all groups, was measured. This distance is a measure of how similar or different each grid cell is to every other grid cell. Grid cells are then iteratively reallocated from one group to another, with the aim of minimizing the total within-group distance. It is necessary to nominate the final number of groups, in this case 30, and to nominate a stopping rule for the iterations, in this case a 2 per cent movement of objects between groups.

The method used in this regionalization is known as Jauncey's variant of Forgy's 'nearest centroid sorting method' (Forgy 1965). This method implicitly minimizes a within-cluster error function and leads to cluster boundaries which are piece-wise linear since the boundaries are equidistant from the nearest centroid. The method is also unaffected by the sequence of data units within the data set.

The general within-cluster error function used was the sum of squared deviations for the kth cluster, given by:

$$E_k = S_i S_j (x_{ijk} - x_{ik})^2$$

where the total sum of squares (E) is defined as:

$$E = S_k E_k.$$

In general, it can be shown that convergence exists and that the error function E is minimized when Euclidean distance is used, but this does not necessarily hold for general error functions. The error function used here provides a distance measure based on the Gower similarity measure (Gower 1971) rather than Euclidean distance. Hence, convergence to a global minimum of the error function cannot be guaranteed.

Having completed a regionalization, the next step is to identify a reserve network, or system of protected areas, within each region.

1. *Locating protected areas within regions*

A number of reserve selection procedures based on the principle of complementarity, collectively called iterative selection procedures, are now available for indicating where reserves should be located so that collectively they maximize the likelihood that regional biological diversity is represented (e.g. Kirkpatrick 1983; Margules and Nicholls 1987; Margules *et al.* 1988; Pressey and Nicholls 1989*a*,*b*; Vane-Wright *et al.* 1991; Rebelo and Siegfried 1992; Nicholls and Margules, in press).

As a first approximation to sampling biological diversity, a reserve network could sample the range of environmental variation in a region. A product of numerical classifications such as the one described above is a quantitative statement of that variation. If a set of reserves, or protected areas, could be positioned so that the range of each attribute was sampled, then the probability that biological diversity has also been sampled, would be maximized, given the obvious limitations of the database.

Conceptually, this is a simple idea, but there are considerable practical difficulties. Each region in Fig. 22.1 consists of a set of grid cells grouped together because they are all more similar to one another than they are to grid cells belonging to other groups. The measure of similarity is one of continuous variation involving a contribution from all 12 variables. Very few, if any, grid cells will be identical according to this measure, so in order to sample the range of variation in a protected area network the within-group variation needs to be determined.

To approximate this variation for each region, it is necessary to determine the distribution of each cell in relation to every other cell within the 12 dimensional space that bounds that region. As a first order estimate it is possible to simplify variation in the data by categorizing the continuous variables.

The nine attributes associated with terrain, temperature, and rainfall are continuous variables, whilst the three soils attributes are already categorized (Table 22.1). The continuous variables were simplified by allocating them to one of two classes; less than or equal to the median value for the region, or greater than the median value.

Region 20 (Fig. 22.1b, the south-western corner of Australia) contains 1672 grid cells. The matrix of categorized and simplified variables reveals 311 combinations. Thus, for a reserve network to represent each combination just once, 311 of 1672 grid cells would be required. Table 22.2 shows that 94 of these are unique; that is, are represented by only one grid cell. A further 48 occur only twice, another 48 occur only three times, 24 occur four times, and so on, with only six combinations being represented by 30 or more grid cells. Thus, even the extreme simplification of dividing continuous variables into two classes reveals an extremely heterogeneous region.

Another approach to determining the within-group variation is to scale down and compile a new database for each region. This might consist of data on species, habitat, or community distributions from a biological survey, or another, lower level classification of environments within regions, or a combination of both. Two examples taken from Margules and Nicholls (1987) and Nicholls and Margules (1993), which are also included in

Table 22.2 The frequency of occurrence of each combination of environmental variables in the region shown in Fig. 22.1b.

Frequency	Number of environmental combinations
1	94
2	48
3	48
4	24
5	14
6	14
7	6
8	11
9	3
10	3
11	4
12	4
13	2
14	2
15	6
16	3
17	3
18	2
19	3
20	1
21	3
22	3
25	1
28	1
29	2
30	1
33	2
35	1
36	1
46	1

another review (Margules and Nicholls, in press), are illustrated below. The first is from a region of coastal hardwood forest in south-eastern Australia. The objects in the database are grid cells over continuous forest and the attributes are the proportion of each grid cell occupied by each of 31 tree communities and 26 environments. The second is from an agricultural region

Table 22.3 The actual percentage area of each environment and each tree community represented in a solution requiring at least 10 per cent of each environment and each tree community.

Environment	% area represented	Community	% area represented
1	10.69	1	10.65
2	27.27	2	45.60
3	11.18	3	10.36
4	12.73	4	17.74
5	17.13	5	21.07
6	14.67	6	14.20
7	10.67	7	10.55
8	100.00	8	33.33
9	14.63	9	10.46
10	28.57	10	26.46
11	24.10	11	20.48
12	23.68	12	18,58
13	60.00	13	15.52
14	13.14	14	39.85
15	100.00	15	14.85
16	100.00	16	16.24
17	20.37	17	10.57
18	11.06	18	18.83
19	100.00	19	12.58
20	100.00	20	19.19
21	10.31	21	14.67
22	10.86	22	32.95
23	50.00	23	13.52
24	33.33	24	17.54
25	25.00	25	11.63
26	14.26	26	13.45
		27	10.38
		28	29.56
		29	17.29
		30	13.07
		31	14.46

of South Australia. The objects are remnant patches of mallee habitat (mallee is a habitat characterized by tall multi-stemmed shrubs or low trees of the genus *Eucalyptus*). The attributes are the probabilities of different mallee communities occurring in the remnant patches. These probabilities were derived from correlative statistical models relating observed occurrences of communities to environmental variables. Full details of the derivation of this database can be found in Margules and Nicholls (1987).

2. *Coastal hardwood forest*

As part of a wider study on the mediation of conflicts over forest resource use (CSIRO, in press), a database was generated consisting of 382 9 km^2 grid cells, and, among other data items, the distribution patterns of the 31 tree communities and 26 environments referred to above. The question was asked 'What is the minimum set of grid cells which together sample at least 10 per cent of each environment and 10 per cent of each tree community?' This set would become the nominal core reserve network for a regional conservation plan to be negotiated with forest industry interests.

(a) The algorithm. The algorithm as programmed has the ten steps listed below. Steps 1 and 9 were not used in this example, but they are included here for completeness and will be referred to in the discussion. The algorithm was first used to identify the minimum set of grid cells which sampled at least 10 per cent of each environment. Then, the tree communities still not sampled at the 10 per cent level were recorded and the extra grid cells required to bring the sample up to 10 per cent were found.

Step 1. This is an optional step at which some grid cells can be included before the selection procedure begins. Examples might be existing reserves or grid cells known to have rare species. It is anticipated that most practical applications of this algorithm will use this first step.

Step 2. Select all grid cells with unique occurrences of environments.

Step 3. Find the next rarest environment and select the grid cell that, when added to those already selected, will represent that environment, plus the greatest number of other environments, at or above the 10 per cent level.

Step 4. If there is a choice, select the grid cell nearest to one already selected.

Step 5. If there is still a choice, select the grid cell that also contributes the largest number of environments not yet represented at the 10 per cent level.

Step 6. If there is still a choice, select the grid cell that will enable the 10 per cent level to be achieved for the rarest group of environments remaining under- or unrepresented.

Step 7. If there is still a choice, select the grid cell that will contribute most to achieving the 10 per cent level of representation of the rarest group of environments remaining under- or unrepresented.

Step 8. If there is still a choice, select the grid cell which either (a) contains the smallest percentage area necessary to achieve the 10 per cent level of representation, or (b) contributes the largest percentage area of that environment if no one grid cell can enable the 10 per cent level of representation to be achieved. If (b) is invoked, the algorithm returns to step 3 so that the last two steps below are used only in the last resort.

Step 9. If there is still a choice, select the smallest (since equal sized grid cells were used for this example, this step was not used).

Step 10. If there is still a choice, select the first one in the list.

(b) Reserve selection. At least 10 per cent of all environments were represented in 34 grid cells but only 24 of the 31 communities were represented at the 10 per cent level. A further three grid cells were required to represent at least 10 per cent of the area of all communities as well as all environments, making a total of 37 grid cells. Figure 22.2 shows the spatial arrangement of this solution. Table 22.3 gives the actual percentage area which each environment and community is represented by. Most environments and communities are represented at a higher level than 10 per cent, some considerably higher. Only three environments and six communities are represented at the lowest level of just over 10 per cent. This result is data dependent and different results will be obtained with different data sets. However, there is no reason to expect that higher levels of representation than the minimum required would not be achieved for most attributes in most cases.

3. *Mallee remnants*

The database was a table of probabilities of each of six mallee vegetation communities occurring on each of 101 remnant vegetation patches in a 15 000 km^2 agricultural region of the western Eyre Peninsula, South Australia (Fig. 22.3). In a vegetation survey of these remnants, only a subset of all remnants was sampled, so probabilities of occurrence in unsampled remnants were estimated with logistic regression models. The availability of data from only a subset of all possible sites of interest may be a common feature. A method for predicting occurrences of species (communities, etc.) on unsampled sites is given in Margules and Nicholls (1987) based on the statistical modelling developed by Austin *et al.* (1984) (see also Margules and Stein 1989; Nicholls 1989; and Austin *et al.* 1990), for elaborations of this method). The algorithm described below is for use with the resultant

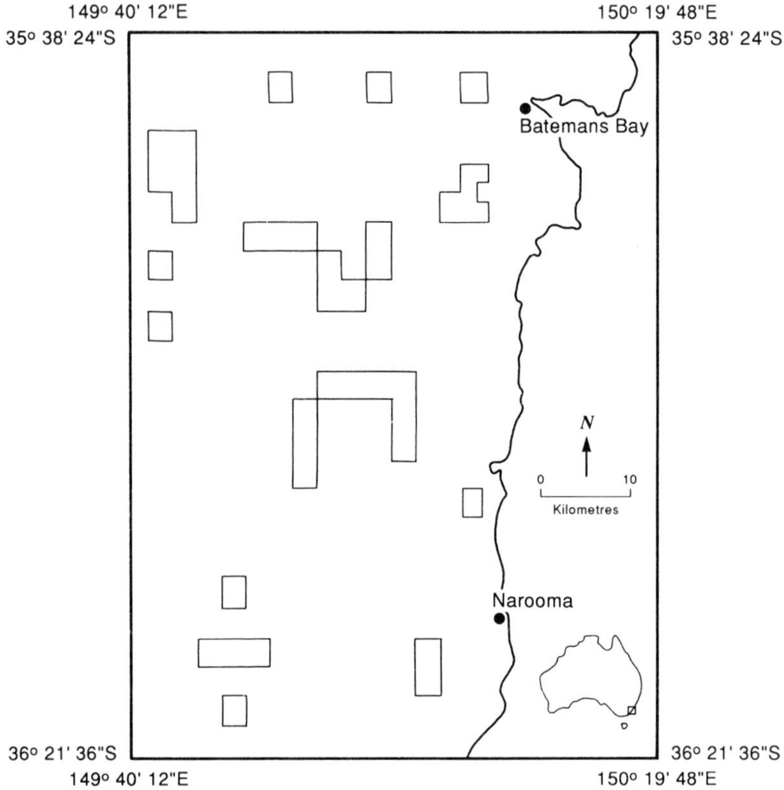

Fig. 22.2 The minimum set of 9 km^2 grid cells needed to sample at least 10 per cent of each environment and 10 per cent of each tree community in a coastal hardwood forest region of New South Wales, Australia.

database, which is a table of probabilities. An abstract of the table of probabilities analysed below is given in Table 22.4.

(a) The algorithm. The objective was to find the minimum set of mallee remnants that, together, had a high chance of sampling all six communities at least once. The level of acceptance has to be specified and in this example it was a 95 per cent chance. The algorithm consists of the following steps.

Step 1. Find the remnant with the highest probability (p) of having community 1.

Step 2. If there is a choice, choose the largest.

Step 3. From the table (the expanded Table 22.4) record the probabilities of the other 5 communities occurring on this remnant.

Table 22.4 An abstract of the table of probabilities of each plant community occurring in each remnant patch.

	Comunity					
Patch	1	2	3	4	5	6
1	0.019	0.517	0.79	0	0	0.053
10	0.035	0.158	0.79	0.272	0	0.050
11	0	0.105	0.79	0	0	0.187
15	0.068	0.067	0.289	0.515	0.67	0.026
22	0.019	0.571	0.79	0	0	0.053
35	0.203	0.157	0.79	0	0	0.189
56	0.068	0.067	0.79	0.515	0.667	0.080
57	0.068	0.067	0.79	0.514	0.667	0.080
65	0.068	0.067	0.79	0.515	0	0.026
76	0.095	0.091	0.79	0.653	0	0
77	0.095	0.091	0.79	0.653		0
78	0.583	0.091	0.41	0	0	0
79	0.583	0.091	0.41	0	0	0
82	0.203	0.571	0.025	0	0	0.715
87	0.203	0.571	0.07	0	0	0.715
93	0.493	0.158	0.41	0	0	0.764
95	0.583	0.158	0.41	0	0	0.764
98	0.203	0.571	0.025	0	0	0.437
99	0.203	0.571	0.025	0	0	0.913

Step 4. Subtract these probabilities from 1 to find the probability of failure (q); i.e. that a community does *not* occur on this remnant.

Step 5. Repeat this procedure for all 6 communities.

Step 6. Find the product of the probability of failure (q) for each community.

Step 7. Repeat step 1 to 6 (in step 1, finding the remnant with the next highest probability, and so on) until the product of the probability of failure falls below the designated level of acceptance; in this case, 0.05 for a 95 per cent chance of success.

Table 22.5 illustrates how this algorithm works. Remnant 95 is the largest remnant with the equal highest probability of having community 1 (0.583). This same remnant also has probabilities of 0.158, 0.410, 0, 0, and 0.764, of having communities 2, 3, 4, 5, and 6, respectively. The probability of remnant 95 not having community 1 is 1 − 0.583, which equals 0.417.

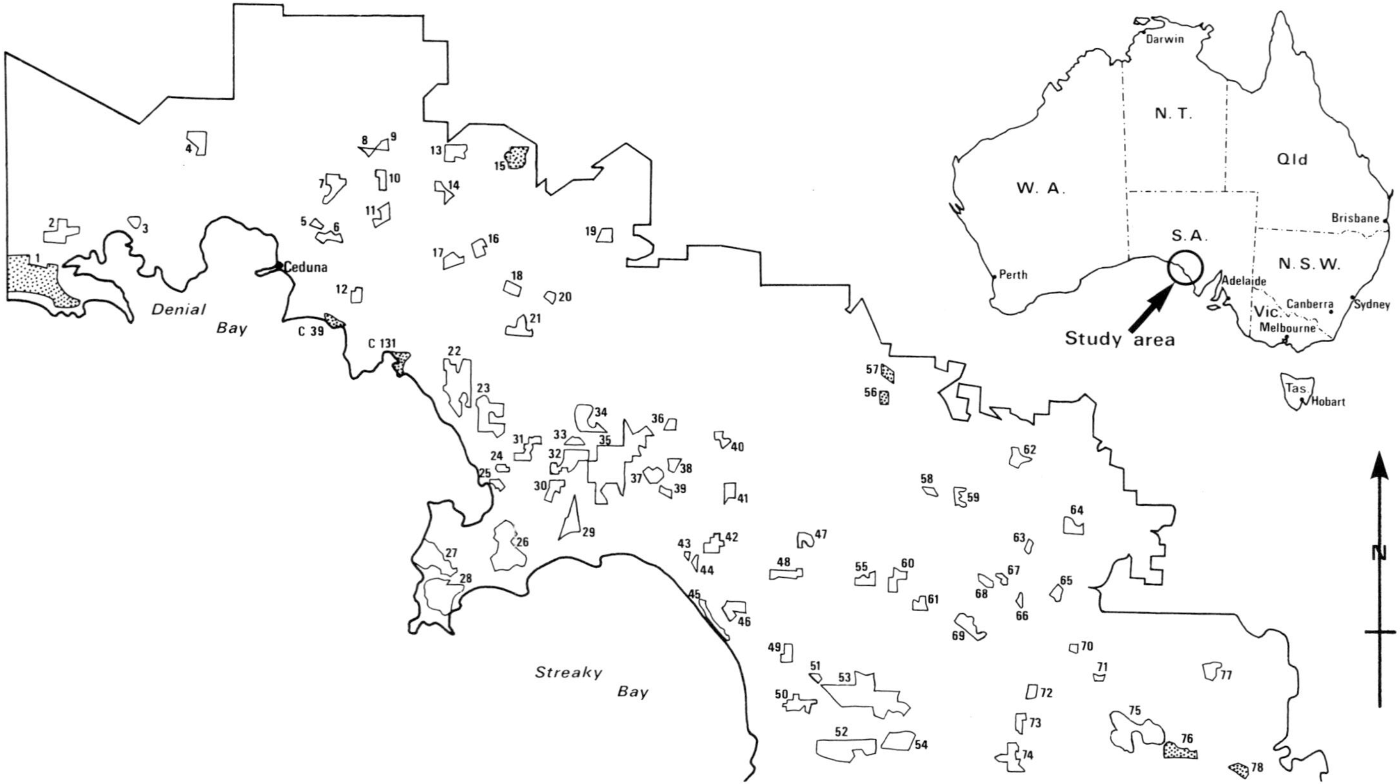

Denial Bay
Ceduna
C 39
C 131
Streaky Bay
1 2 3 4 5 6 7 8 9 10 11 12 13 14 15 16 17 18 19 20 21 22 23 24 25 26 27 28 29 30 31 32 33 34 35 36 37 38 39 40 41 42 43 44 45 46 47 48 49 50 51 52 53 54 55 56 57 58 59 60 61 62 63 64 65 66 67 68 69 70 71 72 73 74 75 76 77 78
Darwin
N. T.
Qld
W. A.
S. A.
Brisbane
N. S. W.
Perth
Adelaide
Canberra
Sydney
Vic.
Melbourne
Study area
Tas.
Hobart
N

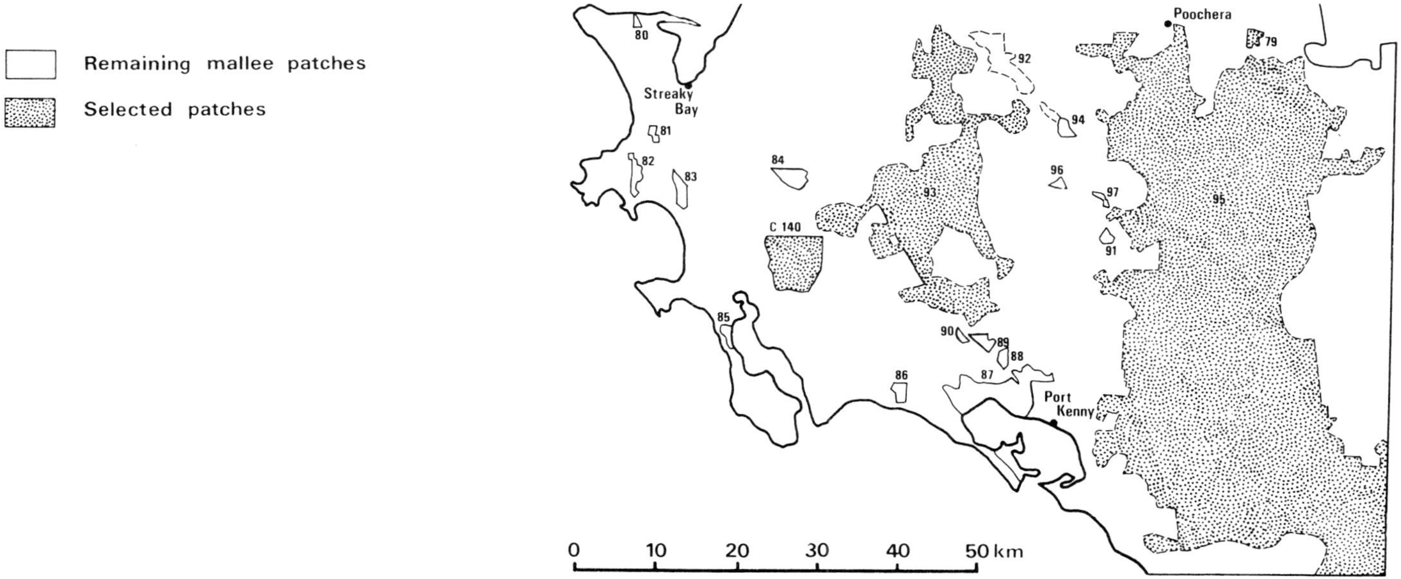

Fig. 22.3 The set of remnant mallee habitat patches which together have a 95 per cent chance of sampling each of six mallee communities at least once, on the western Eyre Peninsula, South Australia.

Table 22.5 The procedure for selecting the minimum set of patches with a 95 per cent chance of representing each community at least once. When patches are otherwise equal, the largest patch is chosen. p is the probability of finding the community and q is the probability of not finding it.

	Community											
	1		2		3		4		5		6	
Patch	p	q	p	q	p	q	p	q	p	q	p	q
95	0.583	0.417	0.158	0.852	0.410	0.590	0	1	0	1	0.764	0.236
99	0.203	0.797	0.571	0.429	0	1	0	1	0	1	0.913	0.087
1	0	1	0.571	0.429	0.790	0.210	0	1	0	1	0	1
76	0	1	0	1	0.790	0.210	0.653	0.347	0	1	0	1
15	0	1	0	1	0.289	0.711	0.515	0.485	0.667	0.333	0	1
product (q)	0.332		0.152		0.019		0.168		0.333		0.021	
79	0.583	0.417	0	1	0.410	0.590	0	1	0	1	0	1
87	0.203	0.797	0.571	0.429	0	1	0	1	0	1	0.715	0.285
77	0	1	0	1	0.790	0.210	0.653	0.347	0	1	0	1
57	0	1	0	1	0.790	0.210	0.515	0.485	0.67	0.333	0	1
Cumulative product (q)	0.100		0.065		0		0.028		0.111		0.006	
78	0.583	0.417	0	1	0.410	0.590	0	1	0	1	0	1
22	0	1	0.571	0.429	0.790	0.210	0	1	0	1	0	1
56	0	1	0	1	0.70	0.210	0.515	0.485	0.667	0.333	0	1
Cumulative product (q)	0.046		0.028		0		0.014		0.037		0.006	

Similarly, the probabilities of remnant 95 not having communities 2, 3, 4, 5, and 6 are 0.852, 0.590, 1, 1, and 0.236 respectively.

Remnant 99 is the largest with the equal highest probability of having community 2 (0.517). It also has probabilities of 0.203, 0, 0, 0 and 0.913 of having communities 1, 3, 4, 5, and 6 respectively. Thus, the probabilities of remnant 99 *not* having communities 1 to 6 are 0.799, 0.429, 1, 1, 1, and 0.087, respectively. In the same way, remnant 1 is the first choice for community 3, remnant 76 is the first choice for community 4, and remnant 15 is the first choice for community 5. There are only five remnants in this first iteration because remnant 99 has the highest probability of having community 6 as well as community 2.

The product of the probability of failure (q) after this first iteration for communities 1 to 6 is 0.332, 0.152, 0.019, 0.168, 0.333, and 0.021 respectively. Thus there is already a 95 per cent chance that communities 3 and 6 are represented at least once in these five remnants because the probability of failure, of not being sampled, is less than 0.05. However, communities 1, 2, 4, and 5 are still not represented with the required probability so the whole procedure is repeated, finding the next best candidate (i.e. the remnant with the next highest probability, or the largest if there is a choice) for those four communities only. After this second iteration, community 4 has reached the required probability of representation. A third iteration is needed for all six communities to fall below a probability of failure of 0.05. Clearly, it is possible to continue the process until all communities have a 99 per cent ($q = 0.01$) or even a 99.9 per cent ($q = 0.001$) chance of being represented.

(b) Reserve selection. Figure 22.3 maps the minimum set of 12 remnant mallee patches needed for a 95 per cent chance of sampling each community at least once. Note that whilst the very largest remnant, and some other relatively large remnants, are included, some small remnants are also required. This is because land clearing differentially favours the best agricultural soils and therefore only small remnants with communities of those soils remain, yet they have high conservation value because they make a contribution to sampling regional biological diversity that the other remnants could not make. Simply choosing, say, the 12 largest remnants, or the 12 remnants with the most species on them, would mean that the nominal reserve network would not have the highest probability of sampling regional biological diversity.

Discussion

The scientific procedure for identifying protected area systems suggested is explicit environmental or biogeographic regionalization, followed by explicit iterative reserve selection procedures within regions. This is designed to maximize the likelihood that systems of protected areas will, collectively, sample biological diversity.

It is better, scientifically, to derive regionalizations numerically as the process is explicit and repeatable. One example is given above, but there are many others. Margules and Belbin (1990) identified 100 climatic zones for Australia using a similar non-hierarchical clustering method, which makes fewer assumptions about the structure of the data (Belbin 1987). That same method was used by Mackey *et al.* (1989) to identify 40 bioclimatic regions of Australia. They were thus able to demonstrate, explicitly, the extent to which these 40 are different from one another. Three of the 40 are encompassed by the (then proposed) Wet Tropics of Queensland World Heritage Property, and this difference was used as part of the successful case for World Heritage listing.

If suitable computing facilities are not available, or if bioclimatic databases are not available at the appropriate scale, then existing regionalizations can be adopted in the interim. Recent examples include a regionalization of the Afrotropical realm (MacKinnon and MacKinnon 1987) and two ambitious attempts at global regionalization by Bailey (1989) and Prentice *et al.* (in press).

The examples of the selection of nominal core reserve networks illustrated here, and other Australian examples in the literature (e.g. Margules *et al.* 1988; Pressey and Nicholls 1989*b*; Bedward *et al.* 1991) use databases that are necessarily incomplete, but are nevertheless real examples of some of the more comprehensive databases available in Australia today.

However, even this level of detail is not available in most parts of Australia, nor in most other parts of the world. Two courses of action suggest themselves, which should be undertaken simultaneously. First, if at all possible, use whatever database is available to derive, via iterative selection procedures, a first approximation of a preferred complementary reserve network. Second, improve the quality of the database. Reserve networks can be supplemented as improved databases become available. The limitations and advantages of iterative selection procedures and ways of improving the quality of databases are discussed below.

1. *Iterative selection procedures*

Nominal core reserve networks identified using iterative selection procedures are certainly not the only options, and are not necessarily the 'best' solutions. Rather, they are heuristic solutions designed to satisfy an objective function (sampling all attributes in the database). A number of decisions have to be made by the user, which strongly influence the outcome. These decisions include which objects (e.g. grid cells) to use and how big they should be, whether to include any sites initially, as allowed for at step 1 of the first algorithm above, and how to choose between two or more equally qualified sites (e.g. choosing the largest or smallest). Iterative selection procedures are not objective. However, they are flexible, efficient and explicit. These features are discussed below.

2. *Flexibility*

Flexibility is inherent in the choices that have to be made, such as those referred to above. There is another aspect of flexibility, and that is that the database can be amended and/or the rules can be changed.

Databases can be amended in two ways. First, the objects, the set of candidate sites, can be manipulated. Many of the sites in any solution are capable of being substituted; that is there are alternatives. For example, sites with high values for exploitative uses could be removed and alternative reserve networks could be derived. The increased costs of these alternatives, in terms of land area required, could be assessed against the value derived from the exploitative use. Conversely, some sites in any solution are irreplaceable; there are no alternatives. Such sites are necessary, non-negotiable components of the reserve system. Fortunately these tend to be fewer in number than negotiable sites so there are opportunites for compromise. Pressey *et al.* (this volume) develop the ideas of irreplaceability and flexibility more fully, and in so doing propose mapping landscapes of conservation value based on complementarity.

Second, the attributes can be upgraded. As new information on, say, species distribution patterns becomes available, it can be added to the database, and new, more comprehensive reserve networks can be derived. The algorithms themselves can also be changed by changing the rules governing choices at each step, or the order of steps. In the first algorithm above, the rule at Step 9 is to choose the smallest, if there is a choice. This rule was used in order to minimize the total area required for the nominal core reserve network in these coastal forests. However, if that did not matter, then there is an ecological argument for choosing the largest, as was, in fact, the case with the second algorithm. It would be a simple matter to change this or any other rule. Similarly, in the first algorithm again, at Step 4 the nearest grid cell is chosen if there is a choice. This rule was included to minimize the likelihood that a set of widely dispersed sites would be chosen; a problem with earlier versions of the algorithm (Bedward *et al.* 1991). However, there is a cost associated with this rule, which is that three more grid cells, or 27 km^2 more forest, is required than would have been the case if that rule had not been used (Nicholls and Margules 1993).

These are just a few examples. There is any number of sensible variations that could be adopted in different situations. Common sense and local ecological knowledge are required for the application of these algorithms.

3. *Explicitness*

The need to be explicit, both in the derivation of biogeographic or environmental regions and in the identification of nominal core reserve networks, has been stressed throughout. The scientific reason is that the results are then repeatable. A practical reason is that the proposed protected area network can be justified more easily and defended more readily. One critic (Graeme Caughley, pers. comm.) argues that this second reason defies the accumulated wisdom of bureaucrats and administrators who find, in fact, that explicit statements are the most difficult to defend. However, conservation competes with alternative exploitative land uses which usually promote quantifiable economic and social benefits (e.g. employment). It can be difficult to argue against them without a clear explicit statement of conservation value which can be easily understood by all parties.

4. *Efficiency*

More efficient protected area networks sample a greater proportion of biological diversity in fewer sites. The minimum set of sites needed to sample all attributes is the most efficient solution of all.

This concept of efficiency can be difficult to justify conceptually because a minimum set of sites that samples all species, communities, or environments will almost certainly not be sufficient to maintain ecological processes in the long run, or to satisfy the legitimate ethical concerns of many people for the likely extinction of many species in the near future. No protected area network, however it was chosen, could maintain ecological processes in perpetuity or prevent the extinction of many species, unless it occupied most, if not all, of the region in question. The role of protected area networks is to help sustain ecological processes and help prevent the extinctions of species. That role can be enhanced if the network encompasses a true sample of biological diversity.

Unless protected area networks are chosen efficiently, the prospect is that they will not adequately sample regional biological diversity before the option to do so is exhausted. Pressey and Nicholls (1989*a*) for example, ranked pastoral leases in the Western Division of New South Wales according to the scores they received from the calculation of 12 indices of conservation value (e.g. rarity, diversity, representativeness, and combinations of these), and then asked how far down the lists it was necessary to go before all attributes, in this case environments called land systems *sensu* Christian and Stewart (1968), were represented a given number of times. They found that between 45 per cent and 98 per cent of the Western Division would have to be set aside, depending on which index was used, to sample each environment at least once, compared to 5.7 per cent using an iterative selection procedure. If 15 per cent of the Western Division could be in a reserve network, and that network was chosen using their iterative algorithm, then all environments could be replicated in the reserve system, some more than twice. Thus, efficiency will have to be a prime consideration if reserve networks are ever to sample regional biological diversity.

5. *Improving databases*

Databases for conservation planning need to be improved in at least three ways (Margules and Nicholls, in press). First, data on the distribution patterns of a wider range of species, and their demographic characteristics, are needed. In one sense, this is a routine task requiring the commitment of funds, staff, and equipment. However, there are also substantial scientific issues to be addressed. Since biological surveys are invariably samples of geographic space, not complete inventories, where precisely should those samples be recorded from? Margules and Nicholls (1987), Austin and Heyligers (1989, 1991), and McKenzie *et al.* (1989) have developed survey designs to sample mallee communities in habitat remnants and assemblages of plants, tree distributions in *Eucalyptus* forests, and vertebrates on the arid Nullarbor Plain, respectively. Their survey design protocols need to be tested and evaluated for other species in those same and other environments.

Second, because it will always be impossible to record all species, there is a need for agreement on which species, or which kinds of species, should be recorded. Concepts such as keystone species and indicator species (e.g. Morris and Brooker 1982; Greenslade and Greenslade 1984), ecological redundancy (Walker 1992), and taxonomic difference (e.g. Vane-Wright *et al.* 1991; Faith 1992) have been suggested. Currently, ease of identification and consistency of sampling seem to be the main criteria.

Third, information on ecological process should be included in databases for conservation planning. Knowledge of why species are where they are and how they respond to change is needed so it can be built into the design of reserve networks and wider regional conservation plans. Databases are usually compiled from surveys done at one time. Monitoring is a form of survey with a temporal stratification. A comprehensive network of monitoring sites could provide valuable information on changes in the spatial distribution patterns of species and their populations. At a global scale, the International Geosphere Biosphere Program (IGBP) is promoting the establishment of a network of stations to monitor global change. Monitoring changes in species distribution patterns at this broad scale would be a start in adopting monitoring programmes for biological diversity generally.

Acknowledgements

Graeme Caughley commented on a draft of the manuscript. Many of the ideas reviewed here were developed in collaboration with Bob Pressey. Richard Thackway kindly allowed the use of the unpublished Fig. 22.1. Kim Malafant supplied a copy of the non-hierarchical clustering algorithm used to produce Fig. 22.1. George Milkovits drew Fig. 22.2.

References

Austin, M. P. (1991). Vegetation: data collection and analysis. In *Nature conservation: cost effective biological surveys and data analysis* (ed. C. R. Margules and M. P. Austin) pp. 37–41. CSIRO, Melbourne.

Austin, M. P. and Heyligers, P. C. (1989). Vegetation survey design for conservation: gradsect sampling of forests in north-eastern New South Wales. *Biological Conservation*, **50**, 13–32.

Austin, M. P. and Heyligers, P. C. (1991). New approach to vegetation survey design: gradsect sampling. In *Nature conservation: cost effective biological surveys and data analysis*. (ed. C. R. Margules and M. P. Austin) pp. 31–6. CSIRO, Melbourne.

Austin, M. P., Cunningham, R. B., and Fleming, P. M. (1984). New approaches to direct gradient analysis using environmental scalars and statistical curve-fitting procedures. *Vegetatio*, **55**, 11–27.

Austin, M. P., Nicholls, A. O., and Margules, C. R. (1990). Measurement of the realized qualitative niche: environmental niches of five *Eucalyptus* species. *Ecological Monographs*, **60**, 161–77.

Bailey, R. G. (1989). Map, and explanatory supplement to ecosystems map of the continents. *Environmental Conservation*, **16**, 307–9.

Bedward, M., Pressey, R. L., and Keith, D. A. (1991). A new approach to selecting fully representative reserve networks: addressing efficiency, reserve design and land suitability with an iterative analysis. *Biological Conservation*, **62**, 115–25.

Belbin, L. (1987). The use of non-hierarchical allocation methods for clustering large sets of data. *Australian Journal of Ecology*, **19**, 32–9.

Belbin, L. (1991) The analysis of pattern in bio-survey data. In *Nature conservation: cost effective biological surveys and data analysis* (ed. C. R. Margules and M. P. Austin) pp. 176–90, CSIRO, Melbourne.

Belbin, L. (in press). Comparing two sets of community data, a method for testing reserve adequacy. *Australian Journal of Ecology*.

Christian, C. S. and Stewart, G. A. (1968) Methodology of integrated surveys. In *Aerial surveys and interpreted studies. Natural resources research*, **6**, pp.233–80. UNESCO, Paris.

CSIRO (in press). *Forest issues: processes and tools for inventory, evaluation, mediation and allocation*. CSIRO Division of Wildlife and Ecology, Canberra, Australia.

Faith, D. P. (1992). Conservation evaluation and phylogenetic diversity. *Biological Conservation*, **61**, 1–10.

Faith, D. P. and Norris, R. H. (1989). Correlation of environmental variables with patterns of distribution and abundance of common and rare freshwater macro-invertebrates. *Biological Conservation*, **50**, 77–89.

Forgy, E. W. (1965). Cluster analysis of multivariate data: efficiency versus interpretability of classifications. *Biometrics*, **21**, 768.

Gower, J. C. (1971). A general coefficient of similarity and some of its properties. *Biometrics*, **27**, 857–71.

Greenslade P. J. M. and Greenslade, P. (1984). Invertebrates and environmental assessment. *Environment and Planning*, **3**, 13–15.

Hills, G. A. (1961). The ecological basis for land use planning. *Research Report 46, Ontario Department of Lands and Forests*, Toronto, Ontario.

Isachenko, A. G. (1973). *Principles of landscape science and physical-geographic regionalisation* (trans. R. J. Zatorski). Melbourne University Press.

Kirkpatrick, J. B. (1983). An iterative method for establishing priorities for the selection of nature reserves: an example from Tasmania. *Biological Conservation*, **25**, 127–34.

McKenzie, N. L., Belbin, L., Margules, C. R., and Keighery, G. J. (1989). Selecting representative reserve systems in remote areas: a case study in the Nullarbor region, Australia. *Biological Conservation*, **50**, 239–61.

Mackey, B. G., Nix, H. A., Stein, J. A., Cork, S. E., and Bullen, F. T. (1989). Assessing the representativeness of the Wet Tropics of Queensland World Heritage Property. *Biological Conservation*, **50**, 279–303.

MacKinnon, J. and MacKinnon, K. (1987). Review of the Protected Areas System in the Afrotropical Region. IUCN/UNEP, Gland, Switzerland.

Margules, C. R. and Austin, M. P. (ed.) (1991). *Nature conservation: cost effective biological surveys and data analysis*. CSIRO, Melbourne.

Margules C. R. and Belbin, L. (1990). Where should climate monitoring stations be? In *Global change regional research centres: scientific problems and concept developments* (ed. A. Breymeyer), Conference Papers 6. Institute of Geography and Spatial Organisation, Polish Academy of Sciences, Warsaw.

Margules, C. R. and Nicholls, A. O. (1987). Assessing the conservation value of remnant habitat 'islands': mallee patches on the western Eyre Peninsula, South Australia. In *Nature conservation: the role of remnants of native vegetation* (ed. D. A. Saunders, G. W. Arnold, A. A. Burbidge, and A. J. M. Hopkins), pp. 89–102. Surrey Beatty, Chipping Norton, NSW, Australia in association with CSIRO and Calm.

Margules, C. R. and Nicholls, A. O. (in press). Where should nature reserves be located? In *Conservation biology in australia and oceania* (ed. C. Moritz, J. Kikkaura, and D. Doley). Surrey Beatty, Chipping Norton, NSW, Australia.

Margules, C. R. and Stein, J. L. (1989). Patterns in the distributions of species and the selection of nature reserves: an example from *Eucalyptus* forests in south-eastern New South Wales. *Biological Conservation*, **50**, 219–38.

Margules, C. R., Nicholls, A. O., and Pressey, R. L. (1988). Selecting networks of reserves to maximise biological diversity. *Biological Conservation*, **43**, 63–76.

Milne, G. (1936). *A provisional soil map of east africa*. Amani Memoirs, East African Agricultural Research Station, Amani, Tanganyika Territory.

Morris, D. L. and Brooker, M. P. (1982). An assessment of the importance of Chironomidae (Diptera). In *Biological surveillance. Chironomidae: ecology, systematics, cytology and physiology* (ed. D.A. Murray), pp. 195–202. Pergamon Press, Oxford.

Nicholls, A. O. (1989). How to make biological surveys go further with generalised linear models. *Biological Conservation*, **50**, 51–75.

Nicholls, A. O. (1991). An introduction to statistical modelling using GLIM. In *Nature conservation: cost effective biological surveys and data analysis* (ed. C. R. Margules and M. P. Austin), pp. 191–201. CSIRO, Melbourne.

Nicholls, A. O. and Margules, C. R. (1993). An upgraded reserve selection algorithm. *Biological Conservation*, **64**, 165–9.

Prentice, J. C., Cramer, W., Harrison, S. P., Monserud, R. A., and Solomon, A. M. (in press). A global biome model based on plant physiology and dominance, soil properties, and climate. *Journal of Biogeography*.

Pressey, R. L. and Nicholls, A. O. (1989*a*). Efficiency in conservation evaluation: scoring versus iterative approaches. *Biological Conservation*, **50**, 199–218.

Pressey, R. L. and Nicholls, A. O. (1989*b*). Application of a numerical algorithm to the selection of reserves in semi-arid New South Wales. *Biological Conservation*, **50**, 263–78.

Rebelo, A. G. and Siegfried, W. R. (1992). Where should nature reserves be located in the Cape Floristic Region, South Africa? *Conservation Biology*, **6**, 243–52.

Spence, N. A. (1968). A multifactor uniform regionalisation of British counties on the basis of employment data for 1961. *Regional Studies*, **2**, 87–104.

Thackway, R. and Cresswell, I. D. (1992). *Environmental regionalistaions of Australia: a user oriented approach.* Environmental Resources Information Network, Australian Parks and Wildlife Service, Sydney.

Trapnell, C. G. (1953). *The soils, vegetation and agriculture of north-eastern Rhodesia.* Government Printer, Lusaka.

Vane-Wright, R. I., Humphries, C. J., and Williams, P. H. (1991). What to protect?—systematics and the agony of choice. *Biological Conservation*, **55**, 235–54.

Walker, B. H. (1992). Biodiversity and ecological redundancy. *Conservation Biology*, **6**, 18–23.

23. New procedures for reserve selection in New South Wales: maximizing the chances of achieving a representative network

R. L. PRESSEY, M. BEDWARD, and D. A. KEITH
New South Wales National Parks and Wildlife Service, PO Box 1967, Hurstville, NSW 2220, Australia

Abstract

The time-honoured approach to reservation is opportunistic and *ad hoc*. This has some important disadvantages, which reduce the likelihood of achieving representative reserve systems and which call for more systematic procedures for selecting reserves. Opportunism and a variety of more systematic approaches for reserve selection are reviewed in relation to three key principles: efficiency, flexibility, and irreplaceability.

Efficiency requires that selected reserves should complement rather than unnecessarily duplicate each other, and is especially important in regions where protected areas will only ever occupy a relatively small proportion of the total area. Flexibility relates to the large number of options for constructing representative reserve networks in most regions, and, the small number of these that are generally considered by conservation planners. Irreplaceability is a measure of the potential contribution of any given site to a fully representative reserve system. The new procedures for reserve selection in New South Wales are explicit, conceptually straightforward, and are underpinned by these three principles. They are applicable to many data bases and any regions, and will help to maximize the chances of conserving biodiversity.

Introduction

Internationally and in Australia, recent policies and statements have stressed the importance of conserving as much biodiversity as possible over the

Systematics and Conservation Evaluation (ed. P. L. Forey, C. J. Humphries, and R. I. Vane-Wright), Systematics Association Special Volume No. 50, pp. 351–73, Clarendon Press, Oxford, 1994.

forthcoming decades (e.g. IUCN 1980; Anon. 1984; Biological Diversity Advisory Committee 1992; WRI, IUCN and UNEP 1992). It is generally recognized that formal reserves will play a critical role in the conservation of biodiversity, although they should be dedicated and managed in the context of a wide range of *in situ* and *ex situ* protection measures (e.g. McNeely *et al.* 1990; Soulé 1991). It follows that reserves should be as representative as possible of natural diversity at all levels, from ecosystems to genetic variation within populations. It also follows that, given their critical role in conservation, they should be selected from the range of alternative areas with great care.

The practice of reservation, however, is different. Reserves rarely have been and rarely still are selected because they are the best sites for representing unprotected natural features. The configuration and content of reserve systems throughout the world have been largely determined by opportunistic and *ad hoc* approaches to reserve dedication. With such approaches, reservation is driven primarily by pragmatic considerations of land availability and political strategy. It is therefore not surprising that the practice of reserve selection has been referred to as 'a chance process' (Sullivan and Shaffer 1975) and 'conservation by default' (Leader-Williams *et al.* 1990), or that there are calls for de-proclamation of poorly sited reserves so that the resources can be better spent on biological conservation elsewhere (e.g. Siegfried 1989). One of the outcomes of many years of opportunistic reservation is a highly uneven representation of natural features in reserve systems with many features, even at the level of ecosystem, missing altogether (e.g. Harrison *et al.* 1982; Mark 1985; Clarke and Bell 1986). Another outcome, discussed below, is that opportunistic reservations can use up conservation options without representing some features at all and therefore actually hinder the development of representative networks (Brown and Hickey 1990; Pressey 1990*a*, 1992).

More systematic and explicit approaches to selecting reserves are necessary if the goal of conserving biodiversity is to be realized. Research by the New South Wales National Parks and Wildlife Service on systematic reserve selection has been underway since 1987. Part of this work has been in collaboration with the CSIRO Division of Wildlife and Ecology.

Despite extensive surveys and research in some regions, reserve planning in New South Wales faces a constraint which is common world-wide: the goal of representativeness must be pursued with a very incomplete knowledge of the natural features to be protected, particularly in the case of species and sub-specific taxa. The first part of this chapter discusses the ways in which this lack of biological inventory is being addressed in New South Wales. After this background is a description of the development of new, systematic procedures to assist planners in making sound decisions on the location, size, and configuration of reserves. The procedures will be effective with any consistent data base and in any region, and are based on three principles: efficiency, flexibility, and irreplaceability. A discussion of these procedures and principles and comparison with other approaches to selecting nature reserves forms the main part of this chapter.

Study regions and data bases

Procedures for systematic reserve selection are being developed with data from two contrasting regions in New South Wales (Fig. 23.1). One is the Western Division of New South Wales, a semi-arid area of some 320 000 km^2, representing 40 per cent of the state. It is predominantly leasehold and used for sheep grazing, with some cropping in the eastern and southern parts which have higher rainfall than the interior. Geology, soils, and vegetation formations are diverse. Much of the original vegetation remains uncleared, although altered by domestic stock and feral herbivores, particularly rabbits and goats. However, clearing for cropping and increased grazing capacity has proceeded rapidly in recent decades in the east and south so that natural habitats on the margins of the region are becoming increasingly fragmented and isolated (Pressey 1990*b*). Conservation reserves occupy about 2.6 per cent of the region and fail to represent many species and natural environments. The results reported in this chapter are for a trial study area

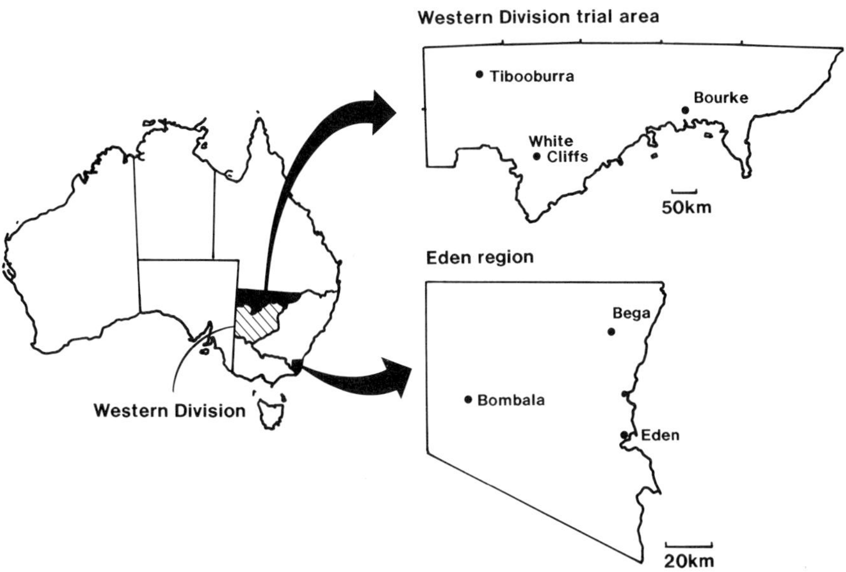

Fig. 23.1 Study areas for the development of reserve selection procedures in New South Wales.

in the north of the Division covering about 140 000 km^2, although continuing work covers the entire Division. The second region occupies about 7 800 km^2 near Eden, on the moist south-eastern coast of the state. This region varies topographically from a low range and undulating terrain in the east, to an inland mountain range and a plateau of low relief at about 1000 m elevation in the west. The complex geology and diverse vegetation formations have been summarized by Keith and Sanders (1990). State forests occupy about 45 per cent of the region and some 36 per cent is cleared, mostly as private agricultural land. Conservation reserves, which are far from representative of the natural environments, make up about 9 per cent of the region.

In neither region is biological site data sufficient as a sole basis for conservation planning. Many taxa in both areas are no doubt undescribed and, of the recognized taxa, few are well known in terms of distribution, abundance, and requirements. Even in the Eden region, where systematic studies on the biota have been undertaken by several institutions for a decade, only a small proportion (0.1 per cent) of the landscape has been directly sampled. This illustrates the need for more information but, more importantly for the purposes of this chapter, it highlights the need to make the best use of scarce biological data. Conservation planners, with the need to make decisions on the other 99.9 per cent of the landscape as well, must therefore rely to a large extent on some form of land classification as a surrogate for biological information and as a framework for spatial extension from samples. Discussion of land classes in this chapter refers to any subdivisions of the landscape used for conservation planning.

The database for research on reserve selection in the Western Division consists largely of digitised land system maps. Land systems are areas throughout which there is a recurring pattern of landform, soils, and vegetation (Christian and Stewart 1968) and are delineated by interpretation of aerial photographs. Mapping of land systems at 1:250 000 by the NSW Soil Conservation Service provides the most consistent and detailed definition of natural environments available for the region. The results reported here for the Eden region are mainly from another interpretation of natural environments, termed environmental domains, derived by numerical classification of grid cells on the basis of climatic, terrain, and soil attributes by Richards *et al.* (1990). A more detailed regional map of floristic vegetation types, involving modelling from more than 1200 quadrats, is being prepared for further research on reserve selection.

Despite their practical necessity, one of the limitations of land classes as a basis for biological conservation was described by Noss (1987). He introduced the metaphor of a 'coarse filter' to refer to land class reservation through which some taxa slip and so lack adequate protection. He advocated a 'fine filter' or taxon-specific approach as a complementary strategy. Extending this idea, at least five limitations of land classes in conservation planning can be identified: (1) relationships between land classes, and spatial and temporal variations in the distribution and abundance of taxa are unclear and difficult to quantify; (2) taxa are patchily distributed within land classes so reserving a piece of a land class might miss many taxa (the 'coarse filter' argument); (3) additional attention is needed for taxa which are threatened or heavily reliant on a particular region, even if they are caught in the coarse filter; (4) land classes do not necessarily delineate or recognize areas of critical resources essential for maintaining populations in times of scarcity; (5) many taxa require a combination of habitats not delineated or recognized by land classification. These limitations and possible ways of minimizing them have been discussed in detail elsewhere (Pressey in press).

The second and third limitations have been addressed for both study regions. In the Western Division, land system maps are complemented by information on selected species of vascular plants and vertebrates which can be placed in several categories of rarity or vulnerability and require specific consideration in the conservation strategy (see Pressey *et al.* 1990*b*; Dickman *et al.* 1993; Sadlier and Pressey, in press; Smith *et al.*, in press). In the Eden region, environmental domains were complemented by information on vascular plant species and plant communities which are rare or biogeographically significant (from Keith 1990; Keith and Sanders 1990; Keith and Ashby 1992), and by opportunistically collected data on fauna. These studies give some *a priori* assessments of the taxa most likely to be missed or inadequately protected by a reserve system based only on land classes.

Research is also attempting to quantify the extent to which rare taxa are missed by reserves based only on land classes (Pressey and Bedward 1991), and to quantify the biological informativeness of land classes (Bedward *et al.* 1992*a*). Current field sampling in the north-west of the state will allow the floristic informativeness of intuitive land classes to be quantified at several hierarchical scales. Such studies give insights into the manner in which land classes relate to the patterns of individual species which in turn help to understand the planning approaches needed for biological conservation.

Efficiency in reserve selection

Efficiency in conservation evaluation refers to the proportional number or area of sites in a region needed to represent all the features to some required level—the smaller the area, the higher the efficiency. It is easily calculated with the formula $1 - (X/T)$, where X is the number or area of sites needed to achieve a representation target and T is the total number or area of sites in the region (Pressey and Nicholls 1989*a*). The idea of efficiency has much in common with that of 'complementarity' (Vane-Wright *et al.* 1991), in that selection of a network of sites which are complementary in terms of the features they contain will lead to efficient representation.

Efficiency is an important consideration for conservation planners for two reasons. First, in most regions, nature reserve proposals have to compete with exploitative land uses such as pastoralism, agriculture, or forestry which are, to varying extents, inimical to conservation goals. It is important to show that the area of new reserves, or their opportunity cost to other land uses, is no more than that required to fulfil conservation objectives. Secondly, very inefficient planning methods will increase the total extent of the system of protected areas without giving a commensurate increase in the representativeness of the system. This leads to the significant danger of reaching a ceiling of acceptable reserve area well before all features are protected. Two examples of this second consideration are given here.

The first example is for opportunistic reservation. Figure 23.2 shows the accumulation of land systems in reserves within the trial study area for the Western Division. Between 1971 and 1988, the total reserve area increased from zero to 4611 km^2 or 3.3 per cent of the study area. The cumulative land system content of the reserves is shown in two ways: for representation simply equal to presence, regardless of the proportional area reserved (Fig. 23.2a), and for representation only recorded when at least 5 per cent of the total extent of each land system was reserved (Fig. 23.2b). The rate of accumulation of land systems in the reserve network is relatively gradual. A selection procedure to represent all the area's land systems with maximum efficiency, using pastoral properties as units of selection, has a much greater average rate of accumulation (Figure 23.2). Using this procedure, the minimum areas needed to represent all land systems at presence-only and 5 per cent levels are 7980 km^2 (5.7 per cent of the region) and 15 312 km^2 (11.0 per cent), respectively. Extrapolation of the average rates of accumulation in the existing reserves indicates total areas of 16 410 km^2 (11.8 per cent of the region) and 29 560 km^2 (21.3 per cent) will be required to represent all land systems at presence only and 5 per cent levels, respectively. The total area required for *ad hoc* reservation at the 5 per cent level is therefore likely to be politically unacceptable, especially when additional areas are included to protect rare species and other features, and to improve reserve viability with corridors and enlargements.

The relative inefficiency of opportunistic reservation can also be demonstrated with a temporal trend (Fig. 23.3). In 1971, before the dedication of reserves in the study area, a minimum of 5.7 per cent of the region would have been required to represent at least one occurrence of each land system (presence only) using pastoral properties as selection units. In 1988, reserves made up 3.3 per cent of the region and the minimum total area to represent all land systems, starting with the existing reserves, would have been 8.3 per cent of the region. The relatively small contribution of the existing reserves to complete representation is clear: the two lines are almost parallel—the reservation goal is retreating almost as quickly as reserves

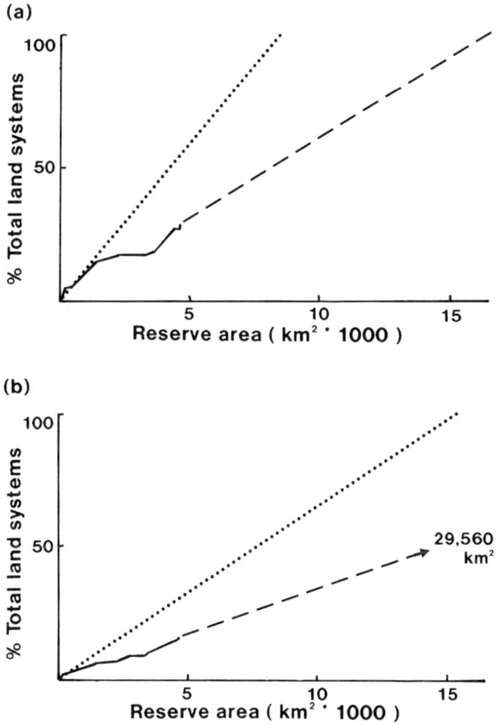

Fig. 23.2 Actual and hypothetical accumulation of land systems in reserves in north-western New South Wales. (a) Accumulation of land systems if representation is based only on the presence of each land system in an existing or hypothetical reserve: solid line—existing reserves; dashed line—extrapolation of average rate of accumulation in existing reserves; dotted line—maximum rate of accumulation using an analysis to maximize efficiency. (b) Accumulation of land systems if representation is based on having at least 5 per cent of the total area of each land system in an existing or hypothetical reserve; solid, dashed, and dotted lines as above; data for maximum rate of accumulation for 5 per cent representation courtesy of A.O. Nicholls, CSIRO Division of Wildlife and Ecology.

advance on it. This is a reflection of the low efficiency with which the existing reserves, regardless of their other values, represent the land systems of the region. A continuation of opportunistic reservation and the upward trend for total area needed involves the serious risk of using up all the conservation 'options' for the region well before the reserve network is fully representative.

The second example of the importance of efficiency as a principle to guide reserve selection concerns systematic scoring. Scoring approaches to conservation evaluation involve recording ratings for one or more criteria for each site (forest fragment, wetland, pastoral holding etc.), and then combining these

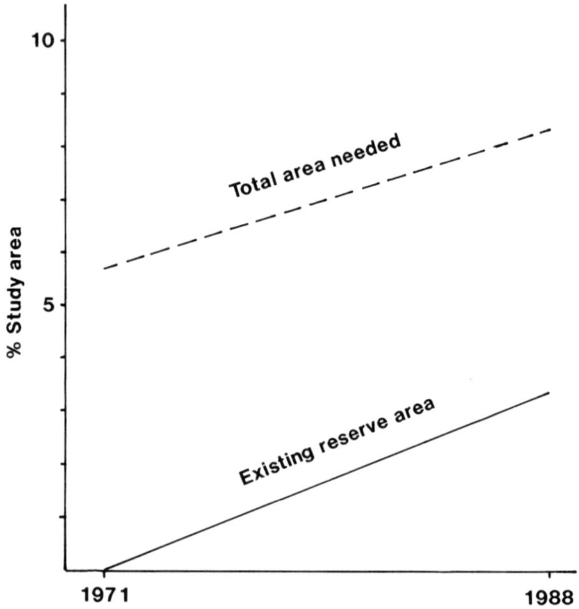

Fig. 23.3 Actual reserve area in north-western New South Wales compared to the minimum total area required to represent all land systems, 1971 and 1988 (after Pressey 1990*a*).

into an aggregate score to indicate value or priority for conservation. This is a procedure which has been widely advocated in the international literature and commonly uses diversity, rarity, representativeness, and size as criteria (see Margules and Usher 1981; Smith and Theberge 1986, 1987 for reviews). Its advantages are explicit and consistent treatment of all sites in a region. Its disadvantage is that, if all the sites in a region are ranked from top to bottom in terms of conservation value, an impossibly large number or area of highest scoring sites can be required to represent all the features in the region (Pressey and Nicholls 1989*a*; Bedward *et al.* 1991; Margules *et al.* 1991). Figure 23.4 shows the percentage number of total sites in the Western Division trial study area needed for one representation of each land system (presence only) if sites are reserved in order of scores for four criteria. The percentages are very large relative to the minimum numbers of properties required for one, two, or three representations of each possible land system (from Pressey and Nicholls 1989*b*). Scoring criteria used in this way are inefficient because they result in unnecessary duplication of some features without targeting sites containing the features needed to complement the existing reserves.

The problem of unnecessary duplication of some features can be overcome by an iterative approach to reserve selection which adds sites to a network in

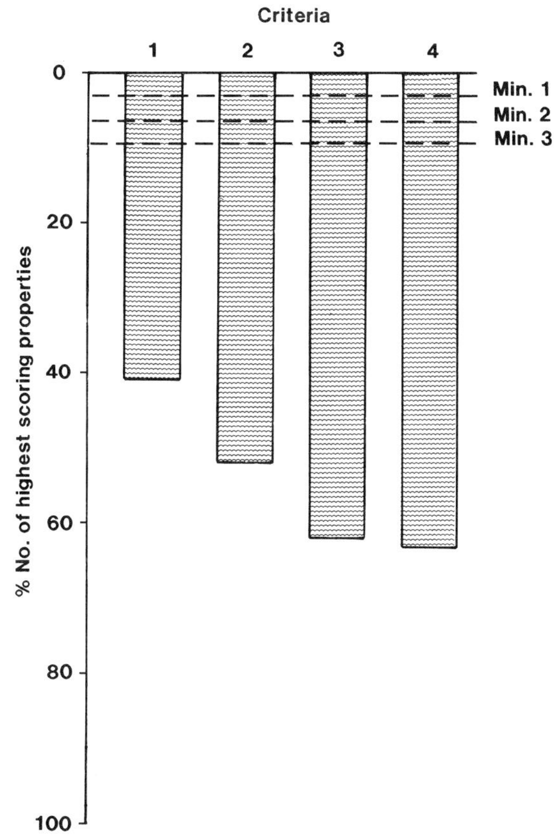

Fig. 23.4 Percentages of total number of properties required to represent all possible land systems at least once (presence only) if properties are reserved in order of scores for four criteria: 1—rarity based on the relative area of land systems in the region; 2—rarity based on relative frequency of land systems; 3—property size; 4—diversity based on number of land systems in each property; note that the relative efficiencies of the criteria vary for other data sets; min1, min2, and min3 indicate the minimum percentages of total property number required to represent all possible land systems once, twice, and three times, respectively.

a stepwise manner according to how well they complement the existing reserves. Such a procedure was first developed and applied in Tasmania by Kirkpatrick (1983), and Kirkpatrick and Harwood (1983). Subsequent applications of an alternative iterative procedure developed by the CSIRO Division of Wildlife and Ecology have dealt with wetlands in north-eastern New South Wales (Margules *et al.* 1988), remnants of mallee vegetation in South Australia (Margules and Nicholls 1987; Margules 1989), pastoral properties in north-western New South Wales (Pressey and Nicholls 1989*b*, 1991; and see Fig. 23.5), and the mallee areas in south-western New South

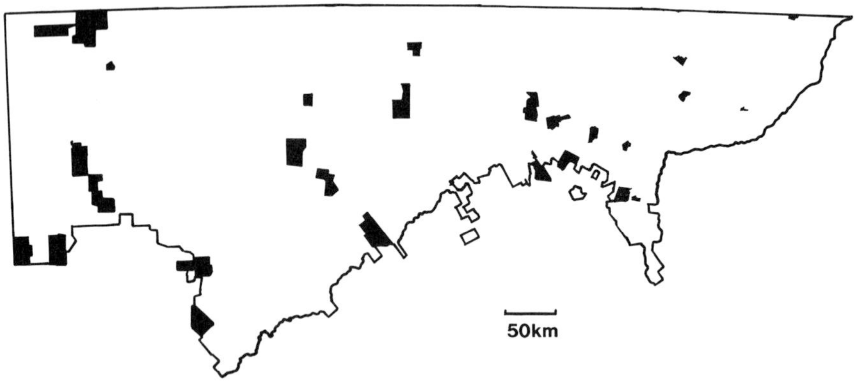

Fig. 23.5 A minimum set of 32 pastoral properties needed to represent all land systems at least once (presence only) in the Western Division trial area.

Wales (Pressey *et al.* 1990*a*). These algorithms are highly efficient and were used to identify minimum requirements for representation in Figs. 23.2, 23.3, and 23.4. Applications of similar procedures in other regions of the world include those of Vane-Wright *et al.* (1991) and Rebelo and Siegfried (1990, 1992).

Such iterative analyses have several advantages for conservation planning. Their high efficiency maximizes the likelihood of achieving the goals of nature conservation in the face of limited availability of land for this purpose. They help to improve decision-making because they clearly identify the minimum requirements for a stated conservation goal. They indicate the threshold number or area of sites above which conservation planners must work to achieve, not only representation of features, but goals related to reserve design and land suitability. Finally, the algorithms are also a valuable aid to decision-making in enabling planners to quickly assess minimum reserve requirements with alternative goals or constraints (e.g. Fig. 23.3).

Flexibility in reserve selection

In most regions, there are many ways of combining sites together to form representative networks of reserves. Two simple examples are in Table 23.1, derived from small environmental provinces in north-western New South Wales delineated by Morgan and Terrey (1990). The number of possible combinations of sites depends on combination size (number of component sites). At any specific combination size, it is possible to identify those combinations which are 'representative' by some criterion. For the examples in Table 23.1, representative combinations of a given size are simply those

Table 23.1 Combinations of properties of different sizes which are possible and those which are 'representative' (containing all land systems in the environmental province at least once).

A. Province 3B (29 properties)

No. properties in combination	No. combinations	No. solutions (and % total)
1	29	–
2	406	–
3	3654	2 (0.05)
4	23751	81 (0.3)
5	118 755	1120 (0.9)
6	475 020	9020 (1.9)
7	1 560 780	50 540 (3.2)
8	4 292 145	214 166 (5.0)
9	10 015 005	718 732 (7.2)
10	20 030 010	1 965 148 (9.8)

B. Province 7A (13 properties)

No. properties in combination	No. combinations	No. solutions (and % total)
1	13	–
2	78	–
3	286	–
4	715	6 (0.8)
5	1287	45 (3.5)
6	1716	49 (8.7)
7	1716	286 (16.7)
8	1287	351 (27.3)
9	715	286 (40.0)
10	286	155 (54.2)

which contain every land system in the province at least once. In both provinces, and in any other data set, there is a minimum number of sites below which all features cannot be represented. This minimum size or minimum set can be identified by the iterative algorithms outlined above and by mathematical programming (e.g. Cocks and Baird 1989). However, an analysis of site combinations shows that there can be more than one minimum-sized solution and hundreds or thousands of solutions only slightly greater in size than the minimum. In large data sets the range of solutions is naturally much larger.

These solutions are, in fact, alternative reserve networks which differ not only in the number of sites they contain but also in overall condition, tenure, cost, extent, and configuration. The more of these alternatives that can be identified and appraised, the more likely planners are to achieve, in addition to representativeness, other important goals such as appropriate reserve design, best condition of component sites, and the most convenient tenure for conversion to nature reserve.

Some of the alternative networks which planners can consider might conflict with other land uses. Other potential networks might be eliminated by clearing, poor condition, or lack of availability of one or more component sites. However, one of the ways in which conservation planners impose constraints on their own flexibility is with their choice of selection procedure. Most systematic procedures for reserve selection do not allow planners to explore the range of options available for constructing a fully representative reserve system. This applies also to the very efficient iterative analyses described above. Iterative procedures can quickly identify an indicative minimum set to represent all natural features in the region, but they do not display the range of alternatives of the same size or slightly larger. Another limitation is that they can achieve their high efficiency of representation by selecting a dispersed set of individual sites as a nominal reserve network (Fig. 23.5), not necessarily optimal for management or the viability of populations within these areas. A new version of an earlier algorithm (Nicholls and Margules 1993) has reduced, but not eliminated, the latter problem. However, the former limitation concerning the options which planners can consider still applies to this new algorithm.

These two limitations have recently been addressed with a new procedure called CODA (Conservation Options and Decisions Analysis) which recognizes the flexibility available for reserve selection in most regions and allows this to be explored. The procedure and its application to the design of new reserves in the Eden region have been described in detail by Bedward *et al.* (1992*b*). Briefly, CODA consists of four main stages:

Stage 1. Identify features of interest and set targets for representation. The extent of representation required for each of the natural features (e.g. vegetation types) in the study region is stated explicitly. This can be expressed as a fixed area of each feature or a fixed percentage of the total extent of each feature. Alternatively, targets can vary between features depending on rarity or status outside the region.

Stage 2. Identify focal areas. These are the fixed, non-negotiable parts of the future reserve system around which the entire system will be designed. Focal areas can include existing reserves and sites which are nominated by planners such as those with endangered species or communities. In this stage, the procedure calculates the extent to which representation targets have been met within focal areas.

Stage 3. Select a preliminary extended reserve network to fully achieve representation targets. A minimum set algorithm selects a set of additional sites to efficiently bring all features up to full desired representation.

Stage 4. Modify the preliminary network to address other conservation objectives. Some of the sites selected in Stage 3 will be contiguous with focal areas, in good condition and immediately suitable for reservation. Others might be problematic because of isolation, disturbance, or excessive cost of acquisition. CODA's support facilities allow the user to find replacements for unsuitable sites and tailor the reserve network to address objectives of design and land suitability. Sites can be taken out of the network, or added to it, and the implications of each change assessed in terms of the achievement of representation targets. CODA also checks for redundant sites in the network, i.e. those that can be removed without any features falling below full representation.

The essence of the CODA system is complete control by the user over the configuration and content of the network by selecting or de-selecting individual sites until, as far as possible, all goals are achieved. It also gives the user the flexibility of placing different priorities on different goals (e.g. representation, land suitability, contiguity) in the consideration of different sites or groups of sites. This level of control is in contrast to a recent approach to reserve selection in Tasmania (Lewis *et al.* 1991) in which flexibility is achieved by changing the weightings given to various criteria, most of which are calculated iteratively. These global settings are very useful in exploring indicative scenarios, but do not allow individual components of the network to be adjusted.

The outcome of CODA's application to the Eden region is shown in Fig. 23.6. Given the flexibility recognized and explored by CODA, other configurations are possible, subject to the constraints imposed by clearing, tenure, condition, and the distribution of natural environments and species. Replacement, in Stage 4, of sites which were selected for the preliminary network in Stage 3 will usually lower the efficiency with which targeted features are represented in the network. Lowered efficiency was also shown by Nicholls and Margules (1993) to come from increased contiguity of

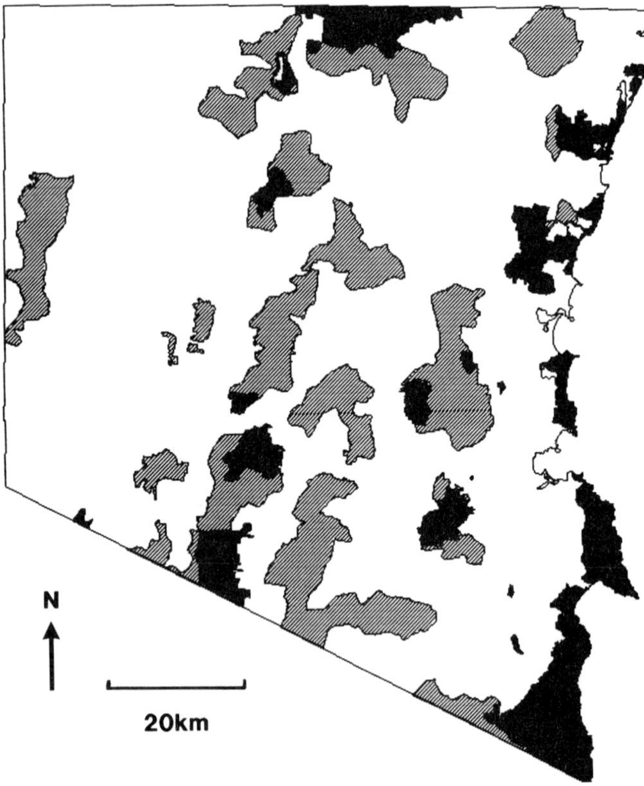

Fig. 23.6 Nominal reserve network for the Eden region designed with the CODA procedure and using catchments of median area 1000 ha as units of selection, except in fragmented areas; catchments have been amalgamated wherever possible; black areas are reserves existing at the start of the process; shaded areas are additional reserves selected with CODA (after Bedward *et al.* 1992).

selected sites in their modified iterative analysis. Lewis *et al.* (1991) showed that loss of efficiency could come from a variety of criteria relating to design and land suitability. This loss of efficiency can be minor and is the cost of meeting conservation objectives other than representation. CODA therefore combines considerations of efficiency with flexibility and allows the user to determine, not only the final extent and shape of the network, but the extent to which efficiency must be reduced to accommodate other goals.

Irreplaceability in reserve selection

Underlying the flexibility of the CODA system is the recognition that some of the sites in a region are essential components of the reserve system to be designed. These are the fixed points which make up the framework of the expanded network and can be identified in a variety of ways and with varying levels of subjectivity. In this sense, irreplaceability is a qualitative attribute of sites—they are either irreplaceable or not. However, irreplaceability can be quantified as well. Sites vary in the number of alternatives that exist to represent the features they contain and so can have varying levels of irreplaceability.

Examples of irreplaceability, in the black-and-white sense, for identifying focal sites in the CODA procedure were existing reserves and areas with endangered species or communities. There is clearly much scope for broadening these considerations. Morton (1990), for example, has argued for the importance of run-on areas in the arid zone as places to which mammal populations, in particular, contract in drought and from which they later expand in good years. The relatively high levels of moisture and nutrients in these patches make them indispensable for the long-term survival of many species. Another example concerns the coastal lowlands of a small part of north-eastern New South Wales which have a high growth index in winter which makes the area an important overwintering area for birds and bats from higher latitudes and altitudes (Nix 1976). Species and communities of plants which flower in the winter in this district can be regarded as sufficiently important for the mobile fauna of a large region to be classed as irreplaceable (Pressey and Griffith 1992). Connections between areas of important habitat can also be seen as critical components of a network. Mwalyosi (1991) discussed the selection and protection of a suitable corridor linking two national parks in Tanzania. The corridor is now essential if some of the migratory animals are to be retained in one of the parks because increasing settlement has prevented the use of the land adjacent to the reserve.

Current research in New South Wales is providing complementary information to assessements such as those above by exploring levels of irreplaceability (Pressey *et al.*, in press). The approach draws on the results of the combinatorial analyses used to derive the data in Table 23.1. When all the alternative representative combinations of sites (potential reserve networks) are identified as in the right-hand column of Table 23.1, it is possible to calculate the frequency with which each site occurs in the range of alternative networks at a given combination size. For example, in Province 3B, there are 1120 combinations of any five sites which are representative. At this combination size one property occurs in all 1120 representative combinations or alternative networks—it is an essential component of any reserve network in the province (Table 23.2). Among the other five levels of occurrence, two sites occur in about half the representative combinations and nine sites in about 2.5 per cent. Different patterns of occurrence are evident in two other data sets.

Table 23.2 Frequencies of occurrence of properties in 'representative' combinations (those containing all land systems at least once); these are equivalent to levels of irreplaceability in a fully representative reserve system; data for 3B—combination size of four, 7A—combination size five, 5BA—combination size six; data from Pressey *et al.* (in press).

Province 3B		Province 7A		Province 4BA	
% occ.	no. props.	% occ.	no. props.	% occ.	no. props.
100.0	1	100.0	2	100.0	4
63.0	1	53.3	2	78.3	1
50.6	2	37.8	3	26.1	1
39.5	1	13.3	6	8.7	5
4.9	15			4.3	12
2.5	9				

These frequencies of occurrence in representative combinations are referred to as levels of irreplaceability. In practical terms, irreplaceability can be defined as the potential contribution of an individual site to a fully representative reserve system. Alternatively, it reflects the extent to which options for achieving a representative network are constrained if a particular site is lost. For example, exclusion of Site X in Fig. 23.7a from consideration as a reserve rules out 63 per cent of the options for constructing a representative network with combinations of any four sites. Exclusion of Site Y narrows the options by only 4.9 per cent. Irreplaceability therefore indicates the extent to which sites should form the framework of the region's reserve system.

Application of this idea at a regional scale requires irreplaceability to be predicted rather than measured directly because of the impossibility of computing all the combinations of sites in large data sets. For representation based only on the presence of features, irreplaceability can be accurately modelled with an index which is easily computed for whole regions (Pressey *et al.* in press). Work is proceeding on analysis and modelling for representation based on areas or population sizes.

Irreplaceability has two important roles in conservation planning. First, it could be used as a basis for the design of whole networks by focusing first on the completely irreplaceable sites, and then making decisions on sites of progressively lower irreplaceabilities. This allows planning to proceed logically from sites with fewest options for replacement in the network to those with most. At each choice of optional sites, the potential contributions of the unreserved sites would change to some extent so the index would be

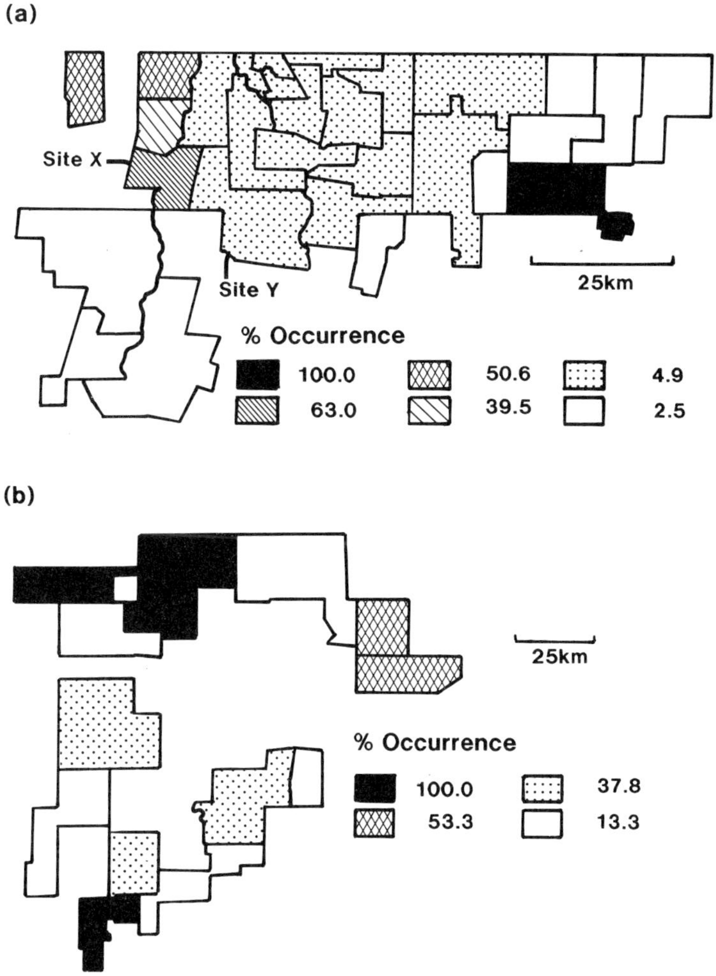

Fig. 23.7 Small environmental provinces in north-western New South Wales (adapted from Morgan and Terrey 1990) showing levels of percentage occurrence in representative combinations of sites; these levels indicate the relative 'irreplaceability' of each pastoral property in constructing a representative reserve system for each province; (a) Province 3B; (b) Province 7A.

recalculated to update values at each step. Second, assessments of specific development proposals and decisions on particular acquisitions, which might involve departures from an initial network design, require some information on the potential value of individual sites. As a guide to such decisions, irreplaceability can be used to provide a map or 'landscape' of conservation significance (Fig. 23.7) in relation to a specific reservation goal. This capability is especially important as an alternative to conservation options being narrowed incrementally, and without any broad view of goals, by the 'tyranny of small decisions' (Odum 1982).

Conservation planning based on levels of irreplaceability leaves maximum options open for developing representative networks of reserves. Opportunistic reservation can do the opposite. Sites with low values for irreplaceability are the great majority in any large region (Pressey *et al.* in press) and so their reservation by opportunistic means is very likely. When this occurs early in the development of a reserve network, few options for efficient representation remain because, for any given size of network, the selected sites occur in few representative combinations. Compounding this problem by reserving multiple sites of low irreplaceability can eliminate all options at a particular combination size and require more sites to be reserved than would have otherwise been the case (see Fig. 23.3). The effect of previous *ad hoc* reservations in increasing the minimum number or area of sites needed to represent all natural features in a region has also been recognized and quantified in several other studies (Margules 1989; Pressey *et al.* 1990; Lewis *et al.* 1991).

Conclusions

Representing the biodiversity of any region in a system of protected areas is a demanding goal. It is demanding in terms of the extent of protected areas, and the political and public commitment required. It is also demanding in terms of the information needed on the elements of biodiversity and the capabilities of procedures needed to support decisions on conservation. Good data are not enough. Good data can be used in inappropriate procedures with little likelihood of success in achieving conservation goals.

Reserve selection procedures should be explicit, systematic, and conceptually straightforward, and recognize the importance of efficiency, flexibility, and irreplaceability. Such procedures maximize the chances of achieving representative networks in several ways:

(1) through their efficiency they increase the feasibility of actually reserving all the sites needed to represent the natural features of a region at the desired level;

(2) they allow planners to appraise a wide range of options for networks in a practical way to achieve goals other than representation and to minimize conflicts with other land uses;

(3) by quantifying irreplaceability they provide a framework for the design of entire networks and for decisions on individual sites which can anticipate their contribution to a reservation goal;

(4) they allow reserve proposals to be explained and defended.

These new procedures are applicable anywhere at the level of land classes, individual taxa, or other features. They can also be used to identify sites appropriate for a variety of protection measures in addition to strict reservation. While CODA and assessments of irreplaceability are practical systems to support real conservation decisions, other earlier analyses still have an important place in planning. For example, minimum set algorithms could continue to play a role in more extended procedures such as CODA, and are valuable for rapid assessments of minimum reserve requirements and the relative costs in reserve area of alternative goals and constraints.

Of course, data and procedures do not, in themselves, create representative reserve networks. For a system of integrated reserves to become a reality, procedures must be applied in the context of social, economic, and political constraints. These will inevitably require compromises in the principles discussed in this chapter. However, for a real chance of conserving biodiversity, conservation should start with the principles and not with the compromises. The legacy of compromise is a world-wide network of reserves which is far from representative and which will make the goal of representativeness harder to achieve. A reserve plan for any region which begins with sound principles will almost certainly need to be modified to accommodate the limitations of real life; but such a plan will be defensible and sound in design and content, it will enable departures from the ideal to be gauged in terms of dollars and loss of natural diversity, and it will maximize the chances of stated goals actually being achieved.

There will always be scope for improving procedures with new ideas, more complex conservation goals, and better capabilities for handling data. It is therefore important that conservation biologists continue to exchange and refine ideas on reserve selection. Ultimately, however, the achievement of conservation goals will depend on convincing politicians and the public that those goals are important enough to justify the costs involved. This will be more easily achieved with the backing of procedures which clearly demonstate the requirements and options for reserve networks.

Acknowledgements

The early ideas reported here, as indicated in the references cited, come from collaborative work with Nick Nicholls and Chris Margules at the CSIRO Division of Wildlife and Ecology. Thanks go to Vicki Logan who, in addition to ably supporting the ongoing research on reserve selection, prepared the figures for this chapter. Attendance by RLP at the Symposium on Systematics and Conservation Evaluation and subsequent collaborative research at The Natural History Museum was greatly assisted by a travel grant from the British Council.

References

Anon. (1984). *A national conservation strategy for Australia.* Australian Government Publishing Service, Canberra.

Bedward, M., Pressey, R. L., and Nicholls, A. O. (1991). Scores and score classes for evaluation criteria: a comparison based on the cost of reserving all natural features. *Biological Conservation*, **56**, 281–94.

Bedward, M., Keith, D. A., and Pressey, R. L. (1992*a*). Homogeneity analysis: assessing the utility of classifications and maps of natural resources. *Australian Journal of Ecology*, **17**, 133–9.

Bedward, M., Pressey, R. L., and Keith, D. A. (1992*b*). A new approach for selecting fully representative reserve networks: addressing efficiency, reserve design and land suitability with an iterative analysis. *Biological Conservation*, **62**, 115–25.

Biological Diversity Advisory Committee (1992). *A national strategy for the conservation of Australia's biological diversity.* Department of Arts, Sport, the Environment and Territories, Canberra.

Brown, M. and Hickey, J. (1990). Tasmanian forest—genes or wilderness? *Search*, **21**, 86–7.

Christian, C. S. and Stewart, G. A. (1968). Methodology of integrated surveys. In *Aerial surveys and integrated studies, proceedings of the Toulouse conference*, pp. 233–80. UNESCO.

Clarke, J. E. and Bell, R. H. V. (1986). Representation of biotic communities in protected areas: a Malawian case study. *Biological Conservation*, **35**, 293–311.

Cocks, K. D. and Baird, I. A. (1989). Using mathematical programming to address the multiple reserve selection problem: an example from the Eyre Peninsula, South Australia. *Biological Conservation*, **49**, 113–30.

Dickman, C. R., Pressey, R. L., Lim, L., and Parnaby, H. A. (1993). Mammals of particular conservation concern in the Western Division of New South Wales. *Biological Conservation*, **65**, 219–48.

Harrison, J., Miller, K., and McNeely, J. (1982). The world coverage of protected areas: development goals and environmental needs. *Ambio*, **11**, 238–45.

IUCN (1980). *World conservation strategy.* International Union for the Conservation of Nature, Gland, Switzerland.

Keith, D. A. (1990). Rare and regionally significant vascular plant species of the Eden region, south-eastern New South Wales: a listing for the 'fine filter' approach. *Proceedings of the Linnean Society of New South Wales*, **112**, 111–32.

Keith, D. and Ashby, E. (1992). Vascular plants of conservation significance in the South-East Forests of New South Wales. *NSW National Parks and Wildlife Service Occasional Paper* 11.

Keith, D. A. and Sanders, J. M. (1990). Vegetation of the Eden region, south-eastern Australia: species composition, diversity and structure. *Journal of Vegetation Science*, **1**, 203–32.

Kirkpatrick, J. B. (1983). An iterative method for establishing priorities for the selection of nature reserves: an example from Tasmania. *Biological Conservation*, **25**, 127–34.

Kirkpatrick, J. B. and Harwood, C. E. (1983). Conservation of Tasmanian macrophytic wetland vegetation. *Proceedings of the Royal Society of Tasmania*, **117**, 5–20.

Leader-Williams, N., Harrison, J., and Green, M. J. B. (1990). Designing protected areas to conserve natural resources. *Science Progress*, **74**, 189–204.

Lewis, A., Stein, J. L., Stein, J. A., Nix, H. A., Mackey, B. G., and Bowyer, J. K. (1991). An assessment of regional conservation adequacy: Tasmania. *Resource Assessment Commission Forest and Timber Inquiry Consultancy Series* (**FTC91/17**).

Margules, C. R. (1989). Selecting nature reserves in the South Australian mallee. In *Mediterranean landscapes in Australia: mallee ecosystems and their management* (ed. J. C. Noble and R. A. Bradstock), pp.398–405. CSIRO, Melbourne.

Margules, C. R. and Nicholls, A. O. (1987). Assessing the conservation value of remnant 'islands': mallee patches on the western Eyre Peninsula, South Australia. In *Nature Conservation: the Role of Remnants of Native Vegetation* (ed. D. A. Saunders, G. W. Arnold, A. A. Burbidge, and A. J. M. Hopkins), pp. 89–102. Surrey Beatty, Chipping Norton, NSW, Australia.

Margules, C. R. and Usher, M. B. (1981). Criteria used in assessing wildlife conservation potential: a review. *Biological Conservation*, **21**, 79–109.

Margules, C. R., Nicholls, A. O., and Pressey, R. L. (1988). Selecting networks of reserves to maximise biological diversity. *Biological Conservation*, **43**, 63–76.

Margules, C. R., Pressey, R. L., and Nicholls, A. O. (1991). Selecting nature reserves. In *Nature Conservation: Cost Effective Biological Surveys and Data Analysis*, (ed. C. R. Margules and M. P. Austin), pp. 90–7. CSIRO, Melbourne.

Mark, A. F. (1985). The botanical component of conservation in New Zealand. *New Zealand Journal of Botany*, **23**, 789–810.

McNeely, J. A., Miller, K. R., Reid, W. V., Mittermeier, R. A., and Werner, T. B. (1990). *Conserving the World's Biodiversity*. International Union for the Conservation of Nature and Natural Resources, World Resources Institute, Conservation International, World Wildlife Fund-US and the World Bank, Gland, Switzerland.

Morgan, G. and Terrey, J. (1990). Natural regions of western New South Wales and their use for environmental management. *Proceedings of the Ecological Society of Australia*, **16**, 467–73.

Morton, S. R. (1990). The impact of European settlement on the vertebrate animals of arid Australia: a conceptual model. *Proceedings of the Ecological Society of Australia*, **16**, 201–13.

Mwalyosi, R. B. B. (1991). Ecological evaluation for wildlife corridors and buffer zones for Lake Manyara National Park, Tanzania, and its immediate environment. *Biological Conservation*, **57**, 171–86.

Nicholls, A. O. and Margules, C. R. (1993). An upgraded reserve selection algorithm. *Biological Conservation*, **64**, 165–9.

Nix, H. A. (1976). Environmental control of breeding, post-breeding dispersal and migration of birds in the Australian region. In *Proceedings of the 16th Ornithological Congress, Canberra 1974* (ed. H. J. Frith and J. H. Calaby), pp. 272–305, Australian Academy of Science, Canberra.

Noss, R. F. (1987). From plant communities to landscapes in conservation inventories: a look at The Nature Conservancy (USA). *Biological Conservation*, **41**, 11–37.

Odum, W. E. (1982). Environmental degradation and the tyranny of small decisions. *Bioscience*, **32**, 728–29.

Pressey, R. L. (1990*a*). Reserve selection in New South Wales: where to from here? *Australian Zoologist*, **26**, 70–5.

Pressey, R. L. (1990*b*). Clearing and conservation in the Western Division of New South Wales. *National Parks Journal (NSW)*, **34**, 16–24.

Pressey, R. L. (1992). Opportunism in acquiring land for reserves: why it's a bad idea. *National Parks Journal (NSW)*, **36**, 19–22.

Pressey, R. L. (in press). Land classifications are necessary for conservation planning but what do they tell us about fauna? *Australian Zoologist.*

Pressey, R. L. and Bedward, M. (1991). Mapping the environment at different scales: benefits and costs for nature conservation. In *Nature Conservation: Cost Effective Biological Surveys and Data Analysis* (ed. C. R. Margules and M. P. Austin), pp. 7–13. CSIRO, Melbourne.

Pressey, R. L. and Griffith, S. J. (1992). Vegetation of the coastal lowlands of Tweed Shire, northern New South Wales: plant associations, species and conservation. *Proceedings of the Linnean Society of New South Wales*, **113**, 203–43.

Pressey, R. L. and Nicholls, A. O. (1989*a*). Efficiency in conservation evaluation: scoring vs. iterative approaches. *Biological Conservation*, **50**, 199–218.

Pressey, R. L. and Nicholls, A. O. (1989*b*). Application of a numerical algorithm to the selection of reserves in semi-arid New South Wales. *Biological Conservation*, **50**, 263–78.

Pressey, R. L. and Nicholls, A. O. (1991). Reserve selection in the Western Division of New South Wales: development of a new procedure based on land system mapping. In *Nature conservation: cost effective biological surveys and data analysis* (ed. C. R. Margules and M. P. Austin), pp. 98–105. CSIRO, Melbourne.

Pressey, R. L., Bedward, M., and Nicholls, A. O. (1990*a*). Reserve selection in mallee lands. In *The mallee lands: a conservation perspective* (ed. J. C. Noble, P. J. Joss, and G. K. Jones), *Proceedings of the National Mallee Conference, Adelaide, April 1989*, pp. 167–78. CSIRO, Melbourne.

Pressey, R. L., Cohn, J. S., and Porter, J. L. (1990*b*). Vascular plants with restricted distributions in the Western Division of New South Wales. *Proceedings of the Linnean Society of New South Wales*, **112**, 213–27.

Pressey, R. L., Johnson, I. R., and Wilson, P. D. (in press). Shades of irreplaceability: towards a measure of the contribution of sites to a reservation goal. *Biodiversity and Conservation.*

Rebelo, A. G. and Siegfried, W. R. (1990). Protection of Fynbos vegetation: ideal and real-world options. *Biological Conservation*, **54**, 15–31.

Rebelo, A. G. and Siegfried, W. R. (1992). Where should nature reserves be located in the Cape Floristic Region, South Africa?—models for the spatial configuration of a reserve network aimed at maximising the protection of floral diversity. *Conservation Biology*, **6**, 243–52.

Richards, B. N., Bridges, R. G., Curtin, R. A., Nix, H. A., Shepherd, K. R., and Turner, J. (1990). *Biological Conservation of the South-East Forests.* Report of the Joint Scientific Committee. Australian Government Publishing Service, Canberra.

Sadlier, R. and Pressey, R. L. (in press). Reptiles and amphibians of particular conservation concern in the Western Division of New South Wales. *Biological Conservation.*

Siegfried, W. R. (1989). Preservation of species in southern African nature reserves.

In *Biotic diversity in southern Africa: concepts and conservation* (ed. B. J. Huntley), pp. 186–201. Oxford University Press, Cape Town. .

Smith, P. G. R. and Theberge, J. B. (1986). A review of criteria for evaluating natural areas. *Environmental Management*, **10**, 715–34.

Smith, P. G. R. and Theberge, J. B. (1987). Evaluating natural areas using multiple criteria: theory and practice. *Environmental Management*, **11**, 447–60.

Smith, P. J., Pressey, R. L., and Smith, J. E. (in press). Birds of particular conservation concern in the Western Division of New South Wales. *Biological Conservation*.

Soulé, M. E. (1991). Conservation: tactics for a constant crisis. *Science*, **253**, 744–50.

Sullivan, A. L. and Shaffer, M. L. (1975). Biogeography of the megazoo. *Science*, **189**, 13–17.

Vane-Wright, R. I., Humphries, C. J., and Williams, P. H. (1991). What to protect?—systematics and the agony of choice. *Biological Conservation*, **55**, 235–54.

WRI, IUCN, and UNEP (1992). *Global biodiversity strategy: guidelines for action to save, study, and use Earth's biotic wealth sustainably and equitably*. World Resources Institute, International Union for the Conservation of Nature and Natural Resources, United Nations Environment Program, Washington DC.

24. Using the Proteaceae to design a nature reserve network and determine conservation priorities for the Cape Floristic Region

A. G. REBELO
Conservation Biology Research Unit, National Botanical Institute, Kirstenbosch, Claremont 7735, South Africa

Abstract

The Cape Floristic Region (CFR), covering 90 000 km^2, comprises one of the world's six floral kingdoms. With 8600 species, of which 68 per cent are endemic, it ranks amongst the richest of temperate and tropical floras, although most of the diversity is concentrated in a few speciose taxa. Some 19 per cent of the area of the CFR is in nature reserves, but by far the majority of the preserved area comprises mountain Fynbos. Fortunately Fynbos vegetation is the richest habitat for Red Data Book (RDB) plant, freshwater fish, amphibian, butterfly, and reptile species in southern Africa. However, only 3 per cent of the original extent of lowland Fynbos is preserved, with the greater Cape Town metropolitan area ranking globally as one of the most urgent conservation priorities. Previously used methods for evaluating priority conservation areas have underrated species-poor areas containing a high proportion of RDB species. By correcting for species richness, a far more realistic picture of threatened areas can be obtained from RDB taxa. Utilizing an iterative selection procedure to identify optimal reserve configurations, based on Proteaceae species richness, suggests that 95 per cent of all vascular plant species in Fynbos can be preserved in 16 per cent of the area. Two null models for evaluating the efficiency of a spatial configuration of reserves are proposed. These results suggest that the ideal approach to designing a reserve configuration is to identify areas of high endemism and richness in distinctive vegetation types within particular biogeographical regions. *A priori* hypotheses based on ecological traits indicate that seed dispersal and regeneration strategies are most strongly correlated with rarity, most specifically with distributional area. The weighting of guilds (based on seed dispersal,

Systematics and Conservation Evaluation (ed. P. L. Forey, C. J. Humphries, and R. I. Vane-Wright), Systematics Association Special Volume No. 50, pp. 375–96, Clarendon Press, Oxford, 1994.

pollination, and regeneration strategies) and taxonomic groups in the Proteaceae does not affect the configuration of nature reserves within the CFR although it alters the sequence of reserve selection. International pressure is required to ensure the effective conservation of this globally important 'hotspot', which should be considered for Biosphere Reserve status.

Introduction

South Africa has a peculiar first- and third-world mix, largely owing to its political development (Huntley *et al.* 1988). As the population of the third-world element increases, the country will undoubtedly become increasingly third-world in character and outlook. This bodes ill for the conservation programme that has developed in a largely first-world framework. Even under the most optimistic of scenarios, the rapid economic development required to uplift the quality of life of all South Africans will have to be at the expense of the environment: dams will have to be built, adequate housing provided, and industrial development fostered (Huntley *et al.* 1988). Rapid political strides, aimed at just dispensation for all South Africans, irrespective of race, are in progress. Perhaps now more than ever is the time to take stock of the situation, to consolidate knowledge gained over the last few decades, and to determine priorities (Soulé and Kohm 1989). Only then can we enter the turbulent political and bioclimatic future confident that we can manipulate the options available so as to minimize the loss of our environmental heritage. This is important as southern Africa harbours a relatively high proportion of the world's biota (Siegfried 1989). Fortunately, much of the floristic diversity, over one-third of the subcontinent's 21 000 plant species, occurs in a small area (4 per cent of the subcontinent): the Cape Floristic Region.

The Cape floristic region

The Cape floristic region (CFR) is one of the world's six Floral Kingdoms (Takhtajan 1986). It has the richest flora of the five Mediterranean climatic regions, which generally have the richest temperate floras, and exceeds that of most tropical and island floras in terms of the proportion of endemic plant taxa (Bond and Goldblatt 1984; Table 24.1). Although it occupies an area of only 90 000 km^2, the CFR contains more than twice as many species (8600) of any island of similar size (Bond and Goldblatt 1984).

Much of this species richness is due to the proliferation of a few taxa (Bond and Goldblatt 1984). The ten largest of the 950 native genera (i.e. those with over 100 species) comprise 20 per cent of the flora—the richest, *Erica*, has 545 species or six per cent of the flora. The 10 largest of the 150 native families contain 58 per cent of the species. However, 193 genera (20 per cent)

Table 24.1 The relative position of the CFR compared to the ten top 'hotspot' areas in tropical forests (Myers 1988 in McNeely *et al.* 1990), ranked by number of endemic plant species.

Area	Original extent (km^2)	Currently untrans-formed (km^2)	*Plant species* Total (A)	Endemics (B)	B/A (%)
Cape Floristic Region	90 000	60 000	8600	5800	68
Atl. Forest, Brazil	1 000 000	20 000	10 000	5000	50
W. Amazonian Uplands	100 000	35 000	20 000	5000	25
Madagascar	62 000	10 000	6000	4900	82
Philippines	250 000	8000	8500	3700	44
Northern Borneo	190 000	64 000	9000	3500	39
Eastern Himalayas	340 000	53 000	9000	3500	39
Western Ecuador	27 000	2500	10 000	2500	25
Colombian Choco	100 000	72 000	10 000	2500	25
Peninsular Malaysia	120 000	26 000	8500	2400	28
New Caledonia	15 000	1500	1580	1400	89

Alternative rankings by:	put the CFR in position no:
Proportion of endemic species	3
Total plant species	7
Proportion of area remaining	10

and seven families of plants are endemic (> 97 per cent of species endemic) to the CFR. Although butterflies and vertebrate taxonomic groups do not share this high species richness, the CFR is a major centre of endemism in southern Africa for butterflies, amphibians, and fish (Rebelo 1992*a*).

Over 60 per cent of the Red Data Book plant species of southern Africa occur in the CFR, which comprises only 4 per cent of the subcontinent (Hall and Veldhuis 1985). Many of the larger mammal species are extinct in the CFR, or have been reintroduced from populations outside the CFR—this is often not reflected in Red Data Books for birds and mammals, as the CFR is a small proportion of the area of the subcontinent (Skead 1980; Rebelo 1992*a*). Together with the remaining tropical rain forests, the CFR must be considered one of the world's top conservation priorities (Macdonald 1989).

Fortunately, the richest element of the flora, Fynbos vegetation, is largely restricted to 'nutrient-poor' soils and is well-protected in the mountains

Table 24.2 Conservation status of vegetation types within the Cape Floristic Region (Rebelo 1992*b*).

Veld Type	Total area (× 10^3 ha)	Area protected (%)	Remaining natural area (%)
Afromontane forest	384.4	4.3	76
Dune thicket	445.3	6.8	76
Renoster shrubland: mountain	475.4	1.3	73
coastal	1528.5	0.3	15
Fynbos (Macchia): coastal	877.0	3.3	53
mountain	3731.0	50.1	93

(Hilton-Taylor and Le Roux 1989; Rebelo 1992*b*). The majority of the endemic bird, reptile, amphibian, fish, and butterfly species occur also in mountains of the Fynbos (Rebelo 1992*a*). However, the lowlands are considerably transformed and urgently require additional protection (Boucher 1981; Jarman 1986), most especially Fynbos and Renoster shrubland vegetation for which only 3.3 and 0.3 per cent of the original extent are currently preserved.

Available data

A basic prerequisite for any conservation strategy is comprehensive knowledge of the region. Measurement of biological diversity is dependent upon the inventory and systematics of species and their distributions in space. Biogeography addresses problems of rarity, habitat requirement, and barriers to dispersal and is thus the cornerstone to any conservation strategy (Soulé and Kohm 1989).

Species taxonomy in the CFR is reasonably complete for vertebrates and plants, but not for invertebrates, except for butterflies. Even for plants, the taxonomy of several of the larger near-endemic families has yet to be completed. However, the major problem is that of the scales used for recording biological information. In southern Africa, species distributional data are mapped routinely at the quarter-degree grid scale (Edwards and Leister 1971), with units measuring *c.* 24 × 27 km^2 in the CFR. However, the CFR consists of a network of mountain versus plain topographies, and nutrient-poor versus nutrient-rich soils and the quarter-degree grid system is too coarse for describing the spatial scale of phytogeographical heterogeneity (Rebelo and Siegfried 1989).

Data at a finer scale (eighth-degree square: 12 × 13 km^2) are available only for the Proteaceae. A *Protea* Atlas Project to be completed in 2001 (Rebelo *et al.* 1986), will provide distributional data to spot localities for analysis using Geographical Information Systems (GIS) technology (Scott *et al.* 1987). The CFR is particularly fortunate in that detailed climatic, pedological, and plant cover data at one second grid and minute resolutions are available from the Department of Agriculture and Water Supply (Macdonald and Pressinger 1989).

An additional source of species data is the many studies of vegetation communities, undertaken throughout the CFR. By 1980, some 122 localities had been studied (Boucher and McDonald 1982). However, the large number of taxonomic reappraisals requires the use of computers merely to compare species lists (Boucher 1990). Comparisons between regions are complicated by different quadrat sizes used, subsampling procedures only capturing a proportion of the total species richness, disparate sizes of study areas, the exclusion of 'rare' and 'infrequent' species from published results, a high (15 per cent) proportion of differential species proving impossible to identify, the uneven geographical coverage, and the lack of an overall vegetation typology (Bond 1981; Campbell 1985). The use of replacement guilds to overcome the high delta diversity between geographical areas is fraught with difficulties (Cowling and Holmes 1992). Furthermore, the high species richness effectively limits the numerical analyses that can be undertaken with existing hardware and software. These data are not considered further herein: they are currently the focus for studies on species turnover and endemism in the CFR (Cowling *et al.* 1992).

A structural community approach (using species-bound characters where detailed structural characters were not known or understood) has been developed to circumvent these problems in the CFR (Campbell 1985). Only three studies utilizing a structural classification of the CFR have been undertaken to date (Rebelo 1992*a*): this approach is not considered further herein.

The Proteaceae and the biogeography of the CFR

The Proteaceae has been used to determine the geographical location of a reserve network as it is the only large taxon for which detailed geographical data exist at a scale of an eighth-degree grid square (Fig. 24.1) (Rebelo and Siegfried 1989, 1992). The CFR comprises 857 such grid squares (some adjacent coastal squares were pooled to maintain a reasonably constant area). Of the 331 taxa (species and distinct subspecies, hereinafter referred to as species), 328 are confined to the gridded region. All Proteaceae species occur in Fynbos which supports the majority of the species in the CFR.

The representativeness of the geographical distribution patterns of the

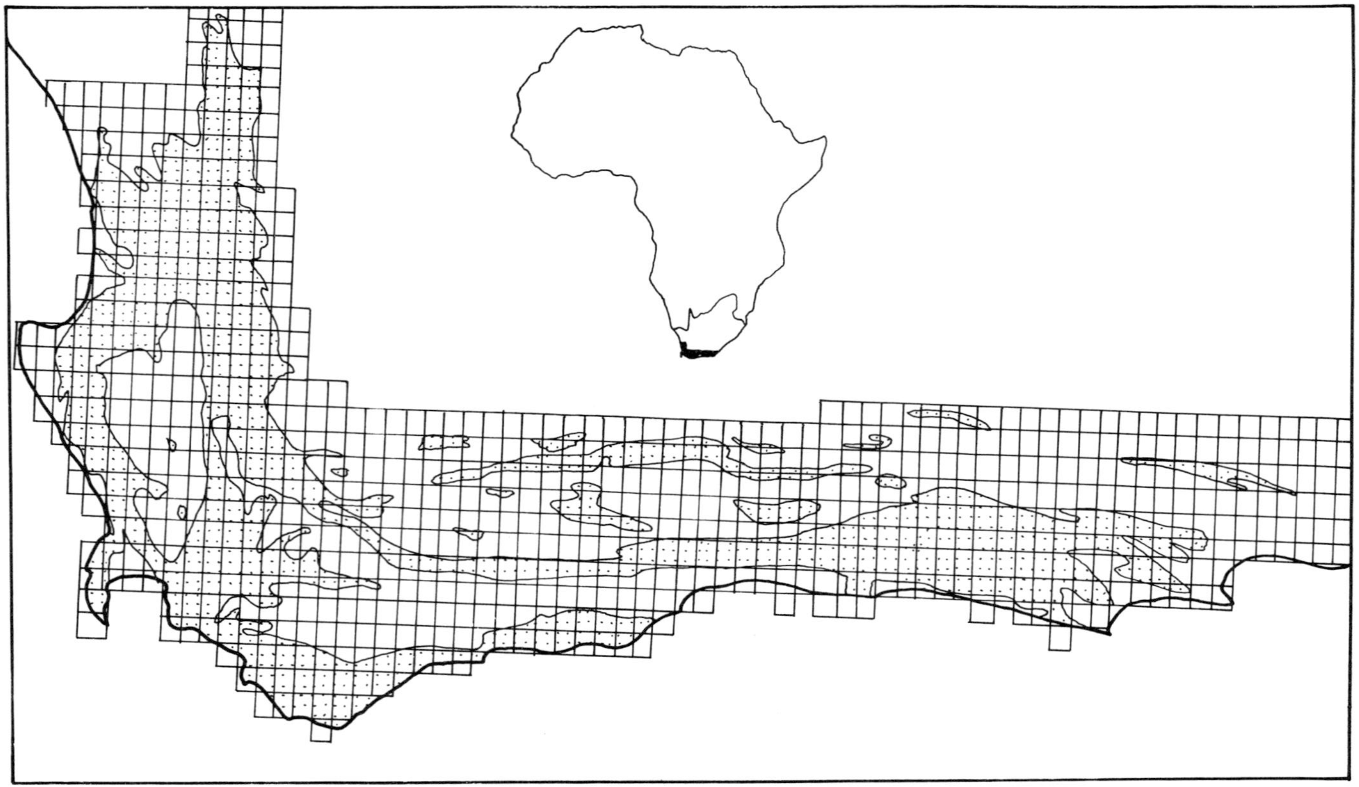

Fig. 24.1 The location of the Cape Floristic Region showing the eighth-degree grid system used and the distribution of Fynbos vegetation (shaded).

Proteaceae was tested by correlating species richness on a quarter-degree grid square with other major CFR families and species. In all cases there is a significant correlation ($P < 0.001$), suggesting that similar factors have influenced speciation and dispersion in the major plant families and genera of the CFR (Rebelo and Siegfried 1989). Due to a lack of data I have assumed that this correlation holds at an eighth-degree scale.

Plant species richness is highest in the south-western Cape, decreasing gradually northwards and rapidly eastwards along the two major fold lines of the CFR mountains. The phytogeographical centres of endemism in the CFR, based on the Proteaceae (Rebelo and Siegfried 1989; Fig. 24.2), show an increase in aerial extent of centres with fewer species to the north and east. Patterns of species richness and endemicity among non-plant taxa in the CFR have not been investigated to date.

Location of a reserve network

1. *Existing reserves*

The location of existing reserves is shown in Fig. 24.2. Some 19 per cent of the CFR (1.78 Mha) is preserved in statutory, private, and contractual nature reserves (Rebelo 1992*b*). By far the largest proportion (79 per cent) of statutory preserved land (1.1 Mha) in the CFR was proclaimed as mountain water catchment areas for an arid South Africa (Rebelo 1992*b*). A recent development (post-1970) has been the proclamation of contractual preservation areas as mountain catchment areas and contractual national parks, which currently comprise 39 per cent of the preserved area. These areas allow grazing, game farming, and flower and plant harvesting on a scale compatible with ecosystem preservation, and are managed as larger units in combination with state water catchment areas (Rebelo 1992*b*).

It is probably true to state that almost all (> 95 per cent) extant bird, non-flying mammal, (> 90 per cent) reptile and amphibian species, and about 93 per cent of mountain Fynbos plant species in the CFR occur in statutory nature reserves (Siegfried 1989; Rebelo and Siegfried 1989).

However, this existing system has excluded the lowlands of the CFR. Thus there is only one Fynbos reserve larger than 10 kha in the lowlands, and a total of 15 kha of Renosterveld vegetation has been preserved (Rebelo 1992b). No estimates of the conservation effectiveness of the lowland reserve system have been made to date.

2. *Priorities: species richness*

(a) Historical prescriptions. The current reserve network is based on Wicht's (1945) prescription, which arose out of concern about the destruction of Fynbos because of its low economic value. He designated five reserves as a

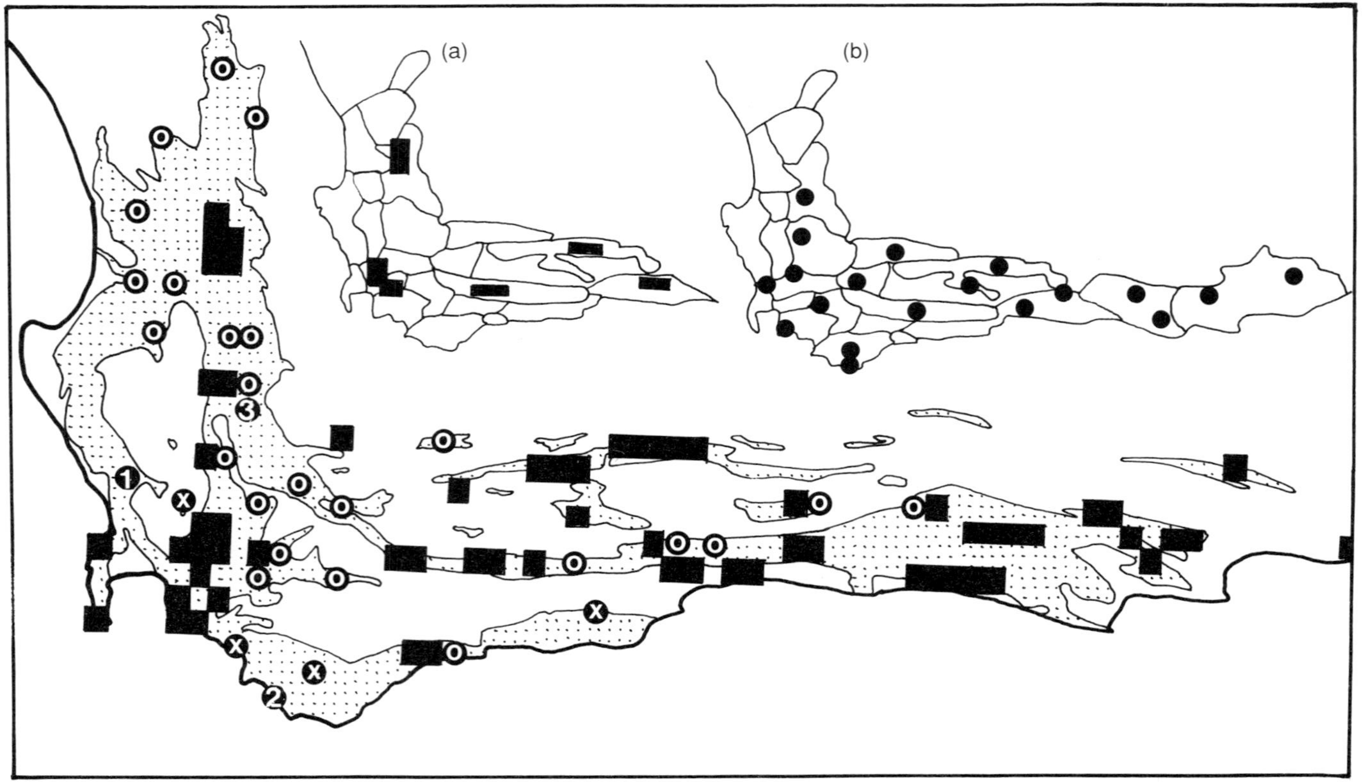

Fig. 24.2 The location of nature reserves (statutary, private, and contractual) in the Cape Floristic Region defined as grid squares containing more than half their area in nature reserves. Additional reserves required to preserve all Proteaceae species in the region are depicted as follows: numerals = reserves required to preserve more than 4 species in order of importance; X = those preserving 3 species; circles: those preserving 1 or 2 species. Inset: prescriptions for reserve configurations by (a) Wicht (1945) and (b) Kruger (1977), superimposed on centres of endemism for the Proteaceae.

core system to preserve Fynbos for its aesthetic and general scientific value by conserving the natural vegetation in 'well selected, representative, relatively large regions, which should be maintained with painstaking care'. This was achieved by selecting large areas of high species richness in each of the five major mountain ranges, apparently ignoring Weimark's (1941) work on centres of endemism. This system theoretically preserves 52 per cent of Proteaceae species in 24 grid squares and requires a total of 72 grid squares to preserve all the species.

Kruger (1977) proposed that each major CFR vegetation type in each of Weimarck's (1941) centres of endemism should be preserved. He did not specify where within the 19 zones identified the reserves should be located. Assuming that the most species-rich grid square in each zone is selected, this system preserves 64 per cent of the Proteaceae species in 19 grid squares and requires a total of 64 grid squares to preserve all species.

Rebelo and Siegfried (1989) mapped species distributions along five transects running along mountain ranges and lowland Fynbos through areas of high species richness. Starting from the grid square with the most endemic species, species turnover was calculated and a 'reserve' designated where species shared with the previous 'reserve' dropped to below 50 per cent, provided that at least 25 per cent additional species above those shared were present in the grid square. This yielded 15 'reserves' containing 69 per cent of the flora. The distribution of the remaining species was plotted and the richest grid square was selected as a reserve. Its species were removed from the map, and the procedure was repeated. This yielded a total of 59 grid squares to conserve the entire flora.

A random selection of grid squares containing more than two species yielded results indistinguishable from both Wicht's (1945) prescription and the existing reserve network. The existing reserve network, selected as those grid squares with more than 55 per cent of the area preserved in statutory nature reserves, preserved 265 species in 66 grid squares, but required 98 grid squares to preserve all the species. Kruger (1977) and Rebelo and Siegfried (1989) performed significantly better than predicted using the random selection.

(b) The ideal reserve configurations. The prescriptions described above beg the question of what the optimal reserve configuration to preserve the entire flora in the minimum number of grid squares might be. Using an iterative approach developed primarily in Australia (Kirkpatrick and Harwood 1983; Thomas and Mallorie 1985; Margules *et al.* 1988; Pressey and Nicholls 1989*a*,*b*), reserves were selected using two algorithms (Rebelo and Siegfried 1992). A species-richness algorithm calculated, at each iteration, the number of species not assigned to a 'reserve'. The richest grid square was selected as a 'reserve'. A rarity algorithm calculated, for each grid square at each

iteration, the sum of the species rarity values (total number of grid squares [857] divided by the number of non-reserve squares occupied). The grid square with the highest score was selected as a 'reserve'.

The species richness and rarity algorithms preserved all the species of Proteaceae in the CFR in 56 and 53 grid squares respectively (Fig. 24.3). However, it is possible to preserve all species in 52 reserves. This inefficiency was due to the inappropriate selection of grid squares when ties occurred (Rebelo and Siegfried 1992).

From these results it is apparent that the spatial configuration of nature reserves is determined primarily by the location of the most geographically restricted species. More widespread species are invariably included within these reserves. The relative number of species shared between grid squares (including surrogate measures, such as Sørenson's coefficient of community) is irrelevant, because endemic species are the focus of efficient reserve location (Rebelo and Siegfried 1992).

For this reason it is not possible to accurately determine the efficiency of any reserve configuration based on the number of species preserved, or the area (number of grid squares) set aside. The only reliable index of the efficiency of a reserve system is the additional area required to preserve all species (Rebelo and Siegfried 1992).

In the absence of detailed data, selection of nodes rich in narrowly endemic species will optimally preserve species richness. Where endemic species have congruent distributions, preservation in a fraction of the region under consideration will be possible. Conservation will be problematic in regions where endemic species have contiguous distribution ranges. There is also not only one ideal reserve configuration for any system. Where endemic species have relatively wide distribution ranges, the number of potential alternative configurations will be greater. Thus in the CFR there are about 5×101^3 alternative configurations which preserve all 331 Proteaceae species in 53 grid squares (Fig. 24.4).

In the CFR, increasing the number of times each species is required to be represented in the reserve system results in a near linear increase in number of grid squares required in the reserve system. Thus to preserve each species twice and four times required 105 and 178 grid squares, respectively. Again the relationship is determined primarily by the dispersion of narrow endemic species: where congruence of endemic species is high, or where endemics are relatively widespread the relationship between number of reserves required per species and total area will be flatter than for contiguous endemics with narrow distribution ranges.

(c) Reality. The existing reserve network is little better than a random selection of grid squares. Although it preserves 80 per cent of Proteaceae species, a further 32 grid squares are required to preserve all the species (Fig.

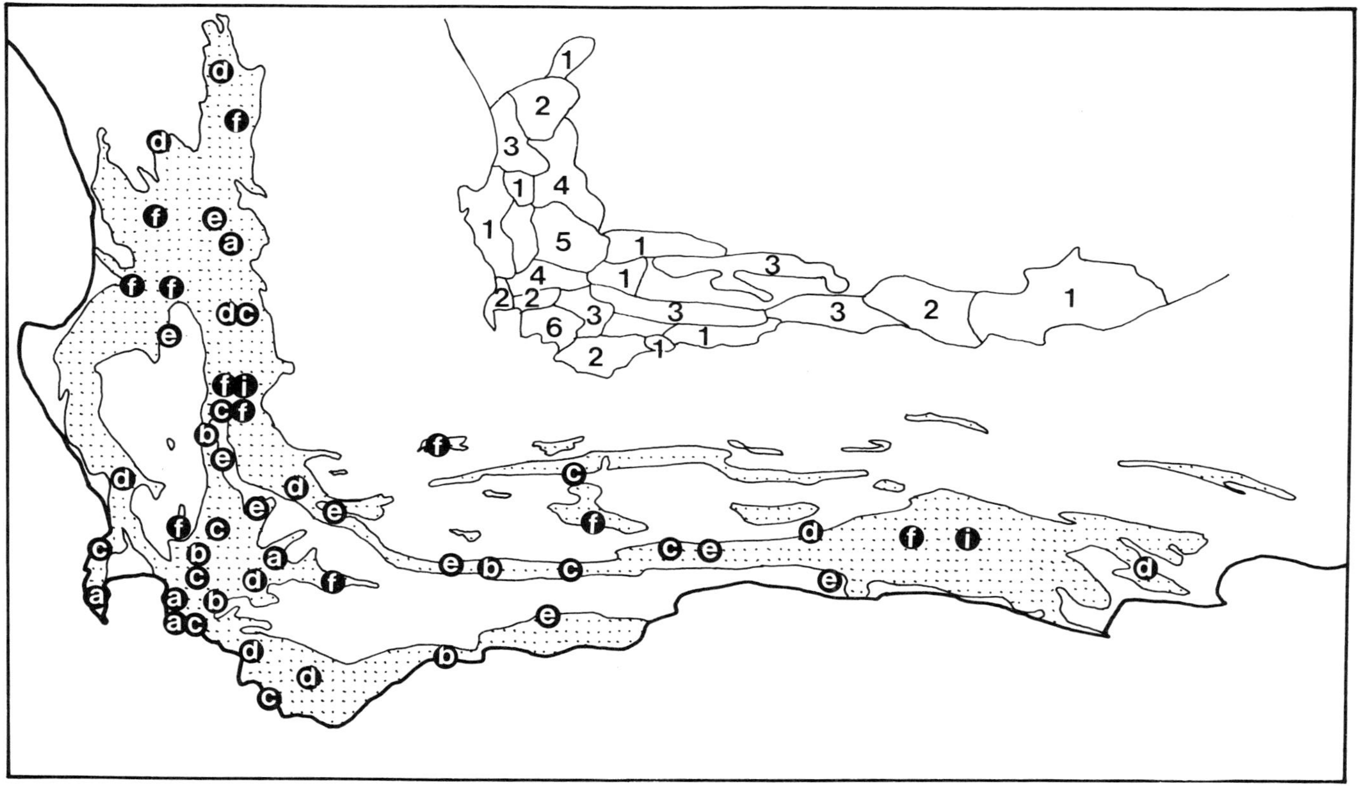

Fig. 24.3 One of 101^3 optimal reserve configurations for preserving all Proteaceae in the Cape Floristic Region in 52 reserves. This configuration additionally protects each species in as many reserves as possible. Inset: Centres of endemism for Proteaceae in the Cape Floristic Region. Numerals are the number of reserves required in each centre for the optimal reserve configuration.

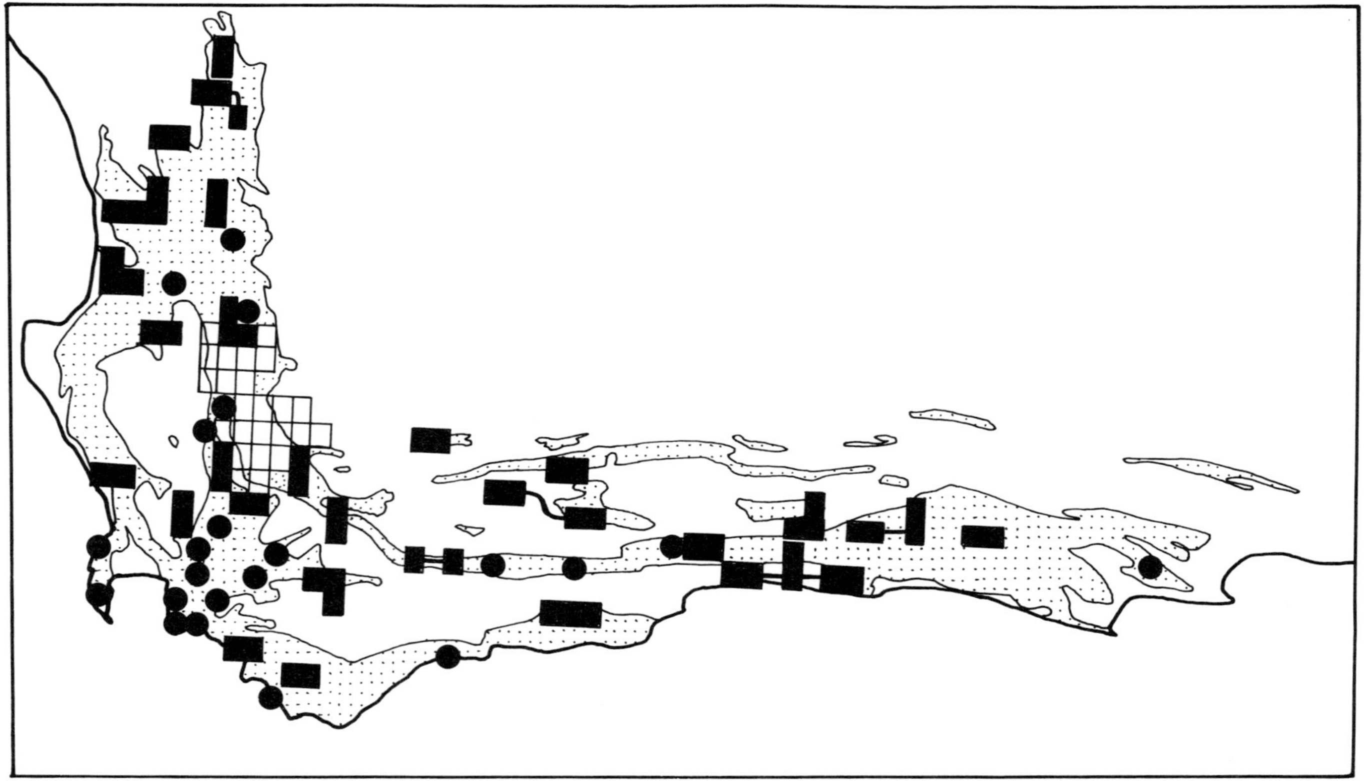

Fig. 24.4 Alternative sitings for each reserve for protecting all species of Proteaceae in the Cape Floristic Region in 52 reserves. Circles are irreplaceable reserves. Contiguous black grid squares and those linked by lines are sets for which any one grid square would equally well achieve the goal of preserving all the species in 52 reserves. The open squares circumscribe an area in which any reserve would probably suffice given inadequacies in existing data.

24.2). Representing all species by building on the existing reserve system thus requires nearly twice (1.87) as many grid squares as the optimal configuration to preserve all the species once.

From the spatial configuration of additional nature reserves required to augment the current reserve system, it is apparent that the lowlands of the CFR are the top priority conservation areas, with the central region also requiring additional conservation status. Three of the required 32 nature reserves preserve more than three target species, and four preserve three target species. These should perhaps be considered for large ecologically-viable reserves. The remaining grid squares preserve only one or two target species and can perhaps be preserved in very much smaller areas. Another major concern is that data are not evenly available throughout the CFR—many remote areas are botanically poorly known. However, provided new species are not found, additional records of species will increase the number of potential reserve configurations without affecting the total number of reserves required. Specifically, as data on poorly-known endemic species increases so does the number of potential alternative configurations for that target taxon. Furthermore, the likelihood of a major centre of endemism having been missed is low, and new species are probably most likely to be found within the existing centres of endemism which also have a high species richness.

3. *Priorities: taxonomic diversity*

Recently, it has been argued that the conservation of biodiversity must take into account the degree of difference between species (Williams *et al.* 1991). Species are not all equal and it is argued that a species in a monospecific family should have a higher conservation requirement than a species with several hundred congeners.

Although the taxonomy of the Proteaceae has been recently appraised, the systematics of the group is poorly known. No detailed cladistic or phenetic studies of the inter-relationships between the genera have been undertaken. Nevertheless, taxonomists have recognized subgenera and sections within genera, which might be used for weighting species so as to ensure that less speciose clades are given a higher weighting.

Some 68 subgenera or sections (hereinafter referred to as subgenera) are recognized within the 13 genera of the Proteaceae (See Rebelo and Siegfried 1989 for references). Some 11 of these subgenera are monospecific, and a further eight have only two species. All 68 subgenera can be preserved in 13 grid squares, a subset of the optimal configuration based on species, although the sequence of selection is quite different (Fig. 24.5). Eight of these feature in the top ten ranking grid squares of the optimal configuration based on species and all are in the top 50 per cent of grid squares.

Only four of the 13 grid squares currently have more than half their area in

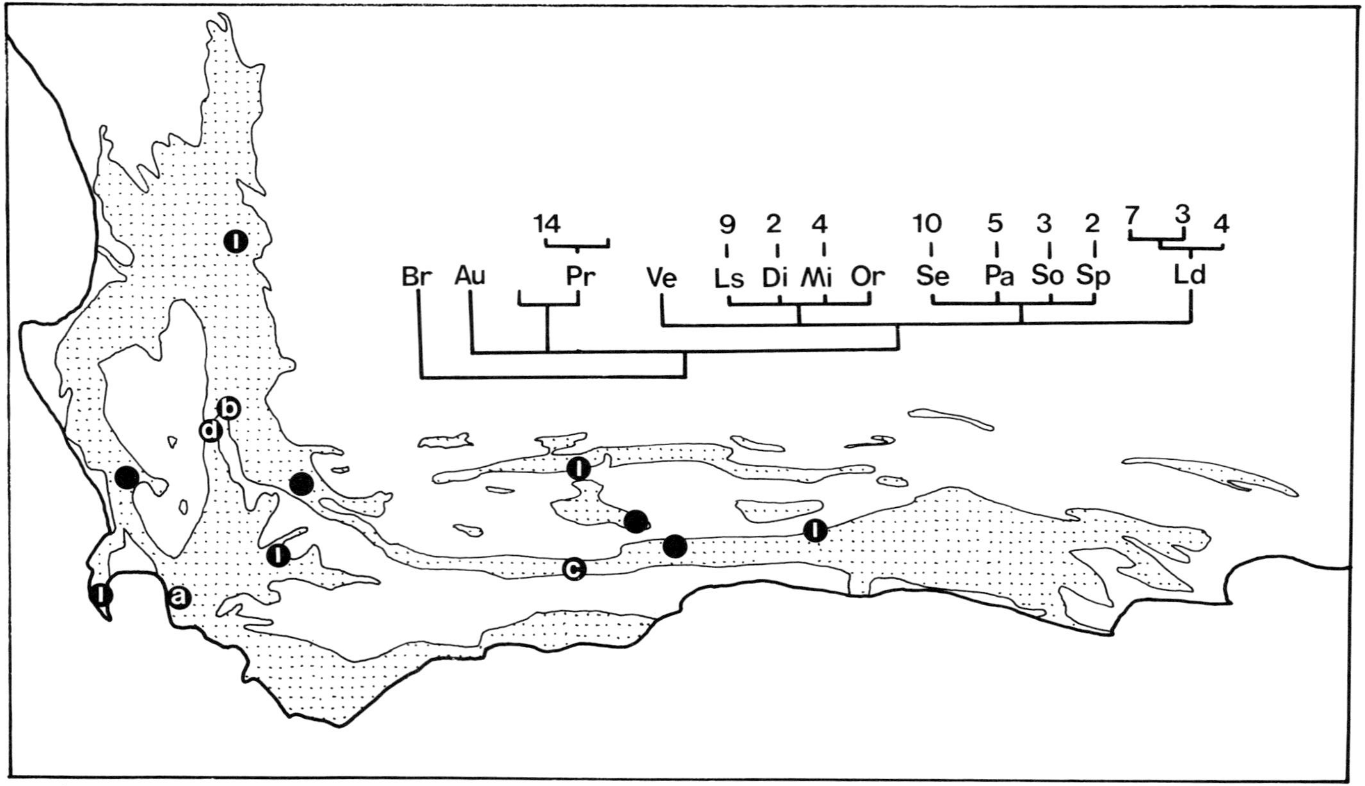

Fig. 24.5 The optimal reserve configuration required to preserve all 68 lineages (see inset) in the Proteaceae of the Cape Floristic Region. The sequence of reserves is shown alphabetically, with reserves preserving only two taxa as ◐ and only one taxon as ●. The relationship between the 13 genera (abbreviated to first two letters), subgenera, and sections (terminal branches above genera) of the family in South Africa are shown with numerals above subgenera (equalling the number of sections per subgenus where greater than 1).

statutory nature reserves. Some 13 subgenera (19 per cent) are currently not preserved: the grid squares containing them feature strongly in the additional nature reserves required to augment the current reserve system.

4. *Priorities: ecological redundancy*

Species are not equal ecologically: some species fundamentally alter the ecosystems in which they occur, whereas others appear to have no measurable influence on ecosystem processes. Walker (1992) has argued that 'redundant' species should be weighted less than keystone and 'driver' species. He suggests that such an assessment could most easily be done by defining guilds, determining the number of species per guild, evaluating the interactions among the species in each guild, and finally considering the relative importance of each guild. Such an approach cannot be taken in the CFR because of lack of data. Guilds within the Proteaceae of CFR mirror taxonomic divisions, with seed dispersal/storage, pollination, breeding system, leaf structure and, to a lesser extent, habitat requirement guilds determining generic and subgeneric affinities. However, within each subgenus, fire survival strategies and leaf-sizes are the plastic variables. Fire is the major perturbation in Fynbos, and masks species interactions to such a degree that multiple redundancy appears to be the major feature of Fynbos vegetation. Certainly, no keystone, mobile link, or 'driver' species has been identified to date in the CFR.

Any reserve configuration based on preserving representatives of either ecological guilds or ecological non-redundant species within such guilds will probably be a subset of the taxonomic diversity reserve system.

Indeed rarity, specifically localized geographical distribution and strict habitat specificity, is strongly correlated with ecological guilds based on regeneration (killed by fire), seed dispersal (myrmecochory), and pollination (insect) strategies in the Proteaceae of the CFR. Furthermore, species richness is also strongly correlated with rarity, with guilds containing many rare species also being the most species rich.

5. *Priorities: Red Data books*

Red Data Books provide data which allow the determining of areas containing an abundance of threatened taxa. A Red Data Book for plants is available for the CFR. Red Data Books for mammals, birds, reptiles, amphibians, fish, and butterflies are available for South Africa (Ferrar 1989). Fortunately, with the exception of birds and mammals, few species are shared between the CFR and the rest of South Africa, but the recording of distribution ranges has not been standardized among the Red Data Books so that comparative data can only be resolved to the scale of a quarter-degree grid square (Rebelo 1992*a*).

As Red Data Books provide data which allow the designation of 'hot spots', or nodes of threatened species, considerable conservation effort should be directed towards these nodes. However, given a geographically-constant threat, areas containing more species are likely to have more Red Data Book species.

Red-Data-Book-Proteaceae species richness in the CFR (Tansley 1988) is significantly correlated ($P < 0.001$) with total Proteaceae species richness at an eighth-degree grid scale, and with total Red-Data-Book species richness at a quarter-degree scale (Rebelo and Tansley 1993). Total Red-Data-Book species richness is also significantly correlated ($P < 0.001$) with total species richness at a quarter-degree scale in the CFR.

Furthermore, the distribution of naturally rare species and threatened species (i.e. vulnerable, endangered, and extinct) are geographically distinct. The distribution of threatened species is in agreement with the findings of the only assessment of conservation priorities (in the lowlands) of the CFR (Jarman 1986) undertaken to date (Fig. 24.6). The greatest threat to plant species in the CFR is urban expansion of Greater Cape Town (Rebelo and Tansley 1993).

However, within grid squares in the greater Cape Town metropolitan area Red-Data-Book Proteaceae richness significantly underestimated total Red-Data-Book species richness. These grid squares also contain local populations of more widespread species, which are not included in the Red Data Book because the threat to them is localized. Thus the threats in the region are underestimated in our analysis using the Proteaceae (Rebelo and Tansley 1993). The results are in concordance with a survey of natural lowland remnants in the greater Cape Town metropolitan area, which revealed 74 threatened species surviving in the remaining 484 ha of natural and semi-natural vegetation, i.e. the arithmetic equivalent of 15 threatened species per km^2 (McDowell *et al.* 1991). This area must rank as one of the top priority conservation areas worldwide, and is primarily impacted by the expansion of the peri-urban areas over much of one of the richer centres of endemism within the CFR.

Size and management of reserves

Once the configuration of reserves in the area has been determined, only then can the regional importance of a reserve be evaluated based on the core species which the reserve must preserve. These species considerations, together with population biology and ecological needs, will prescribe the minimum size of each reserve and its management requirements (Walker 1992).

Even a rudimentary knowledge of the autecology of species in the CFR is confined to higher vertebrates (Rebelo 1992*a*). While some data exist on the importance of species interactions in pollination (Rebelo 1987) and seed

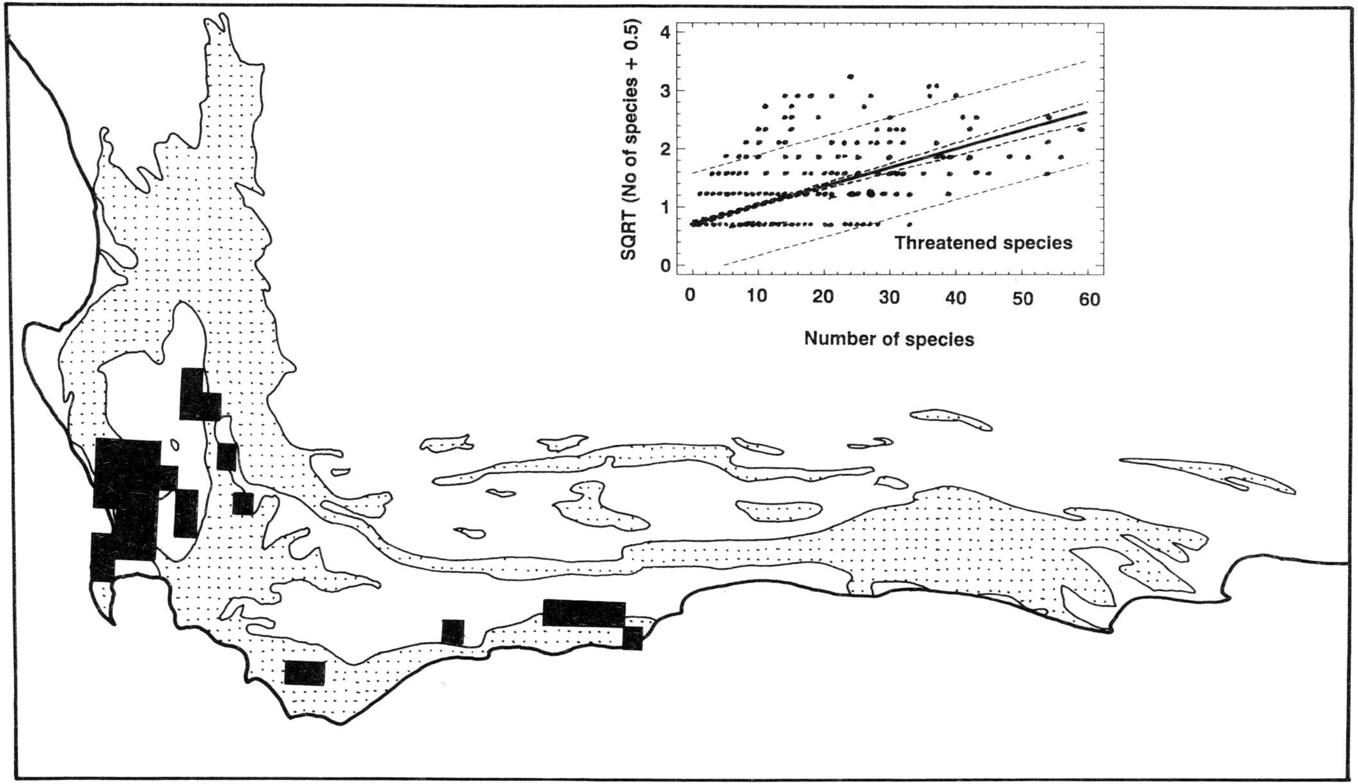

Fig. 24.6 Grid squares containing significantly more threatened (i.e. vulnerable or endangered) species of Proteaceae than predicted by the number of Proteaceae species (inset: r for $\surd$ of both axes = 0.798, $P < 0.001$, $N = 238$). Only two grid squares are not predominantly lowland (i.e. non-mountainous terrain).

dispersal (Bond and Slingsby 1983; Knight 1987), keystone, or mobile link species, if they exist, have not been identified in the CFR.

The high species richness of Fynbos vegetation and the large number of threatened species mitigates against Simberloff's (1988) approach of basing conservation management on the autecology of target species. Even if the autecology of a sizeable proportion of threatened species were known, there is no protocol for extrapolating results to the unstudied species (Bond 1989). Furthermore, the genetic and demographic consequences of small populations may differ between naturally-rare, local endemics, and species with formerly extensive distribution ranges currently reduced to local remnants. Thus ecological studies undertaken in natural populations may not be applicable to threatened species reduced to small, local nature reserves (Bond 1989). Effective conservation strategies for threatened species therefore may involve more intensive management than that required for naturally rare species. We concur with Bond (1989) in advocating an ecosystem approach to conservation in the CFR, with research emphasis placed on identifying keystone species and processes (Soulé and Kohm 1989).

As a rough rule, in Fynbos vegetation fire is the major driving force in the ecosystem. Seed dispersers, pollinators, and predators appear to be largely shared between different vegetation types in the CFR, so that the high diversity of plant species in Fynbos does not seem to be linked to any keystone process. Thus Cowling and Bond (1992) have shown that reserves as small as a few (4–15) ha are potentially viable in Fynbos, provided that the disturbance regimes are maintained. The other major management input in Fynbos is the fight against introduced alien plant species, a threat to which the CFR appears particularly vulnerable (Macdonald and Richardson 1986). Neither of these factors have a major influence on minimum reserve size, although where ecosystem processes are not actively managed it seems that a minimum reserve size in mountain Fynbos is of the order of 600 ha (Bond *et al.* 1988). No data are available for lowland Fynbos, but the area is likely to be much larger given that water tables must be maintained and natural ignition events are much rarer.

Large herbivores and carnivores, which require much larger areas for their effective preservation, were probably not important in Fynbos ecology, due partly to the low carrying capacity of Fynbos (Rebelo 1992*b*). Reserve size considerations based on larger carnivores and herbivores in Fynbos have been reviewed by Rebelo (1992*b*). A reserve of between 10^5 and 10^6 ha would be required to preserve a minimum viable population of 50 pairs of Leopard or Mountain Zebra in Fynbos. Should contractual reserves function adequately as corridors for these top predators and herbivores, then reserves may be as small as 2.2×10^3 in the northwestern CFR and 16×10^3 in the south-west.

It is proposed that additional reserves required to augment the existing reserve system which contain three or more Proteaceae species, should be sufficiently large to maintain ecosystem processes with minimal active management, but additional reserves required for the conservation of one or two species might best be intensively managed in smaller reserves to preserve the target species.

Conclusions

1. Reserve identification software tools exist. Prescriptions exist. Basic research into priorities has been undertaken. The most urgent areas to be acquired to upgrade the CFR nature reserve system have been identified.
2. No significant changes in reserve configuration in the CFR have occurred over the last decade despite the above, although land has been acquired in ecosystems outside the CFR which are under little threat and where land is much cheaper.
3. Given the global importance of the region, it should be accorded the highest conservation status possible. Given that the local population should be involved in this, a Biosphere Reserve is the ideal option.
4. International aid and investment should be conditional on sound conservation strategies being implemented in the CFR. These do not prevent development in the area. Indeed, tourism has been identified as a potential major source of income in the region. The high floristic diversity and a well-developed reserve system with hiking trails and scenic tours are major tourist drawcards. The development of the tourist industry is thus dependent on the efficient preservation of the flora, which is compatible with sustainable utilization of Fynbos for flowers and as a water catchment.

References

Bond, W. J. (1981). Vegetation gradients in the southern Cape mountains. MSc thesis, University of Cape Town.

Bond, W. J. (1989). The dynamic nature of biotic diversity. In *Biotic diversity in southern Africa. Concepts and conservation* (ed. B. J. Huntley), pp. 2–18. Oxford University Press, Cape Town.

Bond, P. and Goldblatt, P. (1984). Plants of the Cape Flora: a descriptive catalogue. *Journal of South African Botany*, Supplement **13**, 1–455.

Bond, W. J. and Slingsby, P. (1983). Seed dispersal by ants in shrublands of the Cape Province and its evolutionary implications. *South African Journal of Science*, **79**, 231–3.

Bond, W. J., Midgley J., and Vlok J. (1988). When is an island not an island? Insular effects and their causes in fynbos shrublands. *Oecologia*, **77**, 515–21.

Boucher, C. (1981). Floristic and structural features of the coastal foreland vegetation South of the Berg river, western Cape Province, South Africa. In *Proceedings of a symposium on coastal lowlands of the western Cape* (ed. E. Moll), pp. 21–6. University Western Cape, Bellville.

Boucher, C. (1990). *SEEK—a Cape species name.* Abstracts to South African Association of Botanists Sixteenth Annual Congress, January 1990, Botanical Reseach Institute, Department of Agriculture and Fisheries.

Boucher, C. and McDonald, A. E. (1982) An inventory of plant communities recorded in the western, southern and eastern Cape Province, South Africa up to the end of 1980. *South African National Scientific Progress Report*, **57**, CSIR, Pretoria.

Campbell, B. M. (1985) Montane vegetation structure in the Fynbos Biome: structural classification and adaptive significance of structural characters. PhD thesis, Utrecht University.

Cowling, R. M. and Bond, W. J. (1992). How small can reserves be? An empirical approach in Cape Fynbos. *Biological Conservation*, **55**.

Cowling, R. M. and Holmes, P. M. (1992). Flora and vegetation. In *The ecology of the Fynbos: nutrients, fire and diversity* (ed. R.M. Cowling), pp. 23–61. Oxford University Press, Cape Town.

Cowling, R. M., Holmes, P. M., and Rebelo, A. G. (1992). Plant diversity and endemism. In *The ecology of the Fynbos: nutrients, fire and diversity* (ed. R. M. Cowling), pp. 62–112. Oxford University Press, Cape Town.

Edwards, D. and Leistner, O. A. (1971). A degree reference system for citing biological records in southern Africa. *Mitteilungen Botanischen Staatsammlung, München*, **10**: 501–9.

Ferrar, A. A. (1989). The role of red data books in conserving biodiversity. In *Biotic diversity in southern Africa. Concepts and conservation* (ed. B. J. Huntley), pp. 136–47. Oxford University Press, Cape Town.

Hall. A. V. and Veldhuis, H. A. (1985). South African red data book: plants—Fynbos and Karoo biomes. *South African National Scientific Programmes Report*, **117**. CSIR, Pretoria.

Hilton-Taylor, C. and Le Roux, A. (1989). Conservation status of the Fynbos and Karoo biomes. In *Biotic diversity in southern Africa. Concepts and conservation* (ed. B. J. Huntley), pp. 202–23. Oxford University Press, Cape Town.

Huntley, B. Siegfried, R., and Sunter, C. (1988) *South African environments into the 21st century.* Human and Rousseau, Cape Town.

Jarman, M. L. (1986). Conservation priorities in lowland regions of the fynbos *biome. South African National Scientific Programmes Report*, **87**, CSIR, Pretoria.

Kirkpatric, J. B. and Harwood, C. E. (1983). Conservation of Tasmanian macrophyte wetland vegetation. *Proceedings of the Royal Society of Tasmania*, **117**, 5–20.

Knight, R. (1987). Aspects of plant dispersal in the southwestern Cape with particular reference to the roles of birds as dispersal agents. PhD thesis, University of Cape Town.

Kruger, F. J. (1977). Ecological reserves in the Cape Fynbos: toward a strategy for conservation. *South African Journal Science*, **73**, 81–5.

Macdonald, I. A. W. (1989). Man's role in changing the face of southern Africa. In *Biotic diversity in southern Africa. Concepts and conservation* (ed. B. J. Huntley), pp. 51–78. Oxford University Press, Cape Town.

Macdonald, I. A. W. and Pressinger, F. M. (1989). *A list of spatio-temporal databases of relevance to the international geosphere-biosphere programme.* Second edition. FitzPatrick Institute, Cape Town.

Macdonald, I. A. W. and Richardson, D. M. (1986). Alien species in terrestrial ecosystems of the fynbos biome. In *The ecology and management of biological invasions in southern Africa* (ed. I. A. W. Macdonald, F. J. Kruger, and A. A. Ferrar), pp. 77–91. Oxford University Press, Cape Town.

McDowell, C. R., Low, A. B., and McKenzie, B. (1991). Natural remnants and corridors in greater Cape Town: their role in threatened plant conservation. In *Nature Conservation 2: the role of corridors* (ed. D. A. Saunders and R. J. Hobbs), pp. 27–39. Surrey Beattie, Chipping Norton, NSW, Australia.

McNeely, J. A., Miller, K. R., Reid, W. V., Mittermeier, R. A., and Werner, T. B. (1990). *Conserving the world's biological diversity.* IUCN, Gland, Switzerland.

Margules, C. R., Nicholls, A. O., and Pressey, R. L. (1988). Selecting networks of reserves to maximize biological diversity. *Biological Conservation*, **43**, 63–76.

Pressey, R. L. and Nicholls, A. O. (1989*a*). Efficiency in conservation evaluation: scoring versus iterative approaches. *Biological Conservation*, **50**, 199–218.

Pressey, R. L. and Nicholls, A. O. (1989*b*). Application of a numerical algorithm to the selection of reserves in semi-arid New South Wales. *Biological Conservation*, **50**, 263–78.

Rebelo, A. G. (ed.) (1987). A preliminary synthesis of pollination biology in the Cape flora. *South African National Science Programmes Report*, **141**. CSIR, Pretoria.

Rebelo, A. G. (1992*a*). Red Data Book species in the Cape Floristic Region: threats, priorities and target species. *Transactions of the Royal Society of South Africa*, **48**, 55–86.

Rebelo, A. G. (1992*b*). Preservation of biotic diversity. In *The Ecology of Fynbos: nutrients, fire and diversity* (ed. R. M. Cowling), pp. 309–44. Oxford University Press, Cape Town.

Rebelo, A. G. and Siegfried, W. R. (1989). Protection of Fynbos vegetation: ideal and real-world options. *Biological Conservation*, **54**: 15–31.

Rebelo, A. G. and Siegfried, W. R. (1992). Where should nature reserves be located in the Cape Floristic Region, South Africa? Models for the spatial configuration of a reserve network aimed at maximizing the protection of floral diversity. *Conservation Biology*, **6**, 243–52.

Rebelo, A. G. and Tansley, S. A. (1993). Using rare plant species to identify priority conservation areas in the Cape Floristic Region: the need to standardize for total species richness. *South African Journal of Science*, **89**, 156–61.

Rebelo, A. G., Cowling, R. M., Gibbs-Russell, G. E., Hockey, P. A. R., Jarman, M. L., Boucher, C., and Hilton-Taylor, C. (1986). Guidelines for the plant atlas of southern Africa: the outcome of a workshop held in Cape Town 5 December 1985. Ecosystem Programmes Occasional Report Series 23.

Scott, J. M., Csuti, B., Jacobi, J. D., and Estes, J. E. (1987). Species richness: a geographic approach to protecting future biological diversity. *Bioscience*, **37**, 782–8.

Siegfried, W. R. (1989). Preservation of species in southern African nature reserves. In *Biotic diversity in southern Africa: concepts and conservation* (ed. B. J. Huntley), pp. 186–201. Oxford University Press, Cape Town.

Skead, C. J. (1980). *Historical mammal incidence in the Cape Province. Vol 1. The western and northern Cape.* Cape Department Nature Environmental Conservation, Cape Town.

Simberloff, D. (1988). The contribution of population and community biology to conservation science. *Annual Reviews of Ecology and Systematics*, **19**, 473–511.

Soulé, M. E. and Kohm, K. A. (ed.) (1989). *Research priorities for conservation biology*. Island Press, Washington.

Takhtajan, A. (1986). *Floristic regions of the World*. University of California Press, Berkeley.

Tansley, S. A. (1988). The status of threatened Proteaceae in the Cape Flora, South Africa, and the implications for their conservation. *Biological Conservation*, **43**, 227–39.

Thomas, C. D. and Mallorie, H. C. (1985). Rarity, species richness and conservation: butterflies of the atlas mountains in Morocco. *Biological Conservation*, **33**, 95–117.

Walker, B. H. (1992). Biodiversity and ecological redundancy. *Conservation Biology*, **6**, 18–23.

Weimarck, H. (1941). Phytogeographical groups, centres and intervals within the Cape Flora. *Lunds Universitets Arsskrift*, **37**. Nr 5, 1–143.

Wicht, C. L. (1945). *Preservation of the vegetation of the Southwestern Cape*. Special Publication Royal Society South Africa, Cape Town.

Williams, P. H., Humphries, C. J., and Vane-Wright, R. I. (1991) Measuring biodiversity: taxonomic relatedness for conservation priorities. *Australian Systematic Botany*, **4**, 665–79.

25. Taxonomy and functional biotic measurement, or, will the Ark work?

STEVEN H. COUSINS
International Ecotechnology Research Centre, Cranfield Institute of Technology, Cranfield, Bedfordshire MK43 0AL, UK

Abstract

Biodiversity is represented differently in systematics and in ecology. This chapter uses hierarchy theory to relate the two approaches. Modifications, based on functional attributes, are made to representations of ecological biodiversity. Taxonomic uniqueness measures are scale independent whereas the functional evaluation tools proposed are different for local, ecosystem, and global scales. Higher taxonomic categories are functionally important for ecosystem analysis and may indicate a shift in the role of taxonomy in ecology. The quantity of life in contrast to the number of types is seen as important at the global scale. The measurement of human interaction strength with the existing biological environment is necessary for conservation evaluation. UK data for birds and mammals are used to exemplify the above points.

Introduction

The two indices of diversity—indeed, the very meaning(s) of the word diversity—are different in ecology and systematics. The mechanisms of extinction may lie squarely in the province of ecology, but we measure extinction taxonomically, squarely within the realm of systematics (Eldredge 1992).

The role of systematics in providing solutions to the global biodiversity crisis has been substantially strengthened by recent work which offers new priorities for conservation (Vane-Wright *et al.* 1991; Faith 1992; Williams and Humphries and others in this volume). This paper integrates these important new developments within a wider framework. As may be inferred

Systematics and Conservation Evaluation (ed. P. L. Forey, C. J. Humphries, and R. I. Vane-Wright), Systematics Association Special Volume No. 50, pp. 397–419, Clarendon Press, Oxford, 1994.

from the quotation above, a different set of priorities may arise from the 'province' of ecology. Given these two views there is a need to explore the relationship between ecology and systematics to see if complementary methodologies for conservation evaluation can be derived. It is suggested here that certain steps towards this are provided by establishing a basis in ecological theory for linking taxonomic and functional ecological measurement derived from Cousins (1985, 1987, 1988), and by making functional measures of biodiversity as it is understood in ecology (Cousins 1991). In spite of improvements to biodiversity measurement in both ecology and systematics, biodiversity itself may not be sufficient to provide a satisfactory basis for conservation evaluation; the quantity of life as well as number of types is important.

Ultimately it is human action which is driving the current extinction process. It is suggested that some independent measure of the strength of human interaction with ecosystems is necessary in order to judge the success or failure of conservation measures which are undertaken, on whatever basis. All these taxonomic and functional measures are relatively new, and require extensive research and testing if these and other solutions are to be realistic and effective in meeting the truly overwhelming task of global biodiversity conservation.

Systematics and ecology

The valuable potential of the new systematics approach is that, by very carefully choosing which parts of the land surface are conserved, a very substantial amount of the world's biodiversity can be contained on a relatively small area, at least in the short term (the principle of efficiency; Pressey and Nicholls 1989). The degree to which this can be achieved depends, in part, on re-assessing the traditional view that all species are equal, and instead prioritising areas according to the uniqueness of biological form that are represented there as well as the absolute number of species found. The rest is achieved by selecting a priority sequence (Williams and Humphries, this volume) or set of reserves (Rebelo, this volume), which contain these species with a minimum of duplication (by application of the complementarity principle of Vane-Wright *et al.* 1991). However, this selection of the most highly complementary pieces of the global biotic landscape does not tell us how to conserve species they include. We cannot presume that a reserve will protect a species for ever and questions of a functional nature immediately arise. How big does the reserve have to be and what would happen if the climate changes, or if a new disease or predator arrives in the reserve? These questions require an ability to understand the functional relationships between species in a strongly biased sample of species-rich ecosystems.

As Eldredge (1992) notes, the process of extinction, or to paraphrase him, of persistence, is an ecological one arising out of the population dynamics and feeding-interactions of locally dispersing and interacting organisms. In contrast, the description of species, and of their relative position vis-a-vis all other species, is a matter for systematics and is achieved by making taxonomic trees showing historical and genealogical relationships.

There are several aspects to this current dichotomy between ecology and systematics. Species diversity in ecology is measured as an index or count of the species found at a particular place or area. More precisely (Eldredge 1992) species diversity, as it is used in ecology, relates to a count of those species represented by at least one organism at a particular location. This concept is spatial and is quantitatively determined by the scale of observation to the extent that the species area curve is fundamental to ecological species diversity measurement. In systematics, species diversity is essentially relational rather than spatial and concerns the number of species within a particular taxon and how those species are related by assumed evolutionary descent. However, because each species tends to be reasonably cohesive in its distribution, we can speak of a spatial distribution of a species as its 'range', the spatial area it occupies on the Earth's surface at any one time. Ecological species diversity measurement is thus concerned with all species at one point, whereas systematic species diversity is concerned with all locations of single or related groups of species. The range of the species is all important here; just one reserve on earth will 'preserve' a species, whereas the range of each species determines how many species are found at each point. From here we can ask what kind of biodiversity does the Rio Convention want: big ranges or little ranges? The result will considerably affect the local biodiversity experienced by the world's human population.

Vane-Wright *et al.* (1991) show how the perceived dichotomy of ecology and systematics carries through into different conservation strategies:

> McNaughton (1989) has observed that we have to 'determine what should be conserved and how it is to be conserved. A critical places strategy . . . could accomplish this objective'. McNaughton is an ecologist, and his 'critical places' refer to representative ecosystems. As systematists we think instead of areas of endemism, or critical faunas and floras for particular taxonomic groups.

However, critical flora and fauna also translate into a critical places strategy, although Vane-Wright's *et al.* set priorities via taxonomic assessment and then require the conservation of all organisms in the preferred location, i.e. the local ecosystem, because the biotic environment in which the critical fauna sits is also necessary to support that fauna.

A theoretical justification for these separate ecological and systematic approaches is provided by Eldredge and others (Grene 1987; Eldredge and Salthe 1984; Vrba 1984; Damuth 1985), who have adopted a model of evolution based on the interactions of a 'dual hierarchy'. This is the hierarchy of systematics going from kingdoms down through taxonomic categories to species, individual genotype to chromosomes and DNA; and the 'economic' or ecological hierarchy of energy and matter transfer which passes from biosphere to ecosystem, down to individual phenotype, cells and subcellular organelles, and so on. These two hierarchies are seen as hierarchies of real objects that can 'do things' and not abstract classes which cannot do anything in the world in their own right.

Taxonomic ecology

One way of establishing a theoretical basis for linking the systematic and ecological hierarchies would be to establish that there is only one hierarchy of objects (that can do things) above the individual organism, and that the other relationships were those of classes, not objects. Although I have put forward such a single hierarchy (Cousins 1988, 1990) I do not propose to reiterate the arguments regarding classes and objects, but instead, I will to try to demonstrate the benefits that accrue from viewing biological systems in this way.

First, we may say that organisms interact to form ecosystems, principally by eating each other or avoiding being eaten. If organisms were all of the same type, that is of the same species, and were of the same size, then feeding interactions would be difficult to sustain. It is the difference between organisms which is at the heart of interactions which form ecosystems. But as we have already seen, it is the task of systematics to describe and relate the difference between organisms. Thus we should not be surprised on this basis that systematics is at the heart of the functional description of ecosystems. The currently perceived mismatch of ecology and systematics comes instead, I suggest, from that old debate of structure versus function and therefore represents a false dichotomy.

Structure is measured in the three dimensions of space, whereas function is measured in four dimensions (i.e. space plus time). Thus, for example, if we view the live mammalian heart at any one time a structure of valves and chambers will be seen which, if we then view at successive times, will change shape and position such that we observe the function of pumping. More generally, energizing a biological structure within a suitable environment generates, when viewed over successive times, function.

A comparable summary of the distinctive features of the systematic (genealogical) and ecological (economic) hierarchies in terms of the dimensions of space and time is given by Eldredge (1992):

> Thus economic biotic systems are inherently spatial—and though they persist for periods of time, their hallmark is definitely moment-by-moment interactions.

Genealogical systems, in contrast, are the by-product of history that act as reservoirs of genetic information.

I interpret these observations simply to mean that the structure of the organism is predominately influenced by genetics, but when this structure (the organism) is viewed moment-by-moment in time, then economic or ecological interactions are observed.

The really interesting question is whether organisms in higher taxonomic classes (structures) when viewed in moment-to-moment interactions create ecosystem function. Snaydon (1973) was perhaps the first to explore this idea, however briefly:

There is still a surprising similarity between taxonomic groupings and ecological behaviour; plants, animals, fungi and bacteria are broadly equivalent to producer, consumer and decomposer levels in ecosystems. Within each of these groups there are also broadly equivalent taxonomic and ecological groupings; for example the taxonomic grouping into algae, bryophytes and angiosperms broadly corresponds to ecological differences and the major taxonomic groupings with the algae broadly corresponds to ecological differences.

A much larger body of knowledge linking the ecological and the taxonomic arises from the study of body size linked phenomena, or biological allometry. Early work related key ecological parameters such as reproductive rate and metabolic rate to body size. In the case of metabolic rate three taxonomically distinct categories were proposed (Fenchel 1974): unicells, multicellular poikilotherms, and multicellular homiotherms, since these three categories radically improved the fit of the data. Over 1000 allometric relationships with ecological relevance were listed by Peters (1983). The abundance of these relationships was caused primarily by refining particular relationships through partitioning the data into particular taxonomic groups. As I wrote at the time (Cousins 1983), the discovery of allometric ecology is the discovery of taxonomic ecology since allometric relationships hold more precisely within taxonomic groups.

Using body-size based parameters for taxonomic categories it is possible to model whole ecosystems on land (Parkin and Cousins 1981) and in the ocean (Platt 1985; Cousins 1985). Size relates the fundamental ecological interactions of who can eat whom, how much food each requires to meet its own metabolic demands, and how much food each represents when eaten; reproductive rates are also likely to be size related within the taxon. The importance of body size in ecology is now well established in the central core of ecological science, trophic ecology (Lawton 1989).

Finally, the difference between a single view of a biotic hierarchy and the dual hierarchy of Eldredge and associated workers amounts to a difference in how organizational levels above the individual organism are treated, since it is assumed (see Eldredge 1992) that the two hierarchies are linked at the level of the organism itself. Hierarchies appear inherently convergent and truncated, which is to say that in a nested hierarchy of parts containing parts, there have to be fewer larger parts at each successive level. Thus categories of parts drop out or truncate at each level. For example, the organs in our own bodies (liver, heart, sexual organs) have a distinct hierarchy of component cell types and cell internal structures. In sum and in interaction they create a whole organism. But the liver has truncated at the organ level; we do not project, except in the arts, a 'liverish' group of adults who form a distinct economic or ecological group. The sexual organs do have a relevance because they commonly form two distinct groups of adults, male and female, but these 'truncate' at the social or reproductive group. Even ecological systems in Eldredge's dual hierarchy truncate biologically at the Earth as a whole, although the solar system plainly has importance in energy transfer to the Earth itself.

In the single hierarchy (Cousins 1990) the ecological object truncates at a much smaller scale than the whole Earth, namely at the area occupied by the social group of the top predator. The ecological object includes all organisms of all taxa found within this area and is named the ecosystem trophic module (ecotrophic module or ETM). The size of the ETM will vary between sites and be dependent on the degree of human interference. There are no larger *ecological* objects than these; there may be larger lakes, oceans, or deserts but these are variously parts of the global climate and water distribution system of weather structures such as cyclones, currents, and watersheds upon the Earth's mantle. ETMs will sit upon or in these structures, but it will not make them any larger as an ecological object. I argue therefore that any biological assemblies larger than the local ETM are aggregates (Rowe 1961) such as forests, breeding colonies of sea-birds, or migrating organisms, but not integrated objects formed by functionally connected and distinct parts (Salthe 1985).

Ecological diversity measurement

So far we have concentrated on showing how systematics and ecology are linked. One of the benefits of the newly developed methods of systematics suggest that it is possible to move away from treating all species as equal, and instead, treat species as being of different relative importance. An equivalent development has been occurring in ecological diversity measurement. These changes are of interest in their own right for the purpose of conservation evaluation. Later they are also examined to see how they might be linked to the systematics approach.

It is noted that mechanism of functional interaction will differ at the global, ecosystem, and local levels of organization. This points to the need for distinctly different conservation evaluation techniques at each scale if the goal of conserving global biodiversity is to be attained. These scale-dependent functional measures may place different priorities for conservation than the scale independent methods arising from measures of taxonomic distinctiveness developed from systematics. Neither is sufficient alone.

Local evaluation

Apart from particularly glamorous species, conservation priorities at the local level are set by some kind of species diversity measure, or by non-specific attributes such as landscape features. We will only consider species diversity measures. The traditional approaches are to make species counts or to use species diversity indices which, as well as being responsive to the number of species in the sample, take into account the relative commonness and rarity of species, that is they include a measure of the abundance of individuals in each species. This was a popular area of research in the 1960s and 1970s which received very full attention from the ecological research community at that time. However, the results have not really met expectations. Some simple modifications to standard practice are possible which may significantly improve the efficacy of species diversity measurement as a conservation evaluation tool at the local level, using selective elements of the now traditional indices. We now consider these.

Species diversity measurement is based around the species number curve. Each time an individual organism is added to a count (N) of organisms in an observed sample, it can be of the same species as organisms already present or it can be a new species in which case (S), the number of species in the observed sample, increases by one also. As N increases S can also increase although it is progressively less likely that, as the number of individuals in the sample increases, the next individual encountered will be of a species not previously met. Functionally, N is proportional to area, such that if the area is doubled within a similar kind of habitat the number of individuals encountered is likely to double also although, for the reasons given above, the number of species increases by a lesser amount.

In summary:

$$N = k.\text{Area} \qquad (25.1)$$

$$S = f(N) \qquad (25.2)$$

$$S = f(\text{Area}) \qquad (25.3)$$

This simple set of relationships has had two basic effects on the measurement of species diversity. While the number of individuals is proportional to the area sampled and measurement achieved by number per unit area, i.e. a number density with units of N/m^2, the approach to measuring species diversity has been twofold. Species diversity is described either as a measure of the number of species per unit area giving a species density measure with units of S/m^2 as typified by species atlases (Sharrock 1976; Dony 1976), or by making measures which are assumed to be sample size independent and are described by the function f as typified by Williams' α (Williams 1964). The frequently used Shannon index is assumed (Pielou 1975) to fall into the class of sample size independent indices although, as will be shown below, this is not the case for samples of the size normally encountered in the ecological literature.

In a study (Cousins 1977) of breeding birds on farms in the United Kingdom using Common Bird Census data from the British Trust for Ornithology, the Shannon index H′, and the related J, were analysed for sample size effects. From the simple relationship which defines H′ and J,

$$H' = J \log_2 S \tag{25.4}$$

we know that S increases with sample size and that therefore, from (4), H′ and J cannot both be constants, and either H′ or J or both are sample-size dependent.

Figure 25.1 shows that for the 85 farm plots (average area 40 hectares), as S increases so does H′. When individual farms are analysed for sample-size effect through the random aggregation of sub-samples, the sample-size dependence is clearly shown in Fig. 25.2 for both S and H′, while J is inversely related to the number of species in the sample and therefore again sample size related, Fig. 25.3. As a conservation evaluation technique the

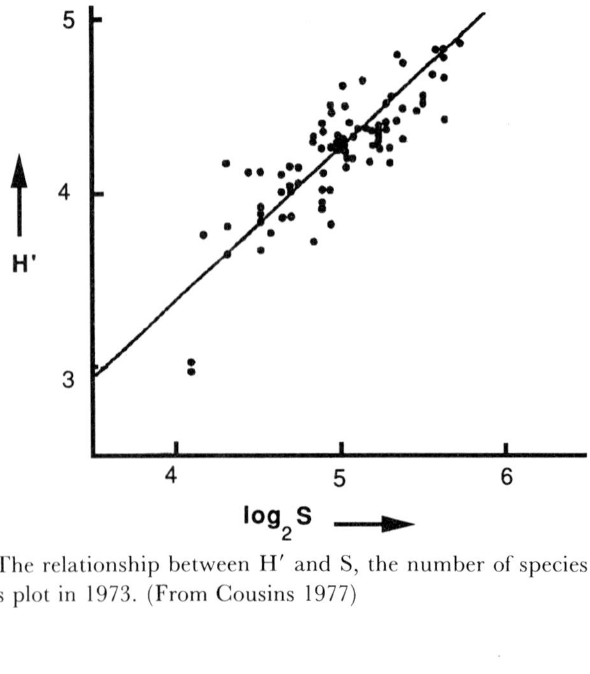

Fig. 25.1 The relationship between H′ and S, the number of species in each BTO farm census plot in 1973. (From Cousins 1977)

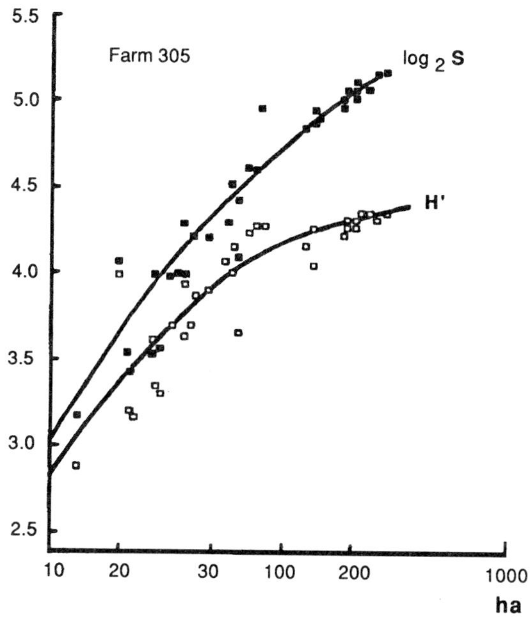

Fig. 25.2 The effect of area sampled on H′ and $\log_2 S$ for a single farm determined by random aggregation of sub-samples. (From Cousins 1977)

Shannon derived indices are therefore very difficult to interpret. At least the area of the sample should be specified. The comparison between samples remains difficult if the sampling is not standardized, or compensated for, as shown in Table 25.1 where three farms are compared.

The highest diversity measured by H′ is 4.28 for farm 209. This is a particularly large farm with the highest total number of individuals (N) and species (S) identified. However, Farm 209 has the lowest evenness value (J). We need to know if this is an artefact of sample size, as would be expected from the relationship of J with S shown in Fig. 25.3, or whether it is really different to the value of J for the other two farms. Clearly the most even distribution, $J = 0.86$, is produced by farm 315, but this similarly could be a result of the low number of species (23) found in that plot. In fact, when we compare the species densities at 10–40 ha, farms 133 and 209 are in effect identical in their species densities. Although J values were not calculated for sub-samples corresponding to these species densities, the J value for farm 209 would be expected to rise if that farm were sampled at the same scale as farm 315. Finally, it is farm 072 which turns out to have a much greater species density than the other two. The lack of sample size independence of H′, $\log_2 S$ and J requires that each of these parameters be plotted against sample area to establish comparability between sites as occurs in Fig. 25.2, although the ratio of H′ to $\log_2 S$ would also require plotting to give J.

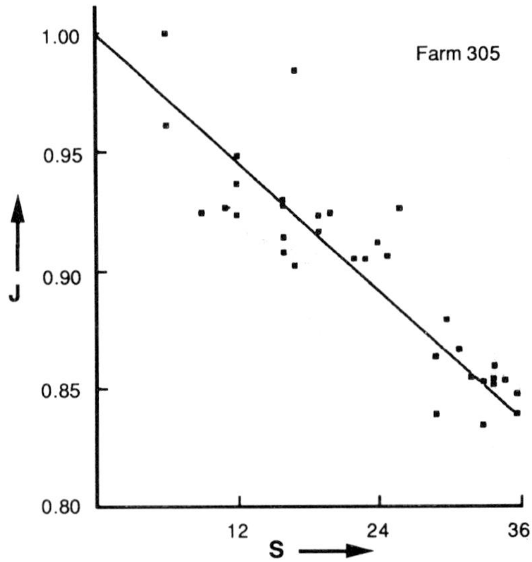

Fig. 25.3 The relationship between J and the number of species present on a single farm determined by random aggregation of sub-samples. (From Cousins 1977)

Table 25.1 Comparison of representations of biodiversity on three farms; S, number of species, N number of individuals. Species densities were obtained from averages of repeated random selection of equal area segments of mapped territory distributions. (After Cousins 1977)

Farm code	315	209	072
N	133	736	250
S	23	41	34
H	3.90	4.28	4.22
J	0.86	0.80	0.83
S/10 ha	9	9	15
S/20 ha	14	15	22
S/40 ha	21	22	30

Given these sample size problems for H′ and J, it is preferable to use clearly understood measures such as species density and the density of individuals to describe the variety and quantity of biota present in an area. A measure of the relative abundance is provided by Williams (1964) alpha,

$$S = \alpha \log_e (1 + N/\alpha) \qquad (25.5)$$

which can be used as a sample size independent measure and, more importantly, can be used to calculate species densities at a common scale by substituting (N′), the number of individuals for the base area, in equation (25.5), where S and N are known for the whole sample. In some habitats strong edge effects of the sample plot need to be removed from the data before calculating the density, N'/m^2 (Cousins 1977). Rarefaction offers an alternative technique for common scaling of diversity (Magurran 1988).

Each of the above methods are cardinal measures of species diversity in which each species is treated as equal and additive, in forming a count or index. A second group of methods is possible, where ordinal measures of species diversity (Cousins 1991) are used and species are ranked in an order and not added together. This difference is important for functional biotic measurement. Earlier it was stated that if species diversity is measured as a species density then this represents in some way the variety of energy paths present in an given area. Members of species of different sizes or of different trophic habit also have different quantitative effects on those energy flows and these can be represented in an ordinal index.

Within the cardinal indices, a sparrow counts the same as an eagle in the count of number of species present. In an ordinal representation, such as the number of species in weight classes, different levels of importance can be attached to species which individually have very different impacts on energy flow.

It is important to make one further relationship explicit between biodiversity and ecosystem function. Biodiversity describes the number of types of organism present in an area; it is the number of individuals of each of these organisms, i.e. the quantity of the biota, which determine the functional consequences of the biodiversity. In the section on taxonomic ecology the importance of body size classes with a taxon has already been identified. This provides an ordinal representation of biodiversity if we plot the number of species against weight, giving the species weight-distribution (S(w)). The plot of numbers of individuals of those weights, gives the individuals weight-distribution (N(w)), which provides the functional description since it is the number of individuals of different weights that are feeding, respiring, and so on. Because it is easier to count the number of types rather than the total number of individuals in an area, it would be useful to know the relationship between N(w) and S(w) for a given sample. For the 85

Fig. 25.4 Mean species weight of the British breeding bird land fauna. Small dots, 115.8–153.0 g; open circles, 153.0–191.3 g; circles with bar 191.3–229.1 g; filled symbols, 229.1–266.9 g. (From Cousins 1989)

farms above, the distributions have correlated medians ($r = 0.70$ $p < 0.001$) and a Chi-squared test showed no difference of medians between the distributions (Cousins 1989). Thus it may be possible to infer functional consequences from biodiversity distributions if the relationship between N(w) and S(w) is known. As an example, the species distribution of body size in UK breeding land birds, Fig. 25.4, would suggest that the distribution of body size of the numbers of individuals of UK breeding land birds also follows that distribution, if the medians of the two distributions are common. This remains to be tested further.

Finally, evidence from a survey of the feeding habits of British insects (Price 1977) shows that species of parasite (72.1 per cent of the fauna) are much more numerous than species of predator (6.3 per cent of the fauna including non-parasitic herbivores). In this case parasites are about 10 times as numerous as predators; the remaining species are saprophytes. Whereas parasites tend to be taxonomically specific in their hosts, foraging theory (Pulliam 1974; Charnov 1976) points to predators including any species in their diet provided that the predator can encounter and capture the prey, and that the rate of energy gain from eating the prey exceeds a threshold value. This comparison of parasitic specificity and predatory opportunism may account for the substantially different number of species in these two categories. Whatever the cause, there is a need to distinguish these two diversities since data on parasites could swamp predator data where they are combined. Warren and Gaston (1992) examine a related issue, the correlation between species richness in different trophic guilds in ecosystems.

This brief review of 'ecological' biodiversity indices has suggested simple alternatives to the more familiar Shannon index. The emphasis on species density, species size, and species as predators, saprophytes, or parasites provides functional descriptors. These labels may also be applied to the relations in a taxonomic hierarchy and used to redefine conservation priorities. A closely related but very differently sized organism may change its ranking in a species priority list for example.

Ecosystem and regional scale evaluation

It has already been noted that one of the differences between ecological and systematic methods for conservation evaluation is that ecological methods are scale specific, while systematic relationships are scale independent. Thus the application of the ecological methods involves different phenomena at the ecosystem level compared to the local scale. By contrast, scale change in systematics requires the introduction of a set of (arbitrary) system boundaries from outside biology, e.g. regional or national boundaries within which to set the priorities for conservation using the new systematic methods.

In terms of the politics of conservation, national or regional boundaries are far from arbitrary and are clearly very important. Funds which are raised locally or nationally are mainly spent locally or nationally. Whereas from the viewpoint of the systematic methods biodiversity priorities are international and predominately located in the tropical and sub-tropical regions, the funds for conservation are sourced and spent disproportionately in Europe and North America. There are, however, larger issues of biological principle present in this problem. There is no doubt that if the systematic methods were applied globally without reference to national interest, and expenditure was made on the basis of where there was the marginally best application, then maximal effect in terms of getting the most diversity of biological form into a minimum number of non-overlapping reserves would be achieved. However, this distribution of reserves may not include a single one in Britain and other large parts of Europe. The question that arises is: when national interest is removed from the analysis, does the 'pure' application of the technique still identify an optimal biological distribution of reserves? Are there functional attributes of the biota which require a different biodiversity strategy?

To gain an insight into this problem let us tackle the question of conservation evaluation at the ecosystem level. The Lindeman (1942) definition of ecosystem, 'the system of physical-chemical-biological processes active within a space time unit of any magnitude' is problematic for conservation evaluation since, whatever change occurs, the 'ecosystem' always remains. The ecosystem object, i.e. the ETM (Cousins 1990), unlike Lindeman's ecosystem concept, is countable and has a characteristic spatial scale. It therefore becomes possible to evaluate particular effects on ecosystems including the loss of an ETM, or its downgrading to a smaller scale, by the extinction of the primitively largest species of predator and its replacement by smaller predators.

In Britain the top predators were, until recent times, brown bears (*Ursus arctos*) and wolves (*Canis lupus*), although their abundance prior to extensive human modification of the landscape is not known. Harting (1894) reports that it was probable that bears were extinct in Britian by the tenth century, whereas the wolf survived until about 1500 in England and Wales, 1740 in Scotland, and 1770 in Ireland. In contemporary Britain the fox is the largest predator; a map of the estimated fox population is shown in Fig. 25.5.

The loss of the primitive top predators from the UK is a total one. In some parts of the world the regional count of ecosystem objects defined by the primitive top predator social groups is still possible, and is an important conservation evaluation technique to assess the status of the natural world. Since these are the largest ecological objects it is interesting to know the biodiversity present at the scale defined by the ETM. Indeed it may eventually be possible to suggest standards for biodiversity that are desirable

Fig. 25.5 Estimated adult fox densities divided into six intervals; light, zero to less than 4 ranging to 20–24 foxes per km^2 for the darkest symbols. (From Macdonald *et al.* 1981)

to provide a 'healthy' ETM akin to the concept of a healthy ecosystem (Cowie 1992) called for elsewhere. Such a 'standard' would need to reflect the latitude and other climatic effects on biodiversity. Each of the ways of classifying biodiversity outlined above could contribute to this; a well-spaced size distribution of species in a full set of expected higher taxa for the location; a balance of predator, saprophyte, and parasite biodiversity; a high biodiversity of biological form as defined by the systematic indices.

Before returning to the relationship between systematics and local biodiversity measurement one further aspect of the ETM needs to be considered. Top predators tend to have extensive species distributions. Thus in the case of the fox in the UK, apart from its absence from some mountainous regions, the fox is ubiquitous. This does not mean, however, that there is only one type of ETM in the UK. The ETM, although defined by the area occupied by the social group of the top predator, includes among its parts all individuals of whatever species present in that area at any given time. Traditionally, plant functional types have defined biomes and at a smaller scale plant taxonomic categories have defined UK ecosystem types (grassland, oak woodland, heather moorland, and so on). In maintaining this tradition, there is a useful parallel to be drawn between the measurement of species biodiversity using the primary variables number of individuals (N), and number of species (S), with the number of ecosystem objects (Ne), and the number of plant-defined types of ecosystem object (Se), found in a given area. Both quantities are potentially useful tools for conservation evaluation. A similar system of plant defined ecological zones is currently under development by English Nature.

The case has been made earlier that systematics and ecology are the other sides of the same coin, that the biological structures, aptly described by systematics, when viewed moment-to-moment in time create the interactions we call ecology. It follows therefore that spatial boundaries which exist in ecosystems also exist for systematic descriptions. Thus when Vane-Wright *et al.*. (1991) use their sytematic methods to determine priority locations for conserving critical faunas, they are in fact selecting a particular, or several ETMs, since they also require that 'once identified, the whole biota of a priority reserve area needs adequate protection as a functional ecological system, or set of such systems'.

The use of political borders to distribute local priorities for conservation using systematic methods will again select particular whole or parts of ETMs within each country. Adding the countries together this approximates to a global collection of ecosystem types and may more usefully be approached, and selected, as such. Because of the relationship to function implicit in the systematic approach this suggests the hypothesis: that the selection of conservation priorities by the representation of ecosystem types offers a minimum area solution within a political boundary, but with the additional benefit of selecting a full range of ecosystems as part of the biodiversity conserved.

The only major functional analysis programme to have occurred at the ecosystem level was the International Biological Programme IBP (carried out 1964–1974). An objective of IBP was to determine a functional description of each of the world's major ecosystem types. IBP was to provide a comparative description of ecosystems using a common methodology, and it was comprehensive in its coverage of the world's biomes. Unfortunately, there were shortcomings in the trophic level methodology (Lindeman 1942; Cousins 1987) and the proposed comparison of systems was not possible at the completion of the study; instead, a trophic-taxonomic analysis was made. This work by Heal and MacClean (1975), must rank as a milestone in taxonomic ecology. However, this solution arose out of the problems with the programme's central methodology. The taxonomic ecology approach if adopted from the start of such a programme would have a very different structure of analysis. It is hard to see how comparative ecology and conservation ecology can proceed without a baseline research description of the world's ecosystems. This is perhaps the most serious gap in our ability to undertake conservation evaluation. The objective which IBP set itself of obtaining a functional description of the world's ecosystems has yet to be fulfilled; the present international geosphere-biosphere programme (IGBP) or global change programmes do not fill this gap left by IBP, and a new initiative is needed.

Human impact assessment

The measurement of biodiversity has itself been commonly used as an index of human impact on the environment. However, for the purposes of conservation evaluation, human impact needs to be seen as an independent variable which is measured and capable of being equated against the results of different conservation strategies. There is no consensus on how definitive measurement of human impact could be achieved, although a number of lines of inquiry are opening up which need to be fully explored. The ecosystem object, the ETM, provides one framework to examine these issues.

The density of top predators forming the social group of the top predator species is the largest organized structure generated by the passage of energy through the food chain. This structure can be disrupted by three primary forms of human impact. The first is direct hunting of the top predators; the second is harvesting other organisms lower in the food chain thereby reducing food availability to the top predators; thirdly, by pollutants reducing the productivity of plants and animals. Data on hunting are available as typified by fur trapping records. Vitousek *et al.* (1986) have made calculations, in a variety of ways, to quantify the proportion of the total incoming solar radiation incident on the land surface that is used by human populations rather than flowing through to top predators, and they calculate that humans may have appropriated 40 per cent of the products of terrestrial photosynthesis. The distribution of pollutants of many kinds is gaining widespread analysis and documentation (UNEP 1991). Agricultural databases form an important measure of human impact on the natural environment. A further group of *ad hoc* measures of human disturbance to natural systems have been developed for specific purposes. A built environment index for the city of London (Sandford 1975) was used as the basis for Fig. 25.6 and produced a negative Spearman correlation (−0.90, p = 0.001) with the species density of land birds (Cousins 1982). If cities are the most dense human distribution, the opposite extreme has been quantified by McCloskey and Spalding (1989) who estimated the areas of the Earth's surface which are wilderness, defined as having no roads or human settlements, in areas of 400 000 hectares or more. Table 25.2 shows how these areas of wilderness relate to biome distributions and may indicate priorities for conservation. Between these extremes of city settlement and wilderness, suitable measures need to be established to measure human impact as it affects biodiversity.

Hunting, appropriation of the products of photosynthesis, and pollution in total reduce the possible density of primitive top predators or ultimately the size of top predator species that can be sustained on an area. Counts of top predators also provide an integrative measure of status of natural ecosystems (Cousins 1990) suitable as a conservation evaluation tool. Again it is not the

Fig. 25.6 An index of the built environment of London derived from Sandford (1975). Four intervals from light, least dense settlements to dark, most dense. Corresponding land bird species densities, light to dark, 43–51 species; 52–60 species; 61–68 species; 69–77 species. (From Cousins 1982)

Table 25.2 Wilderness distribution by biomes. (From McCloskey and Spalding 1989)

Biome	Wilderness distribution by biomes km^2	% total	Count
Tundra Communities	20 047 533	41.7	100
Warm Deserts/Semi Deserts	9 329 531	19.4	389
Temperate Needleleaf Forests	8 799 312	18.3	120
Tropical Humid Forests	3 006 855	6.3	77
Mixed Mountain Systems	1 973 391	4.1	76
Cold Winter Deserts	1 478 494	3.1	51
Tropical Dry Forests	1 424 099	3.0	120
Tropical Grasslands/Savannahs	735 331	1.5	33
Temperate Rainforests	450 215	0.9	15
Temperate Broadleaf Forests	290 646	0.6	20
Temperate Grasslands	272 016	0.6	24
Evergreen Sclerophyllus Forests	170 885	0.4	7
Mixed Island Systems	91 647	0	7
	48 069 951	100.0	1039

number of types of top predator that is important, but the quantity of top predators that provides the evaluation measure.

Each of these techniques capture an aspect of human impact on the existing environment against which the success or failure of conservation management can be evaluated. Since it is the human realm which is driving the reduction in biodiversity, the reasonable economic aspirations of populations, estimated to reach 12 billion by 2150 (Bulatao *et al.* 1990), will generate demands on the natural environment which will need to be anticipated in order to devise appropriate conservation policies. Modestly declining biodiversity may represent a major success against this background.

Global biological function

There is a picture on the front of the paperback edition of *Five Kingdoms* (Margulis and Schwartz 1988) showing the Earth with its climate against a background of stars held in position by the thumb and forefinger of a human hand. On each digit is a symbol representing one of five kingdoms. As well as being a human hand, profiles of the heads of the human male and female are visible as outlines of the forefinger.

This powerful graphic concisely represents the role of life in maintaining the climate of the earth within an otherwise inhospitable void. The influence of life in creating the present proportions of the Earth's atmosphere has been profound (Berkner and Marshall 1965) notably in the creation of an oxygen rich atmosphere with sufficient ozone to substantially reduce the level of ultra-violet reaching the Earth's surface. Without this attenuation of UV, higher plant life would be prevented from existing on land. The present interaction of life and the atmosphere is still closely bound up with the 'five kingdoms'. It is this functional role of life at the planetary level that forms the final case for conservation evaluation.

Here, it is the quantity of life and its appropriate spatial distribution which is important rather than how many species of life there are. Technically, whether the five kingdoms were represented by only five species, or by single local populations of all species, neither case would fulfil the function represented by the hand holding the Earth in space. It is the quantity of active biomass that is determining the physical properties of the Earth's surface and atmosphere. Shukla and Mintz (1982) provided early evidence for the effect on global climate of the removal of existing vegetation from large areas of the Earth's surface. Their model showed that the global distribution of rainfall, temperature, and air motion strongly depend on the land surface evapotransporation, thus confirming the importance of vegetation to climate. A similar call for the conservation of vegetation to preserve global climate has been made by Lovelock (1988).

As global models of climate improve and the role of vegetation in determining climate is better understood, then the consequences, including economic consequences, of anthropomorphic vegetation change upon remote as well as local environments will become calculable. Conservation evaluation on this basis could provide an alternative path to prioritizing land areas for conservation for reasons of the environmental services, such as climate stabilization, that they provide. It is the quantity of life and not its diversity which provide these services. Paradoxically, this approach should justify the continued existence much larger areas of natural and semi-matural land including its inherent biodiversity. The process of selection of priority biodiversity reserves within 'climate reserves' could further be made using the ecosystem–systematics methodology outlined above.

Conclusion

The new systematic methods for prioritising species conservation may be applied at any spatial scale. In this paper the importance of the ecosystem scale, defined by the size of the ecotrophic module, has been stressed as being of particular importance. It is suggested that the systematic methods, in combination with body size and trophic categories of predator, parasite, or saprophyte, ensure a sound functional basis for choice between ecosytem parts or wholes. The case is made for improving local conservation evaluation by using species density measures and Williams' α rather than the Shannon H' and J indices.

There is a profound role for systematics at the heart of functional ecology. This is not through the conventional task of naming species, but relies on the importance of higher taxonomic structure which, when viewed moment-to-moment in time, creates ecosystem function. The congruence should point to a non-problem in the priorities of critical faunas over critical ecosystems, at least for local or regional conservation purposes made, say, within nations.

Globally, the systematic methods provide a fascinating picture of the location of biological form. However, the economy with which the method assembles the ark may not in the end be to the advantage of its inhabitants. In effect, the careful selection of reserves creates very high biodiversity for a small total area and thus 'cheats' the species-area curve. At the global level it is the quantity and distribution of biota, particularly plant life and microbes, that determine the biological influences upon the world's climate. Without relative climatic stability, carefully optimized biodiversity reserves risk losing species and returning eventually to the expected point on the species-area curve for the total area conserved.

The problem of conserving global biodiversity presents an almost overwhelmingly difficult task. An approach to biotic measurement has been proposed which includes the new systematic techniques, but for functional or dynamic reasons, focuses on the ecosystem level. Despite the urgency of this issue, uncertainties mean that major research efforts are needed for practically every aspect if real progress is to be made. The systematic approaches need testing and validating, ecosystems need benchmark and comparative analysis preferably in the taxonomic ecology or ecosystem-systematics framework, human impact measurement needs refining, while climatic and landscape level ecology and economics are yet other planks needed for the ark.

References

Berkner, L. V. and Marshall, L. C. (1965). On the origin and rise of oxygen concentration in the Earth's atmosphere. *Journal of Atmospheric Science,* **22**, 225–61.

Bulatao, R. A., Bos, E., Stephens, P. W., and Vu, M. T. (1990). *World population projections 1989–1990 edition Short and long-term estimates.* World Bank, New York.

Charnov, E. L. (1976). Optimal foraging: the marginal value theorem. *Journal of Theoretical Population Biology*, **9**, 129–36.

Cousins, S. H. (1977). Sample size and edge effect on community measures of farm bird populations. *Polish Ecological Studies*, **3**, 27–35.

Cousins, S. H. (1982). Species size distributions of birds and snails in an urban area.

In *Urban Ecology* (ed. R. Bornkamm, J. A. Lee and M. R. D. Seaward), pp. 99–109. Blackwell Scientific, Oxford.

Cousins, S. H. (1983). An alignment of energy and diversity models of ecosystems. PhD Thesis, The Open University, Milton Keynes.

Cousins, S. H. (1985). The trophic continuum in marine ecosystems: stucture and equations for a dynamic model. *Canadian Bulletin of Fisheries and Aquatic Sciences*, **213**, 76–93.

Cousins, S. H. (1987). The decline of the trophic level concept. *Trends in Ecology and Evolution*, **2**, 312–16.

Cousins, S. H. (1988). Fundamental components in ecology and evolution. In *Ecodynamics* (ed. C. J. Soeder and W. F. Wolff), pp. 60–8. Springer-Verlag, Berlin.

Cousins, S. H. (1989). Species diversity and the energy theory. *Nature*, **340**, 350–1.

Cousins, S. H. (1990). Countable ecosystems deriving from a new food web entity. *Oikos*, **57**, 270–5.

Cousins, S. H. (1991). Species diversity measurement: choosing the right index. *Trends in Ecology and Evolution*, **6**, 190–2.

Cowie, J. (1992). Conserving biodiversity. *Biologist*, **39**, 148.

Damuth, J. (1985). Selection among 'species': a formulation in terms of natural functional units. *Evolution*, **39**, 1132–46.

Dony, J. G. (1976). *Bedfordshire plant atlas*. Luton Museum and Art Gallery, Luton.

Eldredge, N. (1992). Where the twain meet: causal intersections between the genealogical and ecological realms. In *Systematics, ecology and the biodiversity crisis* (ed. N. Eldredge), pp. 1–14. Columbia University Press, New York.

Eldredge, N. and Salthe, S. N. (1984). Hierarchy in evolution. *Oxford surveys of Evolutionary Biology*, **1**, 184–208.

Faith, D. P. (1992). Conservation evaluation and phylogenic diversity. *Biological Conservation*, **61**, 1–10.

Fenchel, T. (1974). Intrinsic rate of natural increase: the relationship with body size. *Oecologia*, **14**, 317–26.

Grene, M. (1987). Hierarchies in biology. *American Scientist*, **75**, 504–10.

Harting, J. E. (1894). *British animals extinct within recent times*. Trubner, London.

Heal, W. O. and MacLean Jr., S. F. (1975). Comparative productivity in ecosystems—secondary productivity. In *Unifying conepts in ecology* (ed. W. H. van Dobben and R. H. Lowe-McConnell), pp. 89–108. Junk, The Hague.

Lawton, J. H. (1989). Food webs. In *Ecological concepts* (ed. J. M. Cherrett), pp. 43–78. Blackwell Scientific Publishers, Oxford.

Lindeman, R. L. (1942). The trophic-dynamic aspect of ecology. *Ecology*, **23**, 399–418.

Lovelock, J. E. (1988). The earth as a living organism. In *Biodiversity* (ed. E. O. Wilson and F. M. Peter), pp. 486–9. National Academy Press, Washington. DC.

McCloskey, J. M. and Spalding, H. (1989). A reconnaissance-level inventory of the amount of wilderness remaining in the world. *Ambio*, **18**, 221–7.

Macdonald, D. W., Bunce, R. G. H., and Bacon, P. J. (1981). Fox populations, habitat characterization and rabies control. *Journal of Biogeography*, **8**, 145–51.

McNaughton, S. J. (1989). Ecosystems and conservation in the twenty-first century. In *Conservation for the Twenty-First Century* (ed. D. Western and M. Pearl), pp. 109–20. Oxford University Press, New York.

Magurran, A. E. (1988.) *Ecological diversity and its measurement.* Croom Helm, London.

Margulis, L. and Schwartz, K. V. (1988). *Five Kingdoms: an illustrated guide to the phyla of life on earth.* Freeman, New York.

Parkin, H. and Cousins, S. H. (1981). Towards a global model of large ecosystems. *ERG Report*, **41**, The Open University, Milton Keynes, UK.

Peters, R. H. (1983). *The ecological importance of body size.* Cambridge University Press.

Pielou, E. C. (1975). *Ecological diversity.* Wiley, London.

Platt, T. (1985). Structure of marine ecosystems: its allometric basis. *Canadian Bulletin of Fisheries and Aquatic Sciences*, **213**, 55–64.

Pressey, R. L. and Nicholls, A. O. (1989). Efficiency in conservation evaluation: scoring vs. iterative approaches. *Biological Conservation*, **50,** 199–218.

Price, P. W. (1977). General concepts on the evolutionary biology of parasites. *Evolution*, **31**, 405–20.

Pulliam H. R. (1974). On the theory of optimal diets. *American Naturalist*, **108**, 59–75.

Rowe, J. S. (1961). The level-of-integration concept in ecology. *Ecology*, **42**, 420–7.

Salthe, S. N. (1985). *Evolving hierarchial systems.* Columbia University Press, New York.

Sandford, H. A. (1975). Habitat overlay. *London Naturalist*, **54**, 72–3.

Sharrock, J. T. R. (1976). *The atlas of breeding birds in Britain and Ireland.* British Trust for Ornithology, Tring, UK.

Shukla, J. and Mintz, Y. (1982). Influence of land-surface evapotranspiration on the earth's climate. *Science*, **215**, 1498–501.

Snaydon, R. W. (1973). Ecological factors, genetic variation and speciation in plants. In *Taxonomy and Ecology* (ed. V. H. Heywood), pp. 1–29. Academic Press, London.

Warren, P. H. and Gaston, K. J. (1992). Predator-prey ratios: a special case of a general pattern? *Philosophical Transactions of the Royal Society of London*, **B 338**, 113–30.

Vane-Wright, R. I., Humphries, C. J., and Williams, P. H. (1991). What to protect—systematics and the agony of choice. *Biological Conservation*, **55**, 235–54.

Vitousek, P. M., Ehrlich, P. R., Ehrlich, A. H., and Matson P. A. (1986). Human appropriation of the products of photosynthesis. *BioScience*, **36**, 368–73.

Williams, C. B. (1964). *Patterns in the balance of nature.* Academic Press, London.

Vrba, E. S. (1984). Evolutionary pattern and process in the sister group Alcelaphini-Aepycerotini (Mammalia:Bovidae). In *Living fossils* (ed. N. Eldredge and S. M. Stanley), pp. 62–79. Springer-Verlag, New York.

UNEP (1991) *Environmental data report.* Basil Blackwell, Oxford.

Index

Systematics Association Publications

1. Bibliography of key works for the identification of the British fauna and flora, *3rd edition* (1967)[†]
   Edited by G. J. Kerrich, R. D. Meikle, and N. Tebble
2. Function and taxonomic importance (1959)[†]
   Edited by A. J. Cain
3. The species concept in palaeontology (1956)[†]
   Edited by P. C. Sylvester-Bradley
4. Taxonomy and geography (1962)[†]
   Edited by D. Nichols
5. Speciation in the sea (1963)[†]
   Edited by J. P. Harding and N. Tebble
6. Phenetic and phylogenetic classification (1964)[†]
   Edited by V. H. Heywood and J. McNeill
7. Aspects of Tethyan biogeography (1967)[†]
   Edited by C. G. Adams and D. V. Ager
8. The soil ecosystem (1969)[†]
   Edited by H. Sheals
9. Organisms and continents through time (1973)[†]
   Edited by N. F. Hughes
10. Cladistics: a practical course in systematics (1992)
    P. L. Forey, C. J. Humphries, I. J. Kitching, R. W. Scotland, D. J. Siebert, and D. M. Williams

[†]Published by the Association (out of print)

Systematics Association Special Volumes

1. The new systematics (1940)
   Edited by J. S. Huxley (Reprinted 1971)
2. Chemotaxonomy and serotaxonomy (1968)*
   Edited by J. G. Hawkes
3. Data processing in biology and geology (1971)*
   Edited by J. L. Cutbill
4. Scanning electron microscopy (1971)*
   Edited by V. H. Heywood
   Out of print

5. Taxonomy and ecology (1973)*
   Edited by V. H. Heywood
6. The changing flora and fauna of Britain (1974)*
   Edited by D. L. Hawksworth
   Out of print
7. Biological identification with computers (1975)*
   Edited by R. J. Pankhurst
8. Lichenology: progress and problems (1976)*
   Edited by D. H. Brown, D. L. Hawksworth, and R. H. Bailey
9. Key works to the fauna and flora of the British Isles and north-western Europe, *4th edition* (1978)*
   Edited by G. J. Kerrich, D. L. Hawksworth, and R. W. Sims
10. Modern approaches to the taxonomy of red and brown algae (1978)*
    Edited by D. E. G. Irvine and J. H. Price
11. Biology and systematics of colonial organisms (1979)*
    Edited by G. Larwood and B. R. Rosen
12. The origin of major invertebrate groups (1979)*
    Edited by M. R. House
13. Advances in bryozoology (1979)*
    Edited by G. P. Larwood and M. B. Abbot
14. Bryophyte systematics (1979)*
    Edited by G. C. S. Clarke and J. G. Duckett
15. The terrestrial environment and the origin of land vertebrates (1980)*
    Edited by A. L. Panchen
16. Chemosystematics: principles and practice (1980)*
    Edited by F. A. Bisby, J. G. Vaughan, and C. A. Wright
17. The shore environment: methods and ecosystems (2 Volumes) (1980)*
    Edited by J. H. Price, D. E. G. Irvine, and W. F. Farnham
18. The Ammonoidea (1981)*
    Edited by M. R. House and J. R. Senior
19. Biosystematics of social insects (1981)*
    Edited by P. E. Howse and J.-L. Clément
20. Genome evolution (1982)*
    Edited by G. A. Dover and R. B. Flavell
21. Problems of phylogenetic reconstruction (1982)*
    Edited by K. A. Joysey and A. E. Friday
22. Concepts in nematode systematics (1983)*
    Edited by A. R. Stone, H. M. Platt, and L. F. Khalil
23. Evolution, time, and space: the emergence of the biosphere (1983)*
    Edited by R. W. Sims, J. H. Price, and P. E. S. Whalley
24. Protein polymorphism: adaptive and taxonomic significance (1983)*
    Edited by G. S. Oxford and D. Rollinson

25. Current concepts in plant taxonomy (1983)*
Edited by V. H. Heywood and D. M. Moore
26. Databases in systematics (1984)*
Edited by R. Alkin and F. A. Bisby
27. Systematics of the green algae (1984)*
Edited by D. E. G. Irvine and D. M. John
28. The origins and relationships of lower invertebrates (1985)[‡]
Edited by S. Conway Morris, J. D. George, R. Gibson, and H. M. Platt
29. Infraspecific classification of wild and cultivated plants (1986)[‡]
Edited by B. T. Styles
30. Biomineralization in lower plants and animals (1986)[‡]
Edited by B. S. C. Leadbeater and R. Riding
31. Systematic and taxonomic approaches in palaeobotany (1986)[‡]
Edited by R. A. Spicer and B. A. Thomas
32. Coevolution and systematics (1986)[‡]
Edited by A. R. Stone and D. L. Hawksworth
33. Key works to the fauna and flora of the British Isles and north-western Europe, *5th edition* (1988)[‡]
Edited by R. W. Sims, P. Freeman, and D. L. Hawksworth
34. Extinction and survival in the fossil record (1988)[‡]
Edited by G. P. Larwood
35. The phylogeny and classification of the tetrapods (2 Volumes) (1988)[‡]
Edited by M. J. Benton
36. Prospects in systematics (1988)[‡]
Edited by D. L. Hawksworth
37. Biosystematics of haematophagous insects (1988)[‡]
Edited by M. W. Service
38. The chromophyte algae: problems and perspectives (1989)[‡]
Edited by J. C. Green, B. S. C. Leadbeater, and W. Diver
39. Electrophoretic studies on agricultural pests (1989)[‡]
Edited by Hugh D. Loxdale and J. den Hollander
40. Evolution, systematics, and fossil history of the Hamamelidae (2 Volumes) (1989)[‡]
Edited by Peter R. Crane and Stephen Blackmore
41. Scanning electron microscopy in taxonomy and functional morphology (1990)[‡]
Edited by D. Claughter
42. Major evolutionary radiations (1990)[‡]
Edited by P. D. Taylor and G. P. Larwood
43. Tropical lichens: their systematics, conservation, and ecology (1991)[‡]
Edited by D. J. Galloway
44. Pollen and spores: patterns of diversification[‡]
Edited by S. Blackmore and S. H. Barnes

45. The biology of free-living heterotrophic flagellates‡
Edited by D. J. Patterson and J. Larsen
46. Plant-animal interactions in the marine benthos‡
Edited by D. M. John, S. J. Hawkins, and J. H. Price
47. The Ammonoidea: environment, ecology, and evolutionary change‡
Edited by M. R. House
48. Designs for a global plant species information system‡
Edited by F. A. Bisby, G. F. Russell, and R. J. Pankhurst
49. Plant galls‡
Edited by M. Williams
50. Systematics and conservation evaluation‡
Edited by P. L. Forey, C. J. Humphries, and R. I. Vane-Wright

*Published by Academic Press for the Systematics Association
†Published by the Palaeontological Association in conjunction with the Systematics Association
‡Published by Oxford University Press for the Systematics Association